Flexible and Stretchable Electronics

Special Issue Editors

Seung Hwan Ko
Daeho Lee
Zhigang Wu

MDPI • Basel • Beijing • Wuhan • Barcelona • Belgrade

MDPI

Special Issue Editors

Seung Hwan Ko
Seoul National University
Korea

Daeho Lee
Gachon University
Korea

Zhigang Wu
Huazhong University of
Science and Technology
China

Editorial Office
MDPI AG
St. Alban-Anlage 66
Basel, Switzerland

This edition is a reprint of the Special Issue published online in the open access journal *Micromachines* (ISSN 2072-666X) from 2016–2017 (available at: http://www.mdpi.com/journal/micromachines/special_issues/flexible_electronics).

For citation purposes, cite each article independently as indicated on the article page online and as indicated below:

Author 1; Author 2. Article title. *Journal Name* **Year**, *Article number*, page range.

First Edition 2017

ISBN 978-3-03842-436-9 (Pbk)
ISBN 978-3-03842-437-6 (PDF)

Table of Contents

About the Special Issue Editors...v

Preface: Technologies Reshaping our Daily Lives vii

Zhigang Wu, Yongan Huang and Rong Chen
Opportunities and Challenges in Flexible and Stretchable Electronics: A Panel Discussion
at ISFSE2016
Reprinted from: *Micromachines* **2017**, *8*(4), 129; doi:10.3390/mi8040129...1

Phillip Lee, Jinhyeong Kwon, Jinhwan Lee, Habeom Lee, Young D. Suh, Sukjoon Hong and Junyeob Yeo
Rapid and Effective Electrical Conductivity Improvement of the Ag NW-Based Conductor by
Using the Laser-Induced Nano-Welding Process
Reprinted from: *Micromachines* **2017**, *8*(5), 164; doi:10.3390/mi8050164...7

Habeom Lee, Jinhyeong Kwon, Woo Seop Shin, Hyeon Rack Kim, Jaeho Shin, Hyunmin Cho, Seungyong Han, Junyeob Yeo and Sukjoon Hong
Large-Area Compatible Laser Sintering Schemes with a Spatially Extended Focused Beam
Reprinted from: *Micromachines* **2017**, *8*(5), 153; doi:10.3390/mi8050153...17

Peng Zhou, Wangnan Li, Tianhui Li, Tongle Bu, Xueping Liu, Jing Li, Jiang He, Rui Chen, Kunpeng Li, Juan Zhao and Fuzhi Huang
Ultrasonic Spray-Coating of Large-Scale TiO_2 Compact Layer for Efficient Flexible
Perovskite Solar Cells
Reprinted from: *Micromachines* **2017**, *8*(2), 55; doi:10.3390/mi8020055...25

Chen-Jui Lan, Song-Ling Tsai and Ming-Tsang Lee
Direct Silver Micro Circuit Patterning on Transparent Polyethylene Terephthalate Film Using
Laser-Induced Photothermochemical Synthesis
Reprinted from: *Micromachines* **2017**, *8*(2), 52; doi:10.3390/mi8020052...37

Ki-Woo Jun, Jong-Nam Kim, Jin-Young Jung and Il-Kwon Oh
Wrinkled Graphene–AgNWs Hybrid Electrodes for Smart Window
Reprinted from: *Micromachines* **2017**, *8*(2), 43; doi:10.3390/mi8020043...47

Jong-Hyun Kim, Qin Zhou and Jiyoung Chang
Suspended Graphene-Based Gas Sensor with 1-mW Energy Consumption
Reprinted from: *Micromachines* **2017**, *8*(2), 44; doi:10.3390/mi8020044...54

Yun Cui, Yuhang Li, Yufeng Xing, Tianzhi Yang and Jizhou Song
One-Dimensional Thermal Analysis of the Flexible Electronic Devices Integrated with
Human Skin
Reprinted from: *Micromachines* **2016**, *7*(11), 210; doi:10.3390/mi7110210...62

Yicong Zhao and Xian Huang
Mechanisms and Materials of Flexible and Stretchable Skin Sensors
Reprinted from: *Micromachines* **2017**, *8*(3), 69; doi:10.3390/mi8030069...74

Minghui Luo, Yanhua Liu, Wenbin Huang, Wen Qiao, Yun Zhou, Yan Ye and Lin-Sen Chen
Towards Flexible Transparent Electrodes Based on Carbon and Metallic Materials
Reprinted from: *Micromachines* **2017**, *8*(1), 12; doi:10.3390/mi8010012 102

Xiaowei Yu, Bikram K. Mahajan, Wan Shou and Heng Pan
Materials, Mechanics, and Patterning Techniques for Elastomer-Based Stretchable Conductors
Reprinted from: *Micromachines* **2017**, *8*(1), 7; doi:10.3390/mi8010007 118

Xuelin Wang and Jing Liu
Recent Advancements in Liquid Metal Flexible Printed Electronics: Properties, Technologies,
and Applications
Reprinted from: *Micromachines* **2016**, *7*(12), 206; doi:10.3390/mi7120206 147

About the Special Issue Editors

Seung Hwan Ko is currently a Professor in the Mechanical Engineering Department of Seoul National University, Korea. He received a B.S. in Yonsei University (2000), his M.S. in Seoul National University (2002), and a Ph.D degree in mechanical engineering at UC Berkeley in 2006. Before joining Seoul National University, he worked as a post-doctoral researcher at UC Berkeley (2006–2009) and an associate professor at KAIST (2009–2013). His research interest is in low temperature process development for flexible, stretchable and wearable electronics, laser assisted nano/micro fabrication process development, laser–nanomaterial interaction and crack assisted nanomanufacturing. He has published over 120 peer reviewed journal papers.

Daeho Lee is currently an Assistant Professor in the Department of Mechanical Engineering, Gachon University, Korea. He received a B.S. (2002) and M.S. (2007) in Seoul National University, a Ph.D degree in mechanical engineering at UC Berkeley in 2012. Before joining Gachon University, he worked as a visiting researcher at Lawrence Livermore National Lab (2013–2014) and as a post-doctoral fellow at UC Berkeley (2012–2014). His research interest is in flexible and stretchable electronics, optoelectronic devices, nanomaterials, low temperature laser processing.

Zhigang Wu is currently a Chutian scholar, distinguished professor in the School of Mechanical Science and Engineering, Huazhong University of Science and Technology (HUST), Wuhan. He received his PhD in Nanyang Technological University, Singapore. Before joining HUST, Dr. Wu was an Associate Professor in Microsystems Technology, Uppsala University, Sweden. His research focused on microscale mixing, flow visualization, soft material-based microprocessing technology and microfluidic devices and systems, especially cell separation chip design and development, as well as stretchable electronic systems.

Preface: Technologies Reshaping our Daily Lives

Electronics are deeply embedded in every corner of our daily lives, be it in factories, offices, apartments or elsewhere. Now they are morphing from a hard and cold machine to a soft and warm intelligence, such as various wearable electronics, which are becoming an essential part of our lives and reshaping our behavior mode and living style. Among them, flexible and stretchable electronics are pivotal from a technical perspective. However, when transiting from the stiff and hard to a flexible and stretchable system, we are facing various technical challenges ranging from general strategy, materials, mechanics, and fabrication processes, through to their applications. Hence, in collaboration with our two colleagues from Korea, and the International Symposium of Flexible and Stretchable Electronics 2016, Wuhan, Micromachines, organized a focus theme issue—Flexible and Stretchable Electronics.

We are happy to see the strong interest expressed by the community on this topic and in this book share selected high-quality papers with you. The Special Issue currently includes 12 papers, 1 editorial, 7 original articles and 4 review papers.

In the editorial, two prestigious scholars, Prof. Yonggang Huang from Northwestern University, USA, and Prof. Yibing Cheng from Monash University, Australia, two rising stars, Dr. Dae-Hyeong Kim from Seoul National University, and Xuanhe Zhao from Massachusetts Institute of Technology, USA, and an excellent researcher, Prof. Haixia Zhang from Peking University, China, shared their eminent opinions, common concerns and perspectives on the opportunities and challenges in flexible and stretchable electronics.

For flexible electronics, Lee, P. et al. demonstrated a way to rapid and effective improvement of the conductance of the Ag nanowired-based conductor by using a laser-induced nano-welding process, while Lee, H. et al. focused on the laser sintering technique for large areal processing using a spatially focused beam. In Zhou, P. et al.'s work, an ultrasonic spray coating was developed for TiO_2 nanoparticles on polymer substrates for the fabrication of flexible perovskite solar cells. In another research article, Lan et al. presented a new and improved approach to the rapid and green fabrication of highly conductive microscale silver structures on low-cost transparent polyethylene terephthalate flexible substrates.

On the subject of stretchable electronics, Jun et al. reported uni- or bi-axially wrinkled graphene–silver nanowire hybrid electrodes comprised of chemical vapor deposition (CVD)-grown graphene and silver nanowires, while Kim et al. presented NH_3 sensing with ultra-low energy consumption for fast recovery and a graphene sheet based on a suspended microheater. In Cui et al.'s work, a one-dimensional analytic thermal model is developed for flexible electronic devices interacting with human skin.

In addition, four in-depth review articles were included: from perspectives of mechanisms and materials, Zhao and Huang further discussed fabrication techniques and representative applications of flexible and stretchable skin sensors. In Luo et al.'s article, the focus was on flexible transparent electrodes based on carbon and metallic materials looking at fabrication techniques, performance improvement, and various applications. Yu et al. discussed several strategies for elastomer-based stretchable conductors. In the last review article, Wang and Liu discussed recent advances in liquid alloy-based flexible and stretchable printed electronics.

In conclusion, research on flexible and stretchable electronics to date has attracted much attention and tremendous advances have been achieved, although it is developing further all the time. We look forward to more innovative work on fabrication development, system integration and diverse applications, which will no doubt become integrated into our daily lives in the near future.

Zhigang Wu

Special Issue Editor

micromachines

MDPI

Editorial

Opportunities and Challenges in Flexible and Stretchable Electronics: A Panel Discussion at ISFSE2016

Zhigang Wu [1,2,*], Yongan Huang [1,2] and Rong Chen [1,2]

[1] State Key Laboratory of Digital Manufacturing Equipment and Technology, Huazhong University of Science and Technology, Wuhan 430074, China; yahuang@hust.edu.cn (Y.H.); rongchen@mail.hust.edu.cn (R.C.)

[2] Flexible Electronics Research Center, Huazhong University of Science and Technology, Wuhan 430074, China

* Correspondence: zgwu@hust.edu.cn; Tel.: +86-27-8754-4054

Received: 10 April 2017; Accepted: 12 April 2017; Published: 18 April 2017

The 2016 International Symposium of Flexible and Stretchable Electronics (ISFSE2016), co-sponsored by the Flexible Electronics Research Center, Huazhong University of Science and Technology (HUST) & State Key Laboratory of Digital Manufacturing and Equipment Technology, National Natural Science and Engineering (NSFC), was successfully held in Wuhan, China, 29–30 June 2016.

A panel of five scholars of international standing led the panel discussion at the conference on important and timely topics including reliability or new functions of flexible and stretchable electronics, advanced materials, flexible electronics manufacturing, ways to link flexible electronics to medical applications, and the roles of organic and inorganic electronics in flexible electronics.

Dr. Yonggang Huang is the Walter P. Murphy Professor of Mechanical Engineering, Civil and Environmental Engineering, and Materials Science and Engineering at Northwestern University. He is interested in the mechanics of stretchable and flexible electronics, and 3D fabrication of complex materials and structures. He is a member of the National Academy of Engineering, USA.

Dr. Yibing Cheng is a professor of Materials Science and Engineering at Monash University. He specializes in inorganic materials. He has particular interests in the development of solution processed solar cells, especially by printing. He is a fellow of the Australian Academy of Technological Sciences and Engineering.

Dr. Dae-Hyeong Kim is an associate professor of chemical and biological engineering at Seoul National University. His research aims to develop technologies for high-performance flexible and stretchable electronic devices using high-quality single crystal inorganic materials and novel biocompatible materials that enable a new generation of implantable biomedical systems with novel capabilities and increased performance.

Dr. Xuanhe Zhao is an associate professor and Noyce Career Development Chair in the Department of Mechanical Engineering, MIT. His current research goal is to understand and design soft materials that possess unprecedented properties and functions. Dr. Zhao is a recipient of the NSF CAREER Award, the ONR Young Investigator Award, and the Early Career Researchers Award from AVS Biomaterial Interfaces Division.

Dr. Haixia Zhang is a professor in the Institute of Microelectronics, Peking University. Her research fields include micro/nano energy harvesting technologies, self-powering systems and active sensors. She has published more than 100 papers in prominent journals, six books/book chapters and 30 patents on micro/nanotechnology.

1. What are the key challenges in flexible electronics? What is the niche market and the killer application for stretchable electronics?

Huang: One killer application of flexible or stretchable electronics is in medicine. Dae-Hyeong, you have done a lot of work in this area, can you make some comments?

Kim: I totally agree with Prof. Huang. The limitation of flexible and stretchable electronics is that the device performance may be lower than conventional rigid electronics. Flexible and stretchable electronics may not be able to compete with rigid electronics in device performances. When we change the substrate from the rigid silicon wafer to plastics, the device performance would be decreased significantly. Therefore, what we need to do is to find out new markets and applications, such as novel medical systems based on flexible electronics. In medical applications, the device should be human-friendly. Individual organs and/or tissues are quite soft, and their shape is curvilinear. So, the device should also be deformable to conform to these biological systems, which is a key property of flexible and stretchable electronics.

Zhao: I fully agree with Yonggang and Dae-Hyeong's points. In the health care industry, I think flexible and stretchable electronics indeed has a niche market. I also want to add some additional points. Flexible and stretchable electronics may address some critical issues in this aging society, in addition to health care; for example, monitoring the well-being of senior people. We will probably not pay a thousand dollars or more to buy a flexible cell phone. But if the cell phone were able to be used as a very comfortable device for senior people and were able to monitor many of their vital signals, we might buy it. Another potential application is in education. We now learn many new things through cell phones. But it would be better if we could have a more conformal and natural way to receive different types of information.

Huang: I totally agree with your points on health care. But I do not really follow your comment on education. I am not sure I understand it yet.

Zhao: Books and rigid devices such as tablet computers have allowed us to learn new information. Flexible electronics, for example, flexible goggles, and different types of virtual reality devices that are more flexible and more conformable, may even do better in this regard. Learning is no longer limited to the classroom; learning is everywhere. I think this kind of conformal devices that can be very comfortably attached to your body may lead to innovative ways of learning and education.

Huang: I fully agree with what you have just said.

Zhang: I want to add some comments here again. So, regarding health care, I agree, we should certainly pay more attention to the wearable market. Another hot topic is artificial intelligence (AI). Recently, AI has become very popular everywhere—such as in famous human-like robots that resemble the most beautiful ladies—and draws a lot of attention. So, if we see this as the future of AI, the future of robots, then soft electronics and conformal electronics are exactly what is needed. If we can put all these on top of our skin, they can detect not only temperature but also many other parameters. Then, these robots will be much smaller in many conditions. I think we should pay attention to these advanced technologies.

2.   Is polydimethylsiloxane (PDMS) the best carrier? Or hydrogel? What is the area of greatest potential for hydrogel?

Huang: I do not think there is one best carrier for everything. There are so many substrates that we have used, such as PDMS, Ecoflex®, and Silbione®, depending on the applications. I do not think we need to identify one single material that suits all the purposes. For example, the elastic modulus of Ecoflex matches well with that of the skin, and is therefore useful for epidermal applications. For some other applications, PDMS may be a good choice.

Zhao: I agree with Yonggang. The choice of material is application-dependent, especially when considering integrating devices with the human body. Different parts of human bodies have different properties, so this is a material design and system design issue. PDMS has been widely used in flexible electronics, while hydrogels have broad applications in biomedicine. But I do not think there is a best material; it will depend on the application.

Huang: Dae-Hyeong uses one special type of substrate, silk, for bio-integrated electronics, which can dissolve inside the human body. This cannot be achieved by PDMS. Therefore, different applications require different substrates.

3.   Shall we focus more on the reliability or new functions?

Cheng: I think it very much depends on what your interest is. If you are interested in applications, I think the reliability is probably more important for consideration. Many excellent research works about different flexible or stretchable electronics have already been reported, but most of these works have not been accepted by the market as commercial products, because most of them failed in reliability or in functionality. For example, I just talked about perovskite solar cells, showing 20% efficiency. But their long-term performance is poor, currently lasting for two weeks or a month. So, from that point of view, I think it would be a huge contribution to the whole field if we can improve the reliability of one or two research outcomes and make them really accepted by the market. However, if you are a young researcher, just coming into this field, and you want to do something exciting or achieve good publications, then focusing on reliability may not be your best choice. This is because the study of reliability is time-consuming and may be difficult to produce many journal papers with very high impact factor. So, if you really want to publish in Nature or Science, you would probably prefer to work on something that is new and more exciting.

Zhang: From my experience in my field, MEMS (Micro-Electro-Mechanical System) is probably not a good field for publication. In most MEMS journals, the impact factor is pretty low, for example, below 2. So, that presents many issues for students, because they want to graduate with very good records. Especially in China, a lot of university academics now asking for high impact publications. So, if you are a student, I strongly suggest that you work on the functions and innovations, and offer some new innovations for good publications. That is not only to make your resume look impressive, it is also very good educational training. At a very young age, you should find something really exciting, not invest your time in something reliable. Reliability research is not for the young students in the lab. So I suggest, if you are working in the lab, and still want to continue academia career as a professor, you should try your best to make something new, try to make something really exciting. After many years, you may move into a big production market, and then you will be able to hire a bunch of people to work on the reliability. I think that is the strategy.

Zhao: For a wide range of applications of flexible and stretchable electronics, reliability is a very import issue. The first light bulb invented by Thomas Edison only lasted a few hours, which probably would not be widely adopted in the society. Edison and his team further improved the design so that it could last over a thousand hours, and then the widespread applications of light bulks made a major impact on our society. For the field of flexible and stretchable electronics, we do need to invent new functions and applications. At the same time, especially for translational and industrial applications, we also need to pay attention to the reliability.

Kim: For students, I think that innovation should be emphasized. It is true that reliability is always important. For example, all the medical devices that we are working on always need to be highly reliable. But I personally think that new ideas and innovations are more important, as we are conducting research at the university. Of course, if you are working at a company, reliability is very important. However, we are working on new ideas, and creating new frontiers and new innovations. So, students should be working more on new, innovative functions.

4. Is it of interest to actuating technology? For what kind of applications?

Kim: Ok, usually I work on soft medical devices, and it is related to sensors and actuators. For the sensing purpose, sometimes we need actuators. If you just want to measure the electrical features, an amplifier might be good enough. But if you want to measure some mechanical properties of specific tissues or organs, then appropriate actuation gives us better sensing results. We need to combine sensors and actuators for a better quality diagnosis. Meanwhile, I think that in the future soft robotics will be a hot field. In the past, robots were rigid. But in the future, we imagine soft, human-like robots. And in that case, we will probably need soft actuators.

Zhang: Actually, it is very important for smart systems. Now, it is very popular to design mini robots to put inside the body, or inject into the body to make detections and perform some surgeries. For that purpose, robots should have internal actuators. I have not worked on these actuators, but I do think they are a very important field and research direction.

Kim: I have something to add. Actuators are important in medical systems. For example, the brain and heart are organs operated by electrical signals. And electrical impulses and stimulations can be used to treat many diseases related to the brain and heart. Also, drug delivery should be controlled by appropriate actuators for controlled drug release.

5. What are the unsolved problems in flexible electronics manufacturing?

Huang: Flexible electronics manufacturing is in its infancy, and there are numerous unsolved problems in this area. In fact, the United States government investigated $70M to form a center on the manufacturing of flexible electronics.

Zhao: Manufacturing of flexible electronics requires different techniques. Is 3D printing a possible technique for the fabrication of flexible and stretchable electronics?

Huang: I will answer this. 3D printing can print polymers and some metals, but it can never print single crystal silicon or other important materials for electronics.

Zhao: How about 3D printing plus transferring?

Huang: This is an interesting idea. Would you like to explain a bit more?

Zhang: 3D printing is good but it is not suitable for mass production. So, if we want to make electronics that are flexible, we must use this kind of traditional technology and basic tools. Therefore, I agree with Yonggang's comments; it is important that fabrication technology tries to use all these existing successful technologies and makes something based on them. This will be easy for mass production and for the actual industry.

Huang: It is important to take advantage of the existing, mature semiconductor fab to develop inorganic, flexible and stretchable electronics.

Cheng: I think printing technologies could be used for flexible electronics manufacturing. 3D printing has potential in the flexible electronics field, such as for health-related products. Many health-related flexible products are associated with individual people. So, for these kinds of products, 3D printing is probably quite suitable. This is a quite new field and we do not need to restrict our minds and initiatives. I guess the point is that while recognizing the existing technologies, such as silicon technology, we should not be afraid of trying and integrating new things.

6.   How can we link flexible electronics to medical applications?

Kim: I think that, at least, we need to develop a completely new device that can solve critical issues which conventional devices cannot solve.

Huang: Or you can make medical electronics flexible so that they can be applied at home.

Kim: Yes, I think that such applications are good candidate markets. We need to persuade medical doctors and patients to use our devices. To do that, the performance of the flexible medical device should be much better than existing commercial technologies. In the medical area, there are many diseases that cannot be cured with existing technologies. In that case, we can create new devices with unconventional functions to address those unmet needs.

Huang: Thank you. In this development, it is really important to work with medical doctors.

Zhang: I think something must be pointed out. My thought is that flexible electronics has great potential for Chinese medicine. Normally, diagnoses are taken by very old doctors. They always follow four ways of diagnosis: look, listen, question and feel the pulse, which are based on their experiences. So, if we can put flexible electronics on the body, if we can monitor these parameters, then a lot of data could be recorded and integrated with these old doctors' experiences. I think that can solve a big problem for Chinese medicine. I think for young students, if they have the chance, they should focus more on these applications. They have a really big future.

Kim: I want to add one point. To translate flexible devices to medical applications, we need to publish articles in medical journals. Although these devices are for general people, the person who needs to decide whether that device can be applied to patients is a medical doctor. And through medical journals, we can let medical doctors know the advantages of our new medical device technology.

Zhao: I want to add another point, regarding the patent. The protection of intellectual property is extremely important in this field. Once a technology is mature enough, we need to consider patenting it. So, when doctors and companies approach you, you or your team will have the intellectual property to proceed with the commercialization of the technology.

7.   What are the roles of organic electronics or inorganic electronics in the field of flexible electronics?

Cheng: In the past, many functional devices were made of inorganic electronics. But in the last ten or fifteen years, there has been an increasing amount of new polymers and new organic materials synthesized and applied to functional electronics. I think there is a catch here. If you favor flexible devices, then you will consider your substrate to be normally of polymer or organic material. If you want to work on organic substrates, then in many cases, it is often a challenge to make devices on organic substrates and some new manufacturing techniques must be developed. For example, you cannot apply high temperature with polymer substrates, which in some cases restricted the application of inorganic materials. Both organic and inorganic materials are important, depending on their functions. But in many cases, they are restricted by processing technology. You may wish to use inorganic materials for flexible devices, however you may not have a suitable processing technology. If new technologies, such as the femto-second laser, can be used, then you can avoid high temperature heating, which can be applied to many organic substrates. As a result, many previously unthinkable inorganic materials may be applicable to polymer substrates.

Huang: Organic and inorganic materials both have their niche applications. Flexible display is an excellent example of organic electronics because one's eyes cannot react in less than 0.1 second anyway, such that the relatively low charge mobility for organics would not be an issue. It is important to develop both organic and inorganic flexible electronics.

8.   Is it possible to utilize the roll to roll technology in stretchable electronics? How?

Cheng: I think roll to roll is an ideal, fast and low-cost technology. However, roll to roll is very much restricted by the substrates. For example, we cannot use roll to roll to print on glass. Of course, on flexible glass maybe, but it is very fragile. Now, if you work on stretchable devices, then that is a great challenge because stretchable means that during the roll to roll processing, materials may be deformed, which would cause enormous problems in roll to roll processing.

Zhang: Just as Prof. Cheng said, this really depends on the material. Another point for roll to roll, is how to make multi-layers. If we make several layers, then roll to roll has a big issue with mismatch; you could not achieve very good results. So, I think it depends on the device and structural design.

Cheng: What is the purpose of roll to roll technology? Roll to roll can be fast, but today, it is not the only technology to achieve fast production and low-cost. So, I don't think that roll to roll technology is an objective; it is only a process. You have many other choices as well. For example, I think that stretchable electronics may not need roll to roll technology in order to make the device cheaper. It is not necessary.

micromachines

MDPI

Article

Rapid and Effective Electrical Conductivity Improvement of the Ag NW-Based Conductor by Using the Laser-Induced Nano-Welding Process

Phillip Lee [1,†], Jinhyeong Kwon [2,†], Jinhwan Lee [2], Habeom Lee [2], Young D. Suh [2], Sukjoon Hong [3,*] and Junyeob Yeo [4,*]

[1] Photo-Electronic Hybrids Research Center, National Agenda Research Division, Korea Institute of Science and Technology (KIST), 5 Hwarang-ro 14-gil, Seongbuk-gu, Seoul 02792, Korea; phillip@kist.re.kr

[2] Applied Nano and Thermal Science (ANTS) Lab, Department of Mechanical Engineering, Seoul National University, 1 Gwanak-ro, Gwanak-gu, Seoul 00826, Korea; jhs0909k@snu.ac.kr (J.K.); mir.ljh@gmail.com (J.L.); habeom.lee@snu.ac.kr (H.L.); youngduksuh@gmail.com (Y.D.S.)

[3] Department of Mechanical Engineering, Hanyang University, 55 Hanyangdaehak-ro, Sangnok-gu, Ansan, Gyeonggi-do 15588, Korea

[4] Novel Applied Nano Optics (NANO) Lab, Department of Physics, Kyungpook National University, 80 Daehak-ro, Bukgu, Daegu 41566, Korea

* Correspondence: sukjoonhong@hanyang.ac.kr (S.H.); junyeob@knu.ac.kr (J.Y.); Tel.: +82-31-400-5249 (S.H.); +82-53-950-7360 (J.Y.)

† These authors contributed equally to this work.

Academic Editor: Zhigang Wu
Received: 2 April 2017; Accepted: 16 May 2017; Published: 19 May 2017

Abstract: To date, the silver nanowire-based conductor has been widely used for flexible/stretchable electronics due to its several advantages. The optical nanowire annealing process has also received interest as an alternative annealing process to the Ag nanowire (NW)-based conductor. In this study, we present an analytical investigation on the phenomena of the Ag NWs' junction and welding properties under laser exposure. The two different laser-induced welding processes (nanosecond (ns) pulse laser-induced nano-welding (LINW) and continuous wave (cw) scanning LINW) are applied to the Ag NW percolation networks. The Ag NWs are selectively melted and merged at the junction of Ag NWs under very short laser exposure; these results are confirmed by scanning electron microscope (SEM), focused-ion beam (FIB), electrical measurement, and finite difference time domain (FDTD) simulation.

Keywords: laser; laser-induced nano-welding; pulse laser; silver nanowires; silver nanowire percolation networks; transparent flexible conductor

1. Introduction

For almost two decades, the flexible transparent conductor and stretchable conductor have received huge interest from many researchers and industry professionals. Since flexible electronics and stretchable sensors are core devices in wearable computer systems, as a next generation electronics platform, manufacturing/fabrication process technologies as well as flexible electronics compatible materials will be more important in the future.

Meanwhile, among the materials for the transparent conductor, fluorine-doped tin oxide (FTO) thin film is the most popular and famous material in research and industry fields. However, FTO thin film is not an appropriate material for the flexible transparent conductor since it is usually delicate and brittle. Hence, instead of FTO thin film, alternative conducting nanomaterials such as carbon nanotube (CNT) [1,2], graphene [3–5], metal nanoparticle (NP) mesh [6–9], metal nanowires (NWs) [10–19], and metal nano-thin film [20,21] are used in the research field for the flexible transparent conductor and stretchable conductor.

Among the nanomaterials for flexible electronics, silver (Ag) NWs have been widely used as a flexible transparent conductor [10–13,16,22–24] and stretchable conductor [15,17,25] due to various advantages such as high electrical conductivity, high transparency, high ductility and simple fabrication process methods. However, a post thermal annealing process is usually required to increase electrical conductivity since Ag NWs are usually too short (up to ~100 μm) to cover the wide area and are generally covered by capping polymer such as polyvinylpyrrolidone (PVP) which hinders electrical conductivity. The conventional thermal annealing process, such as a hot plate and convection oven, is simple and easily applicable to the Ag NW percolation networks for electrical conductivity improvement, while the thermal annealing process has disadvantages such as oxidation problems and long processing time. In particular, the thermal annealing process is not suitable to the flexible polymer substrate due to the low melting temperature of the polymer. Thus, it is important to conduct a post thermal annealing process with low temperature (below 200 °C) to prevent the damages of the flexible polymer substrate.

Recently, the optical NWs annealing process [14,15,18,19,26–28] was introduced to anneal metal NW percolation networks for the improvement of electrical conductivity. Compared to the conventional thermal annealing process, since optical energy in the optical NWs annealing process is irradiated to the sample over a very short time by a ultraviolet (UV) lamp [14], flash light [26] or scanning laser [18], the optical NWs annealing process is very rapid and suitable for the flexible polymer substrate without any macroscopic damages or deformation of the substrate [29,30] under ambient conditions. In addition, oxidation problems on metal NWs during processing are suppressed due to fast processing [18].

Most previous research [15,18,19,26,27,31] focuses on the fabrication of flexible/stretchable electronics and their applications. However, in this study, we attempt to examine mainly the phenomena of Ag NWs' junction and welding properties when a laser is irradiated to the Ag NW percolation networks. In particular, we focus on how the local laser exposure time (from ns to μs) affects the Ag NWs' percolation networks in laser processing. Thus, we compare two different laser-induced nano-welding (LINW) processes as a post annealing process of Ag NWs: the continuous wave (cw) laser scanning system and the nanosecond (ns) pulse laser system. Although the processing mechanism of two LINW processes is basically the same in terms of using laser energy, there are several similar and different results, which are mentioned in the text.

2. Methods

2.1. Experimental Procedure

Figure 1a shows the preparation of an Ag NW-based conductor sample. In order to prepare a film of Ag NW percolation networks on the substrate, Ag NWs are deposited according to the following procedures. Firstly, Ag NWs are synthesized in a solution via the polyol synthesis method [32]. Afterwards, synthesized Ag NWs are filtered out onto the Teflon filter and transferred onto the glass substrate successively. The diameter of the transferred Ag NW percolation networks is ~36 mm due to the size of the Teflon filter. Since transferred Ag NW percolation networks on the glass substrate are weakly bound, the post thermal annealing process, such as a hot plate and furnace, is required to increase the electrical conductivity of the Ag NW percolation networks. Additionally, the LINW process is conducted to the Ag NW percolation networks as an alternative post annealing process for comparison.

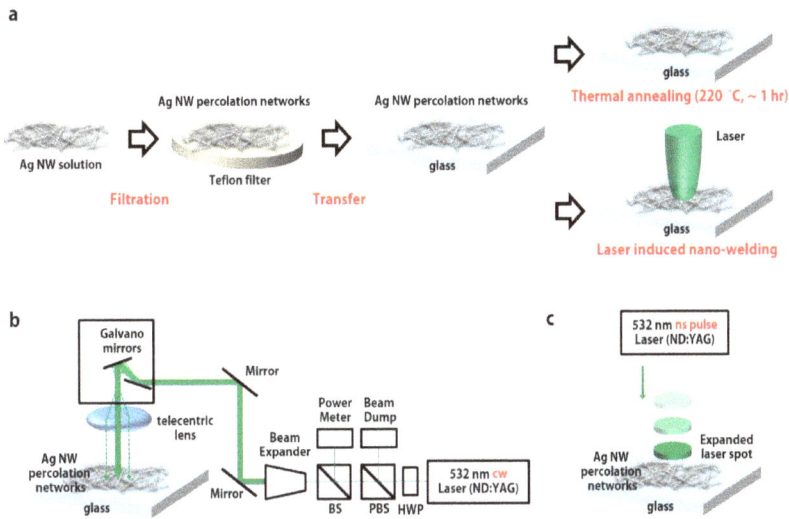

Figure 1. Schematic diagram of the thermal annealing process and laser-induced nano-welding process for Ag nanowires (NWs) percolation networks. (**a**) Sample preparation flow chart. Firstly, Ag NWs are synthesized in a solution by polyol synthesis. Afterwards, they are filtered on the Teflon filter and transferred onto the glass substrate to form Ag NW percolation networks on glass substrate. Finally, the thermal annealing process or laser-induced welding process is applied to Ag NW percolation networks for the improvement of electrical conductivity. The laser-induced nano-welding process through (**b**) the continuous wave (cw) laser scanning system and (**c**) the ns pulse laser system.

2.2. Optical Setup

As we mentioned, two different LINW processes (the cw laser scanning system and the ns pulse laser system) are compared in this study. Figure 1b shows the cw laser scanning system which is combined with Galvano scanning mirrors and a telecentric lens [30,33]. In the cw laser scanning system, a 532 nm cw laser (Millennia 5W, Spectra Physics, Santa Clara, CA, USA) is used on the sample. As shown in Figure 1b, the power of the emitted laser beam is easily controlled through the half wave plate (HWP) and polarized beam splitter (PBS). The beam expander is placed afterwards to enlarge the laser beam for a flat wavefront of laser beam. The angle of the laser beam is deviated by a laser scanner (HurryScan II, Scanlab, Puchheim, Germany) which consists of two electrically driven Galvano mirrors. Afterwards, the laser beam is uniformly focused (10 μm) on the 2D focal plane, without distortion aberration, by a long focal distance f-theta lens (f = 103 mm). The prepared sample that consists of Ag NW percolation networks is placed on the focal plane of the f-theta lens. Figure 1c shows the ns pulse laser system which consists of a 532 nm ns pulse laser (Tempest 300, NewWave, Redwood City, CA, USA) and beam expander. The pulse duration and the repetition rate of the ns laser are 5 ns and 10 Hz, respectively. Since the energy of the applied ns pulse laser is extremely high, only one single shot with proper energy density (mJ/cm^2) is enough for the enhancement of electrical conductivity in Ag NW percolation networks. However, excessive laser energy can ablate Ag NW percolation networks, thus the energy density in the ns pulse laser and the power density (W/cm^2) in the cw laser are carefully controlled by adjustment of the beam waist area through the beam expander and power adjustment, respectively.

2.3. Laser Processing

Firstly, the prepared sample is placed at the focal plane of the applied laser. Afterwards, the laser is irradiated on the sample. In the cw scanning LINW process, the diameter of the prepared sample is ~36 mm. Additionally, the spot size of the focused laser, the laser scanning speed, and the pitch of scanning are 10 μm, 100 mm/s, and 10 μm, respectively. Since the total processing time in the cw scanning LINW process is dependent on sample size and laser scanning speed, fast laser scanning speed is desirable to reduce the processing time. Meanwhile, the laser dwell time, τ ($\tau = 2W_0/v$; τ, W_0, and v are dwell time, beam waist, and scanning speed, respectively), is decreased as the scanning speed increases. Thus, laser energy density (power density by dwell time) and total processing time are in a trade-off relationship, and a laser scanning speed of 100 mm/s is chosen as an optimum scanning speed in this study.

In the case of the ns pulse LINW process, the laser beam is expanded by the beam expander to reduce laser energy density (energy per unit area), since the pulse laser energy is sufficiently high (see Figure 1c). The prepared sample size of Ag NW percolation networks is 6 mm by 6 mm in the ns pulse LINW process, and the extended laser beam illuminates and covers the entire prepared sample area (see Figure S1 in the Supplementary Information).

3. Results and Discussion

Figure 2 shows a photographic image (Figure 2a) and scanning electron microscope (SEM) images (Figure 2b,c) of Ag NW percolation networks after the LINW process. In the case of the conventional thermal annealing process, such as a hot plate [15] and convection oven, the thermal annealing process is often conducted at a low temperature (e.g., below 250 °C) on the flexible substrate due to the low melting temperature of the polymer substrate. Therefore, it is difficult to find meaningful differences in the SEM images of Ag NW percolation networks after the thermal annealing process compared to before the thermal annealing process at a low temperature (below 250 °C).

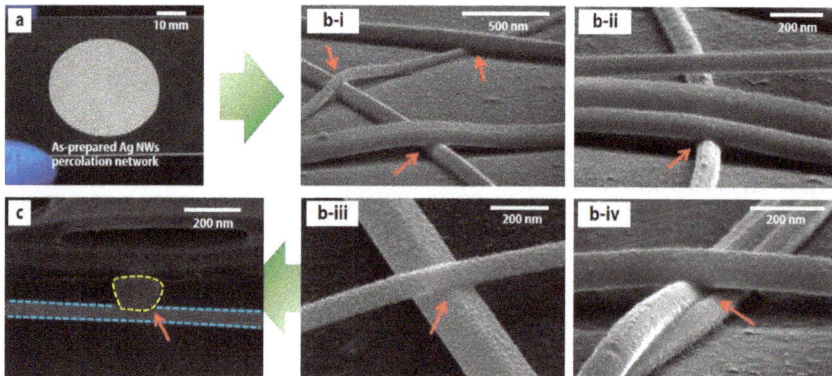

Figure 2. (**a**) Photographic image of the Ag NW percolation networks on the glass substrate (30% transmittance). (**b**) Magnified scanning electron microscope (SEM) images of the Ag NW percolation networks after the laser-induced welding process. (**c**) Cross-sectional SEM image of the junction of the Ag NW percolation networks after the laser-induced welding process. Two Ag NWs are melted and merged at the boundary of two Ag NWs (yellow dots and blue dots). Red arrows represent a junction of Ag NWs in (**b**,**c**).

However, in the case of the LINW process (same for both ns pulse LINW and cw scanning LINW) under ambient conditions (room temperature and atmospheric pressure), melted and merged Ag NWs are found in SEM images after the LINW process, as shown in Figure 2b-i–2b-iv. Since the irradiated laser energy is intensively absorbed to heat up Ag NWs and the laser irradiation time is extremely short (from ns to μs), crossed Ag NWs are only melted and merged at the junction, without damaging the other area of Ag NWs, as shown in Figure 2b.

In order to verify the melting at the junction of Ag NWs, a cross-sectional SEM image at the boundary which is cut by a focused-ion beam (FIB), is examined. It is confirmed that the crossed area of two Ag NWs (yellow and blue dotted lines) are melted and fused at the junction, as shown in Figure 2c. The results for the flexible transparent conductor and stretchable conductor are very noticeable, since melted and merged Ag NW percolation networks will have better electrical/mechanical properties such as electrical conductivity, mechanical elongation, and mechanical strength [13,15,18,19].

Figure 3 shows sheet resistance changes at various times under the thermal annealing process and ns pulse LINW process. The transmittance of prepared Ag NW percolation networks in (a) and (b) are 95% and 96%, respectively. As shown in Figure 3a, the sheet resistance of Ag NW percolation networks is gradually increased and gently dropped below 20 Ω/sq with long processing time (over 1 h) under the thermal annealing process. At first, the sheet resistance is gradually increased due to oxidation formation and the resistance increase of Ag NWs to temperature change, according to their temperature coefficient of resistance [34,35]. Once the temperature of the hot plate reaches 220 °C (~1300 s), the sheet resistance starts to drop due to slight melting at the junction of Ag NW percolation networks. The thermal annealing process ensures stable low sheet resistance in Ag NW percolation networks and easily scales up the sample size. However, long processing time (over 1 h) is generally required to increase the electrical conductivity in the thermal annealing process, since a period of warm-up time for heating is required and only low temperature (below 250 °C) is available for the flexible substrate.

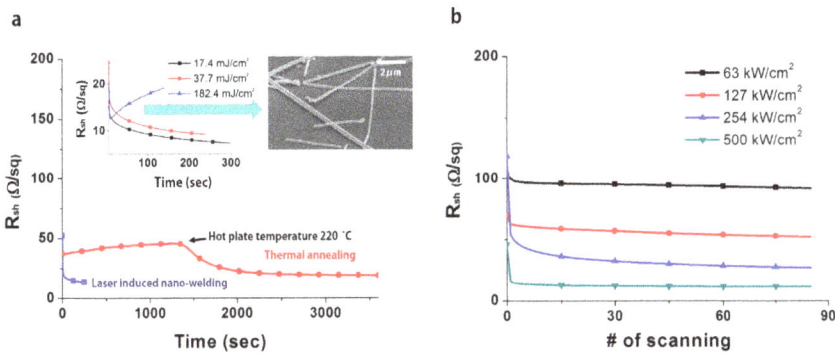

Figure 3. (a) Sheet resistance changes at various times in the thermal annealing process and ns pulse laser-induced nano-welding (LINW) process. The inset graph shows a magnified view of the ns pulse LINW process. SEM images show the laser ablation results of Ag NW percolation networks. (b) The sheet resistance changes with various numbers of laser scans and various laser power levels in the cw scanning LINW process.

In contrast, the sheet resistance of Ag NW percolation networks drops rapidly in the ns pulse LINW process compared to the thermal annealing process. As shown in the inset graph of Figure 3a, the sheet resistance drops immediately after the start of pulse laser exposure with proper energy density (17.4 mJ/cm^2 and 37.7 mJ/cm^2). Even though a single pulse can be enough to improve the electrical conductivity of Ag NW percolation networks, continued laser pulses (10 Hz repetition rate) are employed in the sample to further improve the conductivity. However, extremely high laser energy density (182.4 mJ/cm^2) can ablate/destroy Ag NW percolation networks (SEM image of Figure 3a), resulting in an increase of the sheet resistance of Ag NWs during the laser exposure—so-called "rebound", as shown in the inset graph (light blue line) of Figure 3a. Thus, in Ag NW percolation networks in the ns pulse LINW process, moderate adjustment of the applied laser energy is required.

Similar to the ns pulse LINW process, in the case of the cw scanning LINW process, the sheet resistance of Ag NW percolation networks drops rapidly below 20 Ω/sq with high power density

(500 kW/cm^2), as shown in Figure 3b. The sheet resistance decreases slightly as the number of scans increases, while the sheet resistance drops considerably with high power density. The spot size (2W_0) of the focused laser by the telecentric lens in laser scanning system is ~10 µm. Additionally, the dwell time is 10^{-4} s (10 µm/100 mms^{-1}), thus it is also a very short time compared to the conventional thermal annealing process.

As a result, the LINW process is a more rapid and effective process (due to melting and merging) for improving the electrical conductivity of Ag NW percolation networks than the conventional thermal annealing process. Moreover, it is noticeable that there are no significant differences with respect to the different laser exposure time (from ns to µs) in laser processing. These results are confirmed by SEM images and electrical conductivity measurements. In addition, since the reaction in Ag NWs welding is conducted within an extremely short time—5 ns laser exposure time—this LINW process can be applied to the flexible substrate without any macroscopic damages or deformation of the substrate.

Figure 4 shows SEM images of ablated Ag NWs when the excessive laser energy is applied to the Ag NW percolation networks. Two different LINW processes show fairly different results at excessive laser energy density. Since an extended laser spot is applied to the Ag NW percolation networks in the ns pulse LINW process, entire Ag NWs are heated and melted, resulting in a queue of molten silver micro dots (right SEM images in Figure 4a).

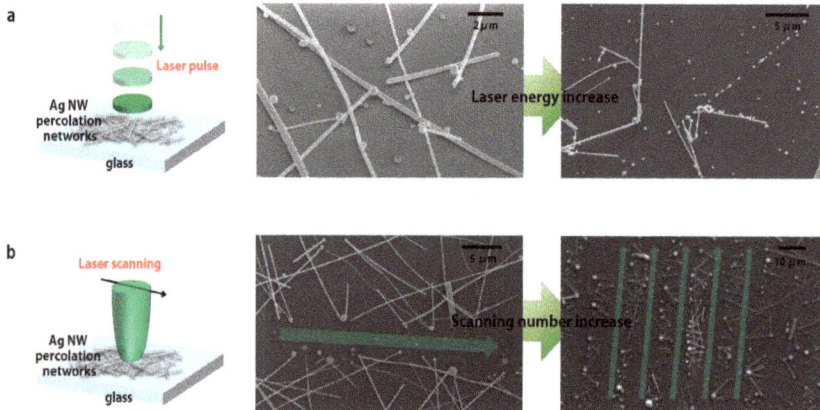

Figure 4. SEM images of Ag NW percolation networks when (**a**) the ns pulse LINW process and (**b**) the cw scanning LINW process are conducted under extremely high power/energy condition.

On the other hand, Ag NWs are selectively melted along the laser scanning direction in the cw scanning LINW process, as shown in Figure 4b (green arrows in SEM images). Since Ag NWs are ablated along the laser scanning direction, the remaining Ag NWs are locally connected to each other, thus this laser ablation technique is applied to fabricate patterned Ag NW mesh for flexible capacitive touch sensors [31].

As shown in previous SEM images, Ag NWs are easily heated and melted in the LINW process when the proper laser energy is irradiated to the Ag NW percolation networks. It is well known that the electromagnetic field enhancements on the surface of Ag NW generate localized thermal heating due to surface plasmon polaritons (SPP) on the surface of Ag NWs [18,36–39]. This behavior can be seen in finite difference time domain (FDTD) simulation (Lumerical), as shown in Figure 5. In this simulation, transverse magnetic (TM) and transverse electric (TE) modes are considered for the two crossed and stacked Ag NWs. In order to simplify the simulation, the diameter size and shape of Ag NWs are fixed at 100 nm and circle, respectively. The complex permittivity of Ag is adopted from Palik, and a simulated pulse covers the wavelength range from 300 to 1000 nm. Total-field/scattering-field (TF-SF), together with perfectly matched layer (PML) formulation, has been employed.

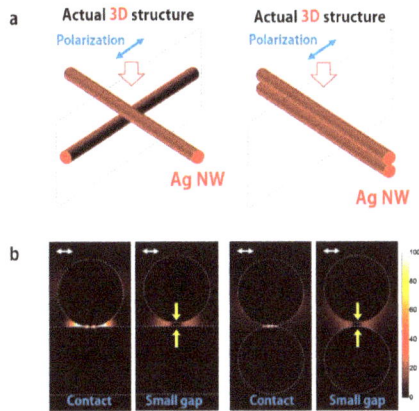

Figure 5. Finite difference time domain (FDTD) simulation at the junction of crossed Ag NWs. (**a**) Simulation layout of two crossed (left) and stacked (right) Ag NWs. (**b**) Electromagnetic field distribution at the junction with various conditions: crossed/contact, crossed/small gap, stacked/contact, stacked/small gap of two Ag NWs. The white arrow is the polarized direction of irradiated light. The yellow arrow indicates a small gap between two Ag NWs.

As shown in Figure 5b, the electromagnetic field enhancements are extremely maximized at the junction of two crossed and stacked Ag NWs. In addition, electromagnetic field enhancement are still maximized near the junction of two Ag NWs, even though two Ag NWs are separated from each other by a small gap. The optical absorption, or the volumetric heat source density generated inside the metal, is calculated by [36]

$$q\left(\overrightarrow{r}\right) = (\omega/2)\mathrm{Im}(\varepsilon)\left|E\left(\overrightarrow{r}\right)\right|^2 \tag{1}$$

As can be seen from the above equation, the optical absorption is directly proportional to the electrical field intensity, and the simulation result in Figure S2 shows that the optical absorption is concentrated at the regions where the field enhancement adjacent to the surface of the nanowire is the largest. This is the reason why Ag NWs are well melted and merged at the junction of Ag NWs, as shown in Figure 2. Additionally, these results are the reason why 5 nanoseconds is enough to improve the electrical conductivity of Ag NW percolation networks.

In summary, two different laser-induced nano-welding processes (ns pulse LINW and cw scanning LINW), as alternative post annealing processes, are investigated to enhance the electrical conductivity of an Ag NW-based conductor in this study. Thus, various phenomena of Ag NWs are examined when the laser irradiates to the Ag NW percolation networks. Through the various characterizations (SEM, FIB, and electrical measurements) and FDTD simulation, it is confirmed that there are no significant differences with respect to the different laser exposure time (from ns to μs). Additionally, the Ag NWs can be selectively melted and merged at the junction of Ag NWs within less than 5 nanoseconds laser exposure.

These results indicate that the LINW process is expected to apply to the flexible polymer substrate without any macroscopic damages or deformation of the substrate due to the rapid processing time and the effect of localized electromagnetic field enhancements. In addition, since melted and merged Ag NW percolation networks will have better electrical/mechanical properties, we expect that the LINW process will be applied to the fabrication of various flexible/stretchable electronics for better performance.

Micromachines **2017**, *8*, 164

Supplementary Materials: The following are available online at www.mdpi.com/2072-666X/8/5/164/s1. Figure S1: The sample preparation of Ag NWs percolation networks for the ns pulse (a) and cw scanning (b) LINW process. The transmittance of the sample is ~91%. Figure S2: Spatial profile of electrical field intensity and the corresponding optical power absorption.

Acknowledgments: This work is supported by the National Research Foundation of Korea (NRF) Grant funded through Basic Science Research Program (NRF-2016R1C1B1014729), Ministry of Trade Industry & Energy (MOTIE) through development program "Development of high drapability of textile type dye-sensitized solar cell materials and outdoor applications (project No. 10052064)", the research fund of Hanyang University (HY-2016-00000002097), and the ICT & Future and the R&D Convergence Program through the R&D program of MSIP/COMPA (Grant No. 2015K000216).

Author Contributions: P.L., J.K., S.H., and J.Y. designed and conducted experiments. J.L. and H.L. synthesized the silver nanowires. J.K. and Y.D.S. prepared sample and conducted various characterization. S.H. conducted FDTD simulation. P.L., J.K., S.H., and J.Y. co-wrote the manuscript.

Conflicts of Interest: The authors declare no conflict of interest.

References

1. Kaempgen, M.; Duesberg, G.S.; Roth, S. Transparent carbon nanotube coatings. *Appl. Surf. Sci.* **2005**, *252*, 425–429. [CrossRef]

2. Delozier, D.M.; Watson, K.A.; Smith, J.G.; Clancy, T.C.; Connell, J.W. Investigation of aromatic/aliphatic polyimides as dispersants for single wall carbon nanotubes. *Macromolecules* **2006**, *39*, 1731–1739. [CrossRef]

3. Eda, G.; Fanchini, G.; Chhowalla, M. Large-area ultrathin films of reduced graphene oxide as a transparent and flexible electronic material. *Nat. Nanotechnol.* **2008**, *3*, 270–274. [CrossRef] [PubMed]

4. Kim, K.S.; Zhao, Y.; Jang, H.; Lee, S.Y.; Kim, J.M.; Kim, K.S.; Ahn, J.-H.; Kim, P.; Choi, J.-Y.; Hong, B.H. Large-scale pattern growth of graphene films for stretchable transparent electrodes. *Nature* **2009**, *457*, 706–710. [CrossRef] [PubMed]

5. Zheng, Q.; Zhang, B.; Lin, X.; Shen, X.; Yousefi, N.; Huang, Z.-D.; Li, Z.; Kim, J.-K. Highly transparent and conducting ultralarge graphene oxide/single-walled carbon nanotube hybrid films produced by langmuir-blodgett assembly. *J. Mater. Chem.* **2012**, *22*, 25072–25082. [CrossRef]

6. Yeo, J.; Hong, S.; Lee, D.; Hotz, N.; Lee, M.T.; Grigoropoulos, C.P.; Ko, S.H. Next generation non-vacuum, maskless, low temperature nanoparticle ink laser digital direct metal patterning for a large area flexible electronics. *PLoS ONE* **2012**, *7*, e42315. [CrossRef] [PubMed]

7. Hong, S.; Yeo, J.; Kim, G.; Kim, D.; Lee, H.; Kwon, J.; Lee, H.; Lee, P.; Ko, S.H. Nonvacuum, maskless fabrication of a flexible metal grid transparent conductor by low-temperature selective laser sintering of nanoparticle ink. *ACS Nano* **2013**, *7*, 5024–5031. [CrossRef] [PubMed]

8. Liu, Y.-K.; Lee, M.-T. Laser direct synthesis and patterning of silver nano/microstructures on a polymer substrate. *ACS Appl. Mater. Interfaces* **2014**, *6*, 14576–14582. [CrossRef] [PubMed]

9. Tsai, S.-L.; Liu, Y.-K.; Pan, H.; Liu, C.-H.; Lee, M.-T. The coupled photothermal reaction and transport in a laser additive metal nanolayer simultaneous synthesis and pattering for flexible electronics. *Nanomaterials* **2016**, *6*, 12. [CrossRef] [PubMed]

10. Lee, J.-Y.; Connor, S.T.; Cui, Y.; Peumans, P. Solution-processed metal nanowire mesh transparent electrodes. *Nano Lett.* **2008**, *8*, 689–692. [CrossRef] [PubMed]

11. De, S.; Higgins, T.M.; Lyons, P.E.; Doherty, E.M.; Nirmalraj, P.N.; Blau, W.J.; Boland, J.J.; Coleman, J.N. Silver nanowire networks as flexible, transparent, conducting films: Extremely high DC to optical conductivity ratios. *ACS Nano* **2009**, *3*, 1767–1774. [CrossRef] [PubMed]

12. Hu, L.; Kim, H.S.; Lee, J.Y.; Peumans, P.; Cui, Y. Scalable coating and properties of transparent, flexible, silver nanowire electrodes. *ACS Nano* **2010**, *4*, 2955–2963. [CrossRef] [PubMed]

13. Lee, J.; Lee, P.; Lee, H.; Lee, D.; Lee, S.S.; Ko, S.H. Very long Ag nanowire synthesis and its application in a highly transparent, conductive and flexible metal electrode touch panel. *Nanoscale* **2012**, *4*, 6408–6414. [CrossRef] [PubMed]

14. Garnett, E.C.; Cai, W.; Cha, J.J.; Mahmood, F.; Connor, S.T.; Greyson Christoforo, M.; Cui, Y.; McGehee, M.D.; Brongersma, M.L. Self-limited plasmonic welding of silver nanowire junctions. *Nat. Mater.* **2012**, *11*, 241–249. [CrossRef] [PubMed]

15. Lee, P.; Lee, J.; Lee, H.; Yeo, J.; Hong, S.; Nam, K.H.; Lee, D.; Lee, S.S.; Ko, S.H. Highly stretchable and highly conductive metal electrode by very long metal nanowire percolation network. *Adv. Mater.* **2012**, *24*, 3326–3332. [CrossRef] [PubMed]
16. Lee, J.; Lee, P.; Lee, H.B.; Hong, S.; Lee, I.; Yeo, J.; Lee, S.S.; Kim, T.-S.; Lee, D.; Ko, S.H. Room-temperature nanosoldering of a very long metal nanowire network by conducting-polymer-assisted joining for a flexible touch-panel application. *Adv. Funct. Mater.* **2013**, *23*, 4171–4176. [CrossRef]
17. Lee, P.; Ham, J.; Lee, J.; Hong, S.; Han, S.; Suh, Y.D.; Lee, S.E.; Yeo, J.; Lee, S.S.; Lee, D.; et al. Highly stretchable or transparent conductor fabrication by a hierarchical multiscale hybrid nanocomposite. *Adv. Funct. Mater.* **2014**, *24*, 5671–5678. [CrossRef]
18. Han, S.; Hong, S.; Ham, J.; Yeo, J.; Lee, J.; Kang, B.; Lee, P.; Kwon, J.; Lee, S.S.; Yang, M.Y.; et al. Fast plasmonic laser nanowelding for a Cu-nanowire percolation network for flexible transparent conductors and stretchable electronics. *Adv. Mater.* **2014**, *26*, 5808–5814. [CrossRef] [PubMed]
19. Song, W.G.; Kwon, H.-J.; Park, J.; Yeo, J.; Kim, M.; Park, S.; Yun, S.; Kyung, K.-U.; Grigoropoulos, C.P.; Kim, S.; et al. High-performance flexible multilayer MoS$_2$ transistors on solution-based polyimide substrates. *Adv. Funct. Mater.* **2016**, *26*, 2426–2434. [CrossRef]
20. Paeng, D.; Yoo, J.-H.; Yeo, J.; Lee, D.; Kim, E.; Ko, S.H.; Grigoropoulos, C.P. Low-cost facile fabrication of flexible transparent copper electrodes by nanosecond laser ablation. *Adv. Mater.* **2015**, *27*, 2762–2767. [CrossRef] [PubMed]
21. Qi, D.; Paeng, D.; Yeo, J.; Kim, E.; Wang, L.; Chen, S.; Grigoropoulos, C.P. Time-resolved analysis of thickness-dependent dewetting and ablation of silver films upon nanosecond laser irradiation. *Appl. Phys. Lett.* **2016**, *108*, 211602. [CrossRef]
22. Lee, H.; Hong, S.; Lee, J.; Suh, Y.D.; Kwon, J.; Moon, H.; Kim, H.; Yeo, J.; Ko, S.H. Highly stretchable and transparent supercapacitor by Ag–Au core–shell nanowire network with high electrochemical stability. *ACS Appl. Mater. Interfaces* **2016**, *8*, 15449–15458. [CrossRef] [PubMed]
23. Suh, Y.D.; Jung, J.; Lee, H.; Yeo, J.; Hong, S.; Lee, P.; Lee, D.; Ko, S.H. Nanowire reinforced nanoparticle nanocomposite for highly flexible transparent electrodes: Borrowing ideas from macrocomposites in steel-wire reinforced concrete. *J. Mater. Chem. C* **2017**, *5*, 791–798. [CrossRef]
24. Moon, H.; Lee, H.; Kwon, J.; Suh, Y.D.; Kim, D.K.; Ha, I.; Yeo, J.; Hong, S.; Ko, S.H. Ag/au/polypyrrole core-shell nanowire network for transparent, stretchable and flexible supercapacitor in wearable energy devices. *Sci. Rep.* **2017**, *7*, 41981. [CrossRef] [PubMed]
25. Hong, S.; Lee, H.; Lee, J.; Kwon, J.; Han, S.; Suh, Y.D.; Cho, H.; Shin, J.; Yeo, J.; Ko, S.H. Highly stretchable and transparent metal nanowire heater for wearable electronics applications. *Adv. Mater.* **2015**, *27*, 4744–4751. [CrossRef] [PubMed]
26. Li, R.Z.; Hu, A.; Zhang, T.; Oakes, K.D. Direct writing on paper of foldable capacitive touch pads with silver nanowire inks. *ACS Appl. Mater. Interfaces* **2014**, *6*, 21721–21729. [CrossRef] [PubMed]
27. Spechler, J.A.; Nagamatsu, K.A.; Sturm, J.C.; Arnold, C.B. Improved efficiency of hybrid organic photovoltaics by pulsed laser sintering of silver nanowire network transparent electrode. *ACS Appl. Mater. Interfaces* **2015**, *7*, 10556–10562. [CrossRef] [PubMed]
28. Hong, S.; Lee, H.; Yeo, J.; Ko, S.H. Digital selective laser methods for nanomaterials: From synthesis to processing. *Nano Today* **2016**, *11*, 547–564. [CrossRef]
29. In, J.B.; Hsia, B.; Yoo, J.-H.; Hyun, S.; Carraro, C.; Maboudian, R.; Grigoropoulos, C.P. Facile fabrication of flexible all solid-state micro-supercapacitor by direct laser writing of porous carbon in polyimide. *Carbon* **2015**, *83*, 144–151. [CrossRef]
30. Yeo, J.; Kim, G.; Hong, S.; Kim, M.S.; Kim, D.; Lee, J.; Lee, H.B.; Kwon, J.; Suh, Y.D.; Kang, H.W.; et al. Flexible supercapacitor fabrication by room temperature rapid laser processing of roll-to-roll printed metal nanoparticle ink for wearable electronics application. *J. Power Sources* **2014**, *246*, 562–568. [CrossRef]
31. Hong, S.; Yeo, J.; Lee, J.; Lee, H.; Lee, P.; Lee, S.S.; Ko, S.H. Selective laser direct patterning of silver nanowire percolation network transparent conductor for capacitive touch panel. *J. Nanosci. Nanotechnol.* **2015**, *15*, 2317–2323. [CrossRef] [PubMed]
32. Lee, J.H.; Lee, P.; Lee, D.; Lee, S.S.; Ko, S.H. Large-scale synthesis and characterization of very long silver nanowires via successive multistep growth. *Cryst. Growth Des.* **2012**, *12*, 5598–5605. [CrossRef]
33. Lee, H.; Hong, S.; Kwon, J.; Suh, Y.D.; Lee, J.; Moon, H.; Yeo, J.; Ko, S.H. All-solid-state flexible supercapacitors by fast laser annealing of printed metal nanoparticle layers. *J. Mater. Chem. A* **2015**, *3*, 8339–8345. [CrossRef]

34. Rathmell, A.R.; Bergin, S.M.; Hua, Y.-L.; Li, Z.-Y.; Wiley, B.J. The growth mechanism of copper nanowires and their properties in flexible, transparent conducting films. *Adv. Mater.* **2010**, *22*, 3558–3563. [CrossRef] [PubMed]
35. Rathmell, A.R.; Wiley, B.J. The synthesis and coating of long, thin copper nanowires to make flexible, transparent conducting films on plastic substrates. *Adv. Mater.* **2011**, *23*, 4798–4803. [CrossRef] [PubMed]
36. Bell, A.P.; Fairfield, J.A.; McCarthy, E.K.; Mills, S.; Boland, J.J.; Baffou, G.; McCloskey, D. Quantitative study of the photothermal properties of metallic nanowire networks. *ACS Nano* **2015**, *9*, 5551–5558. [CrossRef] [PubMed]
37. Kang, T.; Yoon, I.; Jeon, K.-S.; Choi, W.; Lee, Y.; Seo, K.; Yoo, Y.; Park, Q.H.; Ihee, H.; Suh, Y.D.; et al. Creating well-defined hot spots for surface-enhanced raman scattering by single-crystalline noble metal nanowire pairs. *J. Phys. Chem. C* **2009**, *113*, 7492–7496. [CrossRef]
38. Prokes, S.M.; Alexson, D.A.; Glembocki, O.J.; Park, H.D.; Rendell, R.W. Effect of crossing geometry on the plasmonic behavior of dielectric core/metal sheath nanowires. *Appl. Phys. Lett.* **2009**, *94*, 093105. [CrossRef]
39. Lei, D.Y.; Aubry, A.; Maier, S.A.; Pendry, J.B. Broadband nano-focusing of light using kissing nanowires. *New J. Phys.* **2010**, 093030. [CrossRef]

micromachines

MDPI

Article

Large-Area Compatible Laser Sintering Schemes with a Spatially Extended Focused Beam

Habeom Lee [1,†], Jinhyeong Kwon [1,†], Woo Seop Shin [2], Hyeon Rack Kim [2], Jaeho Shin [1], Hyunmin Cho [1], Seungyong Han [3], Junyeob Yeo [4,*] and Sukjoon Hong [2,*]

[1]   Applied Nano and Thermal Science Lab, Department of Mechanical Engineering, Seoul National University, 1 Gwanak-ro, Gwanak-gu, Seoul 151-742, Korea; habeom.lee@snu.ac.kr (H.L.); jhs0909k@snu.ac.kr (J.K.); jayz.shin84@gmail.com (J.S.); augustinus310@snu.ac.kr (H.C.)
[2]   Department of Mechanical Engineering, Hanyang University, 55 Hanyangdaehak-ro, Sangnok-gu, Ansan Gyeonggi-do 15588, Korea; caribou11@hanyang.ac.kr (W.S.S.); kihll1004@hanyang.ac.kr (H.R.K.)
[3]   Department of Mechanical Engineering, Ajou University, San 5, Woncheon-Dong, Yeongtong-Gu, Suwon 16499, Korea; sy84han@ajou.ac.kr
[4]   Novel Applied Nano Optics (NANO) Lab, Department of Physics, Kyungpook National University, 80 Daehak-ro, Bukgu, Daegu 41566, Korea
*   Correspondence: junyeob@knu.ac.kr (J.Y.); sukjoonhong@hanyang.ac.kr (S.H.); Tel.: +82-53-950-7360 (J.Y.); +82-31-400-5249 (S.H.)
†   These authors contributed equally to this work.

Academic Editor: Zhigang Wu
Received: 3 April 2017; Accepted: 3 May 2017; Published: 11 May 2017

Abstract: Selective laser sintering enables the facile production of metal nanoparticle-based conductive layers on flexible substrates, but its application towards large-area electronics has remained questionable due to the limited throughput of the laser process that originates from the direct writing nature. In this study, modified optical schemes are introduced for the fabrication of (1) a densely patterned conductive layer and (2) a thin-film conductive layer without any patterns. In detail, a focusing lens is substituted by a micro lens array or a cylindrical lens to generate multiple beamlets or an extended focal line. The modified optical settings are found to be advantageous for the creation of repetitive conducting patterns or areal sintering of the silver nanoparticle ink layer. It is further confirmed that these optical schemes are equally compatible with plastic substrates for its application towards large-area flexible electronics.

Keywords: laser sintering; metal nanoparticle; micro lens array; cylindrical lens; flexible electrode

1. Introduction

A metal electrode layer, either patterned or in a thin-film form, is an inevitable component of an electronic device. Fabrication of a metal electrode layer based on conventional photolithographic means with vacuum evaporation techniques has achieved tremendous success with silicon-based electronics thus far, however, it has often been found to be improper for flexible and stretchable electronics. New types of conductive layers based on metal nanoparticle (NP) ink have provided a possible solution to this problem. As the melting temperature of metal NPs decreases according to its diameter due to the size effect, [1,2] a metal electrode layer is easily created on an arbitrary substrate at a mild temperature in ambient conditions through the direct coating and sintering of metal NP ink. However, despite its advantages, the minimum feature size of the resultant electrode prepared by metal NPs was relatively coarse due to the limited resolution of the conventional printing techniques which were employed for the selective deposition of metal NP ink [3]. At the same time, the sintering scheme remained questionable for a number of heat-sensitive substrates, as it was experimentally verified that the conventional sintering step with substantial sintering time can damage the underlying substrate [4].

As an alternative to the conventional sintering scheme, selective laser sintering of metal NPs was introduced to conduct patterning and sintering of metal NP ink simultaneously using a focused laser beam [5,6]. The focused laser beam is employed as a localized heat source based on the photothermal reaction, [7] and the area subject to heating is mainly determined by the spatial intensity distribution of the focused laser beam. Yeo et al. [8] demonstrated that the feature size of the resultant electrode can be easily reduced down to several micrometers using a laser beam focused by a telecentric lens module, which is difficult to achieve with other printing techniques. In addition, it was confirmed that the thermal damage on the underlying substrate can be minimized owing to the reduced heat-affected zone generated by a focused laser. These results suggest that laser sintering can be a convenient technique for the creation of metal patterns on flexible substrates. At the early stage, noble metal NPs such as gold (Au) [5,6,9–15] and silver (Ag) NP ink [1,2,4,8,16–23] were employed as the target materials for laser sintering, but recent studies demonstrate that the application of laser sintering can be extended to other oxidation-sensitive metal NPs such as copper (Cu) [24–26], even in ambient conditions, by reducing the local heating time through rapid scanning of the focused laser beam.

Laser sintering, however, is often considered to be inappropriate for mass production since it is a direct writing method in principle. As the electrode pattern becomes denser, the time required for the scanning increases linearly. At the same time, a thin-film electrode is another form of electrode that is difficult to be manufactured by the laser sintering method. Laser sintered thin-film electrodes not only require raster scanning of the entire area, but also often show imperfect electrical conductivity due to the discontinuities in electrical path originated from separate scanning steps. In this study, we introduce extended laser sintering schemes with a spatially modified focused beam to assist in the facile production of the electrodes with denser patterns or thin-film metal layers. In detail, a focusing lens is substituted by a micro lens array (MLA) and a cylindrical lens, designated for the fabrication of densely patterned electrodes and thin-film metal layers, respectively. It is also confirmed that the proposed optical schemes are still compatible with plastic substrates for their application in large-area flexible electronics.

2. Materials and Methods

Ag NP ink is firstly synthesized with the two-phase method, as reported in previous studies [8,17]. The resultant Ag NP is ~5 nm in diameter and encapsulated with self-assembled monolayer (SAM) to prevent agglomeration between NPs. It is confirmed that the Ag NP ink experiences melting when the temperature reaches ~150 °C, which is significantly lower than its bulk counterpart, due to the melting point depression from the size effect [2]. The synthesized Ag NP ink is coated on arbitrary substrates by spin-coating at 1000 rpm, and dried in a convection oven at 70 °C for 3 min to evaporate the excessive solvent. The target substrate can be either a rigid substrate such as a silicon wafer or a slide glass, or a flexible substrate. Polyimide (PI) thin film with a thickness of 150 μm is selected as the flexible substrate throughout the study.

A thin-film composed of Ag NPs is prepared on the substrate after the coating and drying steps, as shown in Figure 1a. Although Ag NPs are closely packed together, the as-prepared NP ink layer does not exhibit good electrical conductivity since the Ag NPs exist as separate entities. These Ag NPs can be transformed into a continuous conductive layer once a focused laser beam is scanned along the designated path, as shown in Figure 1b. The local temperature of the area subject to the laser irradiation increases rapidly and the Ag NPs experience melting and solidification steps as a consequence. The Ag NPs after the laser scanning subsequently show different physical properties from the as-deposited area, such as reflective color and high electrical conductivity.

An optical system is required to focus the laser beam on a designated spot, and an objective lens is frequently employed for the focusing, as shown in Figure 2a. The scanning path is controlled by a programmable motorized stage. Instead of moving the stage, a galvano-mirror together with a telecentric lens can be employed to achieve fast scanning of the laser beam over a large area [8,17]. Together with single focusing scheme, an MLA and cylindrical lens have been exploited in this study as new optical schemes for large-area compatible laser sintering, as depicted in Figure 2b,c. For single focusing, a green wavelength continuous wave laser (Millenia V, Spectra-Physics, Santa Clara, CA, USA) is scanned by a 2D galvano-mirror scanning system (hurrySCAN II, Scanlab GmbH, Puchheim, Germany) while the laser is focused by an f-theta telecentric lens with f = 100 mm. The laser scanner system is controlled by a computer with CAD software (SAMLight, SCAPS GmbH, Oberhaching, Germany) to draw arbitrary patterns. For the generation of multiple beamlets, an MLA with 400–900 nm anti-reflective coating is employed together with a 5× objective lens. More detailed information on the optical system is included in the Results and Discussions section. The MLA has a pitch of 300 μm and a focal length of 18.6 mm. For a focal line, the MLA is replaced by a plano-convex cylindrical lens with f = 50 mm. The laser scanning speed is fixed at 5 mm/s in every case.

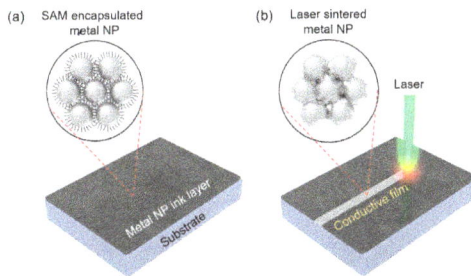

Figure 1. Schematic illustration of metal nanoparticle (NP) ink layer on a substrate (**a**) before and (**b**) after laser sintering.

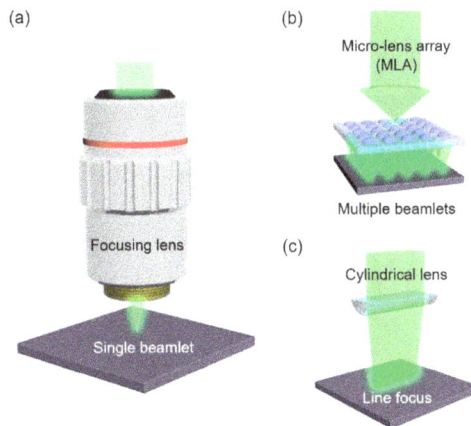

Figure 2. Optical schemes for selective laser sintering with (**a**) a single beamlet, (**b**) multiple beamlets, and (**c**) a focal line.

3. Results and Discussion

Figure 3 shows the sintering results created by the 2D galvano-mirror scanning system with the single focusing beamlet on a glass substrate in terms of their optical, scanning electron microscope (SEM) and atomic force microscopy (AFM) images. More information on the optical setting can be found in the previous studies [8,17]. Laser sintering is firstly conducted on an Ag NP ink layer according to the irradiated laser power. Figure 3a shows the optical microscope image of the Ag NP ink layer after the scanning, whereas the left and the right columns are the bright and dark field images of the same region, respectively. From the bright field image, it is noticeable that the optical transmission starts to change at the laser power of ~20 mW, while the dark field image shows no difference until the power reaches ~60 mW. Electrical measurement reveals that the resultant conductor line does not exhibit substantial electrical conductivity until it becomes reflective in the dark field image. We therefore anticipate that the change in optical transmission in the bright field image is due to the evaporation of the trapped solvent, or from the incomplete sintering between a small fraction of Ag NPs. The width of the area affected by the laser sintering is slightly bigger in the bright field image, and it is highly dependent to the drying condition.

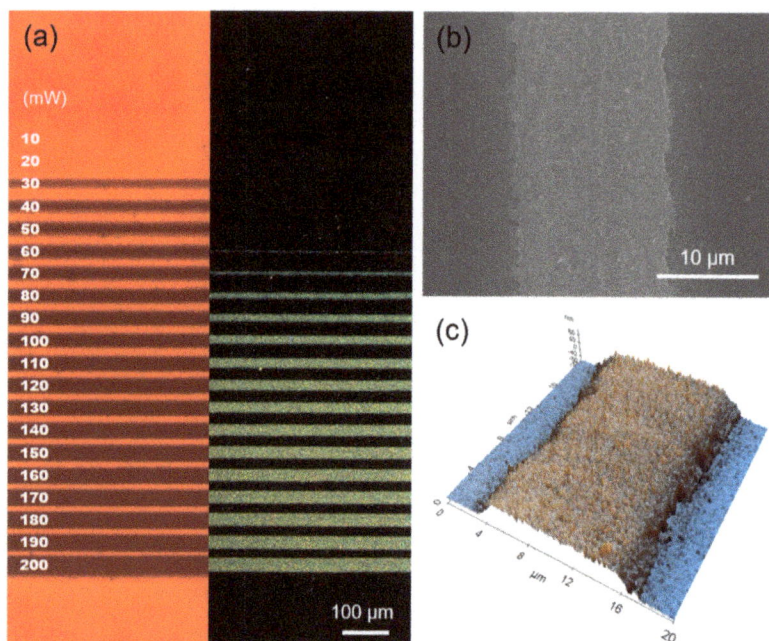

Figure 3. Laser sintering results by a single beamlet. (**a**) Optical images of Ag NP ink sintered at different laser power. (Left) Bright field; (Right) Dark field images. (**b**) Scanning electron microscope (SEM) image of the laser-sintered Ag NP at 100 mW laser power, and (**c**) its atomic force microscopy (AFM) image.

Figure 3b is the SEM image of the laser-sintered Ag NP at 100 mW laser power after the removal of remaining Ag NP ink from the substrate. The area irradiated by the laser remains on the substrate as a thin electrode while the other Ag NPs are washed away. More detailed information about its 3D morphology can be confirmed from its AFM image in Figure 3c. The height of the resultant conductor is ~ 120 nm, and this value can be further controlled by changing the wet processing conditions including spin-coating speed and drying conditions [17].

In order to generate a spatially extended focused beam, we first modified the optical setting, as shown in Figure 4a. Since our objective is to split the incident beam into multiple beamlets or to extend the focus into a focal line, incident laser intensity becomes an important issue. Two achromatic lenses with different focal lengths have been added in a Keplerian telescope configuration in order to reduce the beam size for sufficient laser intensity. At the same time, two adjustable slits are installed in the x- and y- directions to cut the incident laser beam, since a flat-top intensity profile is demanded in large-area applications instead of a TEM$_{00}$ (Fundamental Transverse Mode) Gaussian beam to ensure spatial uniformity in processing. The truncated laser beam shows a relatively flat-top intensity profile, but it is not quite perfect and other optical components such as a homogenizer will be required for further advancement of the proposed technique [18].

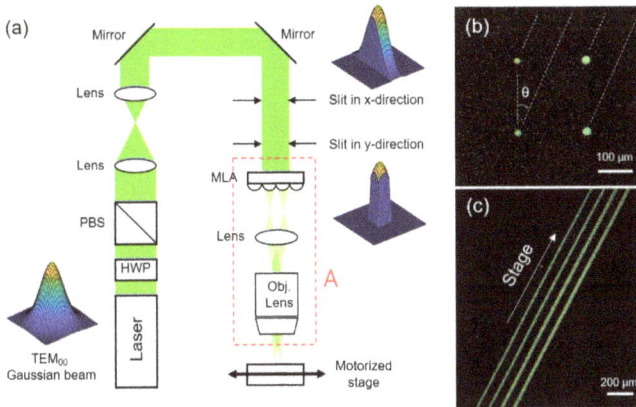

Figure 4. Laser sintering results by multiple beamlets. (**a**) Optical setting for multiple beamlets and a focal line (HWP: Half-wave plate, PBS: polarized beam splitter, MLA: Micro lens array). (**b**) Optical image of Ag NP ink by sintering with the multiple beamlets at a single exposure and (**c**) after moving the stage at a slanted angle.

Figure 4b is the optical image of the Ag NP ink layer after a single exposure to the multiple beamlets. The multiple beamlets are generated by opening each slit at a width of 600 μm, so that the incident beam covers 2 × 2 = 4 cells in the 300-μm pitch MLA. Different from the single beamlet case, four distinct spots are transformed into the conductive electrode with a single exposure as confirmed from the reflective color on the dark field optical image. Their relative positions can be further controlled by changing the parameters in the optical setting. Using the multiple beamlets, repetitive patterns can be easily created. As a representative example, parallel conductive lines, which are basic components for various applications including wire-grid polarizer and grid-type transparent conductors, [17] are created by moving the stage at a slanted angle. Figure 4c shows the resultant parallel lines produced by a single translational movement of the motorized stage.

For the sintering of metal NP ink over the entire area as a thin-film conductor without any patterns, a focal line is more suitable compared to the multiple beamlets. The optical components in Figure 4a (denoted as "A") are replaced by a cylindrical lens to create a focal line. In the current configuration, slit width in the y- direction (L_y) determines the laser intensity at the focus, while the slit width in the x-direction (L_x) controls the length of the focal line, as shown in Figure 5a. In this experiment, L_y has been fixed at ~300 μm throughout the study, while L_x is altered from 50 μm to 250 μm. Although the intensity of the incoming laser remains the same, it is found that the sintering characteristics of the resultant Ag NP ink are different in each case, as shown in Figure 5b. We estimate that the inconsistency comes from the different heat transfer conditions. For an extended focal line, an equal amount of heat is generated along the line and hence the major heat dissipation should occur towards a perpendicular direction to the focal line. However, as the length of the focal line becomes smaller, heat starts to dissipate in every lateral direction so that the resultant temperature reached by a shorter focal line should be lower at L_x = 50 μm. Figure 5b shows that the Ag NP ink at a width of >150 μm is successfully sintered by a single scanning with the cylindrical lens. It is worth mentioning that the resultant conductive layer does not have apparent boundaries which can be found in the thin-film conductors produced by raster scanning of a single beamlet, as shown in the inset.

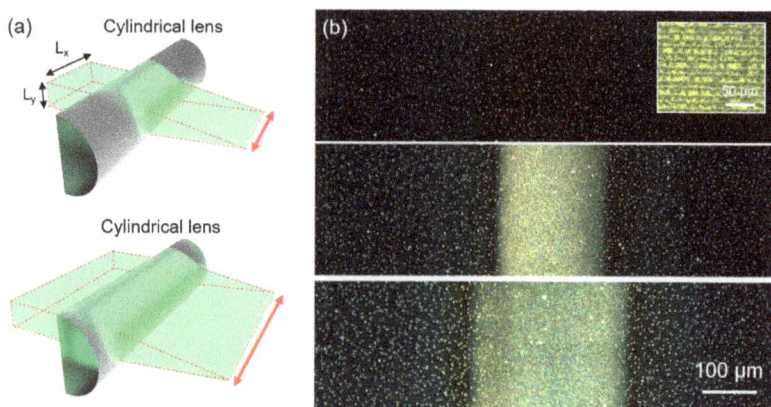

Figure 5. (**a**) Schematic illustration of the laser sintering by a focal line. (**b**) The optical image of the resultant Ag NP ink after scanning at (Top) $L_x = 50$ μm, (Middle) $L_x = 150$ μm and (Bottom) $L_x = 250$ μm (Inset: Ag NP thin-film conductor created by raster scanning of a single beamlet).

In order to verify that the proposed optical schemes are equally compatible with flexible substrates, parallel Ag NP lines are created on a flexible PI substrate with the modified optical setting. It is apparent from Figure 6a,b that consistent lines are created on the PI substrate without any apparent thermal damage. The current-voltage (IV) curve of the resultant conductor is measured to ensure that their electrical characteristics are suitable for the application of flexible electronics. The IV curve in Figure 6c shows that the sintered lines exhibit ohmic behavior and the calculated resistivity is estimated to be 7~8 times higher than bulk Ag, while the as-prepared NP film displays insignificant electrical conductivity.

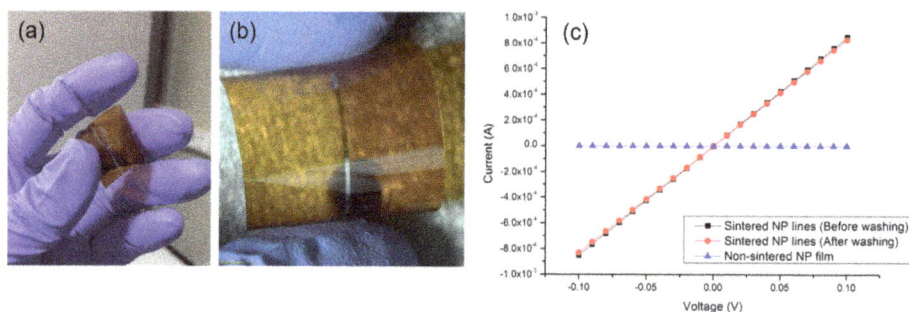

Figure 6. Application of the proposed optical schemes for laser sintering of Ag NP ink on a flexible substrate. Optical image at (**a**) low magnification and (**b**) high magnification; (**c**) current-voltage (IV) curve of the resultant Ag electrode on the flexible substrate.

In summary, we propose two forms of a spatially extended focused beam intended for the fabrication of Ag NP conducting layers with dense patterns or in thin-film configuration. It is demonstrated that the modified optical setting can produce multiple beamlets for repetitive patterns or focal lines for thin-film conductors. Since these optical schemes are equally compatible with flexible substrates, it is expected that the proposed techniques will supplement fabrication of large-area flexible electronics.

Acknowledgments: This work is supported by the National Research Foundation of Korea (NRF) Grant funded through the Basic Science Research Program (NRF-2016R1C1B1014729) and the research fund of Hanyang University (HY-2016-00000002097).

Author Contributions: S.H. and J.Y. designed the experiments; J.K. and H.L. synthesized the material; J.K., H.L. and S.H. conducted the experiments; W.S.S. and H.R.K. assisted in analyzing the experimental result; J.S. and H.C. built the optical setting; S.H., J.Y., J.H. and H.L. wrote the manuscript.

Conflicts of Interest: The authors declare no conflict of interest.

References

1. Son, Y.; Yeo, J.; Ha, C.W.; Lee, J.; Hong, S.; Nam, K.H.; Yang, D.-Y.; Ko, S.H. Application of the specific thermal properties of Ag nanoparticles to high-resolution metal patterning. *Thermochim. Acta* **2012**, *542*, 52–56. [CrossRef]

2. Son, Y.; Yeo, J.; Moon, H.; Lim, T.W.; Hong, S.; Nam, K.H.; Yoo, S.; Grigoropoulos, C.P.; Yang, D.-Y.; Ko, S.H. Nanoscale electronics: Digital fabrication by direct femtosecond laser processing of metal nanoparticles. *Adv. Mater.* **2011**, *23*, 3176–3181. [CrossRef] [PubMed]

3. Cummins, G.; Desmulliez, M.P.Y. Inkjet printing of conductive materials: A review. *Circuit World* **2012**, *38*, 193–213. [CrossRef]

4. Yeo, J.; Kim, G.; Hong, S.; Kim, M.S.; Kim, D.; Lee, J.; Lee, H.B.; Kwon, J.; Suh, Y.D.; Kang, H.W.; et al. Flexible supercapacitor fabrication by room temperature rapid laser processing of roll-to-roll printed metal nanoparticle ink for wearable electronics application. *J. Power Sources* **2014**, *246*, 562–568. [CrossRef]

5. Chung, J.; Bieri, N.R.; Ko, S.; Grigoropoulos, C.P.; Poulikakos, D. In-tandem deposition and sintering of printed gold nanoparticle inks induced by continuous Gaussian laser irradiation. *Appl. Phys. A* **2004**, *79*, 1259–1261. [CrossRef]

6. Bieri, N.R.; Chung, J.; Haferl, S.E.; Poulikakos, D.; Grigoropoulos, C.P. Microstructuring by printing and laser curing of nanoparticle solutions. *Appl. Phys. Lett.* **2003**, *82*, 3529–3531. [CrossRef]

7. Bäuerle, D.W. *Laser Processing and Chemistry*; Springer Science & Business Media: Berlin, Germany, 2013.

8. Yeo, J.; Hong, S.; Lee, D.; Hotz, N.; Lee, M.-T.; Grigoropoulos, C.P.; Ko, S.H. Next generation non-vacuum, maskless, low temperature nanoparticle ink laser digital direct metal patterning for a large area flexible electronics. *PLoS ONE* **2012**, *7*, e42315. [CrossRef] [PubMed]

9. Choi, T.Y.; Poulikakos, D.; Grigoropoulos, C.P. Fountain-pen-based laser microstructuring with gold nanoparticle inks. *Appl. Phys. Lett.* **2004**, *85*, 13–15. [CrossRef]

10. Chung, J.; Ko, S.; Bieri, N.R.; Grigoropoulos, C.P.; Poulikakos, D. Conductor microstructures by laser curing of printed gold nanoparticle ink. *Appl. Phys. Lett.* **2004**, *84*, 801–803. [CrossRef]

11. Ko, S.H.; Pan, H.; Grigoropoulos, C.P.; Luscombe, C.K.; Fréchet, J.M.J.; Poulikakos, D. Air stable high resolution organic transistors by selective laser sintering of ink-jet printed metal nanoparticles. *Appl. Phys. Lett.* **2007**, *90*, 141103. [CrossRef]

12. Chung, J.; Ko, S.; Grigoropoulos, C.P.; Bieri, N.R.; Dockendorf, C.; Poulikakos, D. Damage-free low temperature pulsed laser printing of gold nanoinks on polymers. *J. Heat Transf.* **2005**, *127*, 724–732. [CrossRef]

13. Ko, S.H.; Pan, H.; Grigoropoulos, C.P.; Luscombe, C.K.; Fréchet, J.M.; Poulikakos, D. All-inkjet-printed flexible electronics fabrication on a polymer substrate by low-temperature high-resolution selective laser sintering of metal nanoparticles. *Nanotechnology* **2007**, *18*, 345202. [CrossRef]

14. Ko, S.H.; Chung, J.; Pan, H.; Grigoropoulos, C.P.; Poulikakos, D. Fabrication of multilayer passive and active electric components on polymer using inkjet printing and low temperature laser processing. *Sens. Actuators A Phys.* **2007**, *134*, 161–168. [CrossRef]

15. Pan, H.; Hwang, D.J.; Ko, S.H.; Clem, T.A.; Fréchet, J.M.J.; Bäuerle, D.; Grigoropoulos, C.P. High-throughput near-field optical nanoprocessing of solution-deposited nanoparticles. *Small* **2010**, *6*, 1812–1821. [CrossRef] [PubMed]

16. An, K.; Hong, S.; Han, S.; Lee, H.; Yeo, J.; Ko, S.H. Selective sintering of metal nanoparticle ink for maskless fabrication of an electrode micropattern using a spatially modulated laser beam by a digital micromirror device. *ACS Appl. Mater. Interfaces* **2014**, *6*, 2786–2790. [CrossRef] [PubMed]

17. Hong, S.; Yeo, J.; Kim, G.; Kim, D.; Lee, H.; Kwon, J.; Lee, H.; Lee, P.; Ko, S.H. Nonvacuum, maskless fabrication of a flexible metal grid transparent conductor by low-temperature selective laser sintering of nanoparticle ink. *ACS Nano* **2013**, *7*, 5024–5031. [CrossRef] [PubMed]

18. Ko, S.H.; Pan, H.; Ryu, S.G.; Misra, N.; Grigoropoulos, C.P.; Park, H.K. Nanomaterial enabled laser transfer for organic light emitting material direct writing. *Appl. Phys. Lett.* **2008**, *93*, 151110. [CrossRef]

19. Son, Y.; Yeo, J.; Ha, C.W.; Hong, S.; Ko, S.H.; Yang, D.-Y. Fabrication of submicron-sized metal patterns on a flexible polymer substrate by femtosecond laser sintering of metal nanoparticles. *Int. J. Nanomanuf.* **2013**, *9*, 468–476. [CrossRef]

20. Lee, H.; Hong, S.; Kwon, J.; Suh, Y.D.; Lee, J.; Moon, H.; Yeo, J.; Ko, S.H. All-solid-state flexible supercapacitors by fast laser annealing of printed metal nanoparticle layers. *J. Mater. Chem. A* **2015**, *3*, 8339–8345. [CrossRef]

21. Suh, Y.D.; Jung, J.; Lee, H.; Yeo, J.; Hong, S.; Lee, P.; Lee, D.; Ko, S.H. Nanowire reinforced nanoparticle nanocomposite for highly flexible transparent electrodes: Borrowing ideas from macrocomposites in steel-wire reinforced concrete. *J. Mater. Chem. C* **2017**, *5*, 791–798. [CrossRef]

22. Lee, M.-T.; Lee, D.; Sherry, A.; Grigoropoulos, C.P. Rapid selective metal patterning on polydimethylsiloxane (PDMS) fabricated by capillarity-assisted laser direct write. *J. Micromech. Microeng.* **2011**, *21*, 095018. [CrossRef]

23. Yang, M.; Chon, M.-W.; Kim, J.-H.; Lee, S.-H.; Jo, J.; Yeo, J.; Ko, S.H.; Choa, S.-H. Mechanical and environmental durability of roll-to-roll printed silver nanoparticle film using a rapid laser annealing process for flexible electronics. *Microelectron. Reliab.* **2014**, *54*, 2871–2880. [CrossRef]

24. Kwon, J.; Cho, H.; Eom, H.; Lee, H.; Suh, Y.D.; Moon, H.; Shin, J.; Hong, S.; Ko, S.H. Low-temperature oxidation-free selective laser sintering of cu nanoparticle paste on a polymer substrate for the flexible touch panel applications. *ACS Appl. Mater. Interfaces* **2016**, *8*, 11575–11582. [CrossRef] [PubMed]

25. Suh, Y.D.; Kwon, J.; Lee, J.; Lee, H.; Jeong, S.; Kim, D.; Cho, H.; Yeo, J.; Ko, S.H. Maskless fabrication of highly robust, flexible transparent Cu conductor by random crack network assisted Cu nanoparticle patterning and laser sintering. *Adv. Electron. Mater.* **2016**, *2*. [CrossRef]

26. Michael, Z.; Oleg, E.; Amir, S.; Zvi, K. Laser sintering of copper nanoparticles. *J. Phys. D Appl. Phys.* **2014**, *47*, 025501.

micromachines

MDPI

Article

Ultrasonic Spray-Coating of Large-Scale TiO$_2$ Compact Layer for Efficient Flexible Perovskite Solar Cells

Peng Zhou [1], Wangnan Li [2], Tianhui Li [1], Tongle Bu [1], Xueping Liu [1], Jing Li [1], Jiang He [1], Rui Chen [1], Kunpeng Li [1], Juan Zhao [3,*] and Fuzhi Huang [1,*]

[1] State Key Laboratory of Advanced Technology for Materials Synthesis and Processing, Wuhan University of Technology, Wuhan 430070, China; whutzhp@gmail.com (P.Z.); litianhui@whut.edu.cn (T.L.); tonglebu@whut.edu.cn (T.B.); xuepingliu@whut.edu.cn (X.L.); lijing123@whut.edu.cn (J.L.); milijiang@whut.edu.cn (J.H.); Rui-Chen@whut.edu.cn (R.C.); likunpeng@whut.edu.cn (K.L.)
[2] Hubei Key Laboratory of Low Dimensional Optoelectronic Materials and Devices, Hubei University of Arts and Science, Xiangyang 441053, China; wangnan.li@yahoo.com
[3] School of Automotive Engineering, Wuhan University of Technology, Wuhan 430070, China
* Correspondence: juan.zhao@whut.edu.cn (J.Z.); fuzhi.huang@whut.edu.cn (F.H.); Tel.: +86-27-8716-8599 (J.Z. & F.H.)

Academic Editors: Seung Hwan Ko, Daeho Lee and Zhigang Wu
Received: 31 December 2016; Accepted: 26 January 2017; Published: 14 February 2017

Abstract: Flexible electronics have attracted great interest in applications for the wearable devices. Flexible solar cells can be integrated into the flexible electronics as the power source for the wearable devices. In this work, an ultrasonic spray-coating method was employed to deposit TiO$_2$ nanoparticles on polymer substrates for the fabrication of flexible perovskite solar cells (PSCs). Pre-synthesized TiO$_2$ nanoparticles were first dispersed in ethanol to prepare the precursor solutions with different concentrations (0.5 mg/mL, 1.0 mg/mL, 2.0 mg/mL) and then sprayed onto the conductive substrates to produce compact TiO$_2$ films with different thicknesses (from 30 nm to 150 nm). The effect of the different drying processes on the quality of the compact TiO$_2$ film was studied. In order to further improve the film quality, titanium diisopropoxide *bis*(acetylacetonate) (TAA) was added into the TiO$_2$-ethanol solution at a mole ratio of 1.0 mol % with respect to the TiO$_2$ content. The final prepared PSC devices showed a power conversion efficiency (PCE) of 14.32% based on the indium doped tin oxide coated glass (ITO-glass) substrate and 10.87% on the indium doped tin oxide coated polyethylene naphthalate (ITO-PEN) flexible substrate.

Keywords: ultrasonic spray; titanium oxide; flexible perovskite solar cells; low temperature; large area

1. Introduction

With the global energy consumption increasing, cheap, green and environmentally friendly energy sources are in urgent demand due to the reduction of fossil fuels. Photovoltaic technology is an ideal solution to alleviate the energy crisis and environmental pollution problems. Organic-inorganic lead halide perovskite solar cells (PSCs) have attracted great interest due to their rapid increase in power conversion efficiencies (PCE) from 3.8% to 22.1% within only seven years [1–3]. These great improvements are mainly attributed to the excellent photo-electronic properties of perovskite materials, such as high light absorption properties, direct bandgaps, high charge-carrier mobility and a long electron–hole exciton transport distance (more than 1 μm) [4–6]. Compared to the commercial silicon-based solar cells, PSCs show great advantages with a simplified architecture and a low-cost solution-processed technology, which give them great potential for the future photovoltaic industry [7].

Organic-inorganic lead halide perovskite ($CH_3NH_3PbI_3$) was first used as a sensitizer in dye-sensitized solar cells (DSSCs) by Kojima in 2009 [8] with a PCE of 3.8%, but the performance decreased rapidly due to the dissolution of the perovskite in the liquid electrolytes. Two years later, the solid-state hole-transporting material 2,2′,7,7′-tetrakis-(N,N-di-p-methoxypheny-lamine) 9,9′-spirobifluorene (spiro-OMeTAD) was introduced by Park and Grätzel, and achieved a reported PCE of 9.7% [9]. Since the all solid-state-type PSCs were fabricated, the photo-electronic performance improved rapidly with the use of different electronic/hole transporting semi-conductive materials, such as TiO_2, ZnO, SnO, PCBM, P_3HT, etc. [10]. Depending on the different architectures, the PSCs can be generally classified into mesoscopic, meso-superstructured and planar heterojunction types [11]. Among these types, the planar PSCs have a simplified architecture and are easily produced by a solution process. In a planar solar cell, the photoactive layer, $CH_3NH_3PbI_3$, is sandwiched between an electron-transporting layer (ETL) and a hole-transporting layer (HTL), which is suitable for large-scale commercial manufacturing layer-by-layer. The ETL plays an important role as it allows the transport of electrons while blocking the holes. Thus, it influences the carriers' injection, collection, transportation, recombination, and then the overall performance of the PSCs [12]. Anatase TiO_2 is the most widely used ETL for planar PSCs, but it still requires a high temperature sintering of the TiO_2 compact layer to achieve a high efficiency. In general, this compact layer is prepared by spin-coating or spray pyrolysis of a TiO_2 precursor solution with subsequent sintering at 500 °C to transform the amorphous oxide layer into the crystalline phase (anatase), which provides good charge transport properties [13]. The involvement of the high temperature sintering process of TiO_2 has limited the development of flexible PSCs fabricated on plastic substrates, such as polyethylene terephthalate (PET) and polyethylene naphthalate (PEN).

Since the first flexible perovskite solar cell was reported [14], more efforts have been devoted to it and have achieved efficiencies of over 10% on polymer substrates [15,16], demonstrating that efficient perovskite solar cells can be fabricated at low temperatures with a "regular" design. Atomic layer deposition [17,18], microwave sintering [19] and inductively coupled plasma (ICP)-assisted DC magnetron sputtering [20] have been used for deposition of the TiO_2 compact layer at low temperatures. A photonic-cured compact TiO_2 layer has been used on a PET substrate with a high efficiency of 11.2% by Xiao [21]. Snaith et al. realized flexible PSC with an efficiency of 15.9% on a low-temperature processed TiO_2 compact layer by spin-coating the TiO_2 precursor on fluorine doped tin oxide coated glass (FTO-glass) [22]. Yang et al. developed a process to fabricate a very dense amorphous TiO_2 using DC magnetron sputtering at room temperature and achieved 15.07% PCE based on a flexible PET substrate [23]. Sanjib et al. used a combination of ultrasonic spray-coating and a low-thermal-budget photonic curing technology for the first time to fabricate a flexible PSC with an efficiency of 8.1% [24]. Supasai et al. fabricated compact layers of crystalline TiO_2 thin films using aerosol spray pyrolysis on FTO-glass for PSCs, achieving the best efficiency of 6.24% [25]. Kim et al. fabricated a mesoporous TiO_2 ETL with a large area of 10 cm × 10 cm using electro-spray deposition (ESD) for the first time, which resulted in an optimized PCE of 15.11%, higher than that (13.67%) of the PSC with spin-coated TiO_2 films [26]. Both the aerosol spraying and ESD methods demonstrated great potential in the large-scale fabrication of PSCs. However, the TiO_2 films deposited by the above two spraying methods require high temperature sintering, which is not suitable for the preparation of flexible PSCs. The use of these technologies has allowed for the development of all-low-temperature processed PSCs on flexible substrates, but at the same time, has made the process more complicated, resulting in the increase of the manufacturing cost.

Here, we report a low temperature fabrication (<150 °C) of a compact layer composed of highly crystalline small nanoparticles of anatase TiO_2 (diameter <10 nm) dispersed in ethanol. The ultrasonic spray-coating method was employed for the deposition of the TiO_2 compact layer, demonstrating the capability to precisely and reliably deposit thin and uniform layers. Various parameters have to be considered in the process of ultrasonic spraying, such as the flow rate of the TiO_2 ethanol solution, the gas flow pressure which carries the sprayed droplets to the substrate, the distance between the spray nozzle and the substrate, as well as the moving speed of the nozzle during spraying [27,28]. Through the optimization of these parameters, two technological regimes named "wet-film" (w-film) and "dry-film" (d-film) are compared in our work. The former approach results in a dense TiO_2 layer of different thicknesses (30 nm to 150 nm) by changing the concentration of the dispersion. In addition, a small amount of titanium diisopropoxide *bis*(acetylaceto-nate) (TAA) (1.0 mol % with respect to the TiO_2 content) is added to the colloidal TiO_2 dispersion, resulting in the highest PCE of 14.32% based on ITO-glass and 10.87% based on ITO-PEN, both with an active area of 0.4 cm × 0.4 cm. Through the same method, a large-area flexible PSC of 5 cm × 5 cm is fabricated.

2. Materials and Methods

2.1. Materials

Unless specified otherwise, all the chemicals were purchased from either Alfa Aesar or Sigma-Aldrich and used as received. 2,2',7,7'-tetrakis-(N,N-di-p-methoxypheny-lamine) 9,9'-spirobifluorene (spiro-OMeTAD) was purchased from Shenzhen Feiming Science and Technology Co., Ltd. (Shenzhen, China), and MAI (CH_3NH_3I) was purchased from Lumtec, Taiwan.

2.2. Substrates

Indium doped tin oxide coated glass (ITO-Glass) and polyethylene naphthalate (ITO-PEN) were etched by zinc powder and hydrochloric acid, then followed by ultrasonic cleaning in detergent, pure water and ethanol for 15 min, respectively. The substrate (ITO-Glass or ITO-PEN) was cut into suitable size and plasma cleaned for 5 min to remove any organic material on the surface. Especially, the flexible ITO-PEN substrate was mounted onto glass micro-slides before using.

2.3. Synthesis of TiO_2 Nanoparticles

The anatase titanium oxide nanoparticles were synthesized according to previously reported method published elsewhere [29]. Briefly, 0.5 mL of anhydrous $TiCl_4$ (99.9%, Aladdin, Shanghai, China) was added dropwise into 2 mL of anhydrous ethanol (Sigma-Aldrich, Shanghai, China), stirring for 5 min till the mixed yellow liquid was cooled down to room temperature. Then the whole content was transferred into a vial containing 10 mL anhydrous benzyl alcohol (Aladdin, Shanghai, China), with the color of the clear solution changing to light yellow. The solution was heated to 80 °C and reacted for 9 h. After the reaction the solution was cooled down to the room temperature, and a translucent dispersion of very fine TiO_2 nanoparticles was obtained. Then add 36 mL of diethyl ether to 4 mL of the above solution to precipitate the TiO_2 which was centrifuged at 4000 rpm for 5 min, washed with ethanol and diethyl ether. The above steps were repeated three times and the final precipitation was redispersed in 10 mL of anhydrous ethanol, resulting in a colloidal solution of approximately 12 mg TiO_2/mL ethanol but easily aggregated. The TiO_2 ethanol solution was further diluted into 0.5 mg/mL, 1.0 mg/mL and 2.0 mg/mL respectively for the late ultrasonic spraying. In order to disperse the TiO_2 nanoparticles and enhance the adhesion as well as the connection between nanoparticles, a small amount of TAA was added into the diluted dispersion with a mole ratio of 1.0 mol %, 2.0 mol % and 3.0 mol % respectively. The solution was left to stand for at least 2 h before spraying, but could be stable for months.

2.4. Deposition of the TiO_2 Compact Layer

The low-temperature processed TiO_2 compact layer was deposited by ultrasonic spraying the colloidal dispersion of anatase particles in anhydrous ethanol, formulated with TAA, followed by treating at 135 °C for 1 h to remove ethanol solvent. Figure 1a shows the schematic representation of setup for the spray coating. Figure 1b showed the spray nozzle moving path during the spray. Spray coating was carried out in an Exacta Coat Ultrasonic Spraying System (Sono-Tek Corpration, Milton,

NY, USA) equipped with an AccuMist nozzle. The thickness of the compact layer was tuned by the concentration of TiO_2 nanoparticles (0.5–2.0 mg/mL).

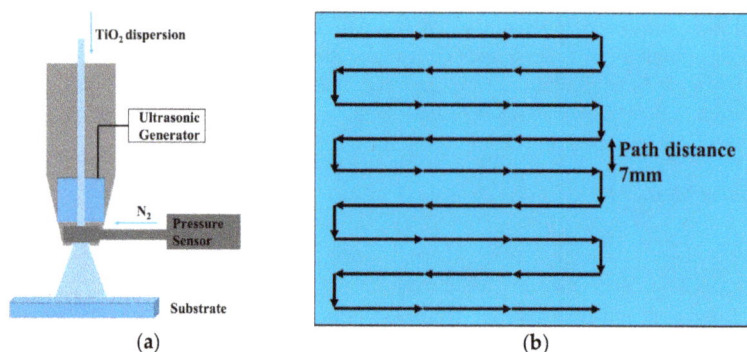

Figure 1. Schematic representation of (a) setup for the spray coater and (b) the spray nozzle moving path.

2.5. Device Fabrication

After deposition of the TiO_2 compact layer, the substrate with appropriate size (1.5 cm × 1.5 cm for small devices and 5.0 cm × 5.0 cm for large area devices) was transferred into a N_2 filled glove box. The perovskite and hole conductor solutions were prepared in a N_2 glove box before deposition. The perovskite was deposited by spin coating 25 µL of a 1.25 mol/L solution of CH_3NH_3I and PbI_2 (molar ratio 1:1) in DMF at 6500 rpm for 30 s using the gas-assisted method [30], meanwhile a 60 psi dry N_2 gas stream was blown onto the film for 8 s from the third second of spinning. For the fabrication of flexible devices with large area, 200 µL perovskite solution was coated on TiO_2 compact layer and spin-coated at 5000 rpm for 30 s, followed by dropping 200 µL chlorobenzene (CBZ) at the fifth second of spinning. The films were subsequently annealed on a hot plate at 100 °C for 10 min in the glove box. After letting the films cool for 5 min, 20 µL spiro-OMeTAD/CBZ solution (68 mM spiro-OMeTAD, 150 mM tert-butylpyridine and 25 mM lithium *bis*(*tri*-fluoromethanesulphonyl)imide) was spin-coated at 3000 rpm for 30 s. Promptly after the hole transport material deposition, a gold counter electrode (60 nm) was evaporated under high vacuum to complete the device.

2.6. Characterizations

The morphologies and microstructures of the prepared low-temperature processed TiO_2 compact and the cross-sectional structure of the perovskite solar cells were investigated using a field-emission scanning electron microscopy (FE-SEM, Zeiss Ultra Plus). To examine the surface roughness, the films were characterized by BY2000 atomic force microscopy (AFM). The TiO_2 anatase phase was tested by an X-ray diffractometer (XRD, D8 Advance). The photocurrent density-voltage curves of the PSCs were measured using a solar simulator (Oriel 94023A, 300 W). The intensity (100 mW/cm^2) was calibrated using a standard Si-solar cell (Oriel, VLSI standards). All the devices were tested under AM 1.5 G sun light (100 mW/cm^2) using a metal mask of 0.16 cm^2 with a scan rate of 0.01 V/s. Some parameters would be mentioned to evaluate the performance of PSCs, such as open circuit voltage (V_{oc}), short circuit current (J_{sc}), fill fact (FF), and power conversion efficiency (PCE).

3. Results and Discussion

Anatase-type TiO_2 nanoparticles were synthesized through a method published elsewhere [29], using titanium tetrachloride (TiCl$_4$) as the precursor, and anhydrous ethanol and benzyl alcohol as solvents. The TiO_2 nanoparticles synthesized via this method are highly crystalline, shown in Figure 2a,b. In the XRD pattern of the TiO_2 powder prepared by drying the as-synthesized nanoparticles at 135 °C to evaporate the ethanol solvent, three typical diffraction peaks occur at 25.5°, 38.5° and 48°, which respectively belong to the (101), (004) and (200) planes of the anatase TiO_2 crystal. The as-synthesized TiO_2 nanoparticles could easily be dispersed in ethanol, which can be used for ultrasonic spraying directly. After the deposition of the TiO_2 layer, the films had to be treated at 135 °C for 1 h, replacing the high-temperature sintering process. The size of the TiO_2 nanoparticles was around 5 nm, as shown in the TEM image (Figure 2a), which can also be calculated by the Scherrer equation using the (101) diffraction peak through the formula reported elsewhere [31]. Figure 2c shows the XRD patterns of ITO-glass with/without the TiO_2 layer, in which just a weak diffraction peak of TiO_2 occurs at 25.5° due to the thin thickness of the TiO_2 layer.

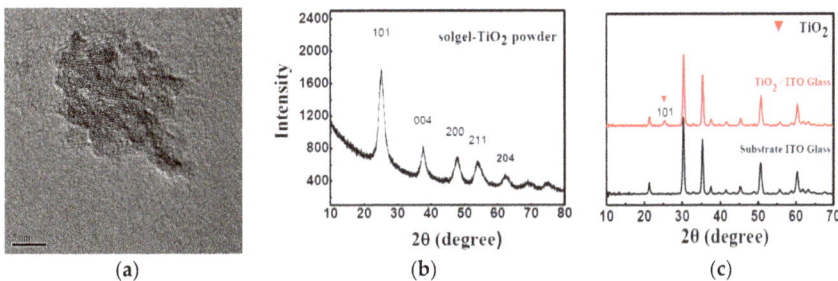

Figure 2. (**a**) Transmission electron microscopy (TEM) image of the TiO_2 nanoparticles dispersed in ethanol at a concentration of 1.0 mg/mL; (**b**) XRD pattern of the TiO_2 powder drying at 135 °C; and (**c**) XRD patterns of ITO-glass with/without the TiO_2 compact layer coated by the ultrasonic spraying.

In order to obtain uniform and dense films of TiO_2 compact layers, two technological regimes of ultrasonic spraying were built. During ultrasonic spraying, many parameters influence the uniformity, roughness, and coverage of the TiO_2 film, such as the flow rate of the TiO_2 ethanol solution, the gas flow pressure which carries the sprayed droplets to the substrate, the distance between the spray nozzle and the substrate, as well as the moving speed of the nozzle. The ultrasonic spraying technology was employed in the deposition of polymer films such as polyvinylpyrrolidone (PVP) by Sanjukta Bose et al. in 2013 [27]. Parts of the parameters have been discussed, resulting in uniform organic films with small roughness compared to the thickness of the films. Based on the work mentioned above, we optimized the process of spraying, and employed it to deposit inorganic TiO_2 film. Figure 1a shows the schematic representation of the setup for the spray-coating. Figure 1b shows the ultrasonic spraying trajectory over the substrate. Snaith et al. realized low-temperature processed PSCs with power convention efficiencies (PCE) of up to 15.9% using a spin-coating method, demonstrating the optimum thickness of TiO_2 to be approximately 45 nm [22], while the thickness of TiO_2 prepared by high-temperature sintering at 500 °C is around 30 nm [7]. Based on these reports, we tuned the thickness of ultrasonic spraying TiO_2 compact layer ranging from 30 nm to 100 nm by changing the spraying parameters and the concentration of the TiO_2 ethanol solution.

During the spraying, two different technological regimes were termed "wet-film" (w-film) and "dry-film" (d-film), depending on the drying type of the ethanol solvent. The former regime means that the ethanol solvent evaporates after the droplets are sprayed onto the substrate, forming a very uniform thin liquid membrane layer before the ethanol evaporates. A few seconds later, the solvent begins to evaporate, resulting in a uniform thin compact layer of TiO_2. The latter regime means that the processing of ethanol evaporation occurs mainly before the droplets are sprayed onto the substrate, without forming the liquid membrane. The TiO_2 compact layer prepared by the former regime was influenced by the "coffee-ring" effect, which is described by Deegan et al. as the result of capillary flow [32], where liquid evaporates faster from the pinned contact line of a deposited solution and is replenished by additional liquid from inside; however, this did not happen in the latter regime. During the spraying of the w-film regime, the pressure of the N_2 flow was set to 0.3 psi, the flow rate of solution was set to 1.2 mL/min, and the speed of the moving nozzle was set to 50 mm/s, which can result in a thin liquid membrane on the surface of the substrate before the solvent evaporates; during the d-film regime, the three parameters were respectively set to 1.0 psi, 0.3 mL/min, and 20 mm/s. Figure 3 showed the AFM surface images of the TiO_2 compact layers by ultrasonic spraying using 1.0 mg/mL of the TiO_2 (without TAA) ethanol solution. The TiO_2 film prepared by the w-film regime presented a root-mean-square (RMS) surface roughness of around 3.3 nm, while it increased to 44.6 nm rapidly with the d-film regime. It was attributed to the big holes between the TiO_2 nanoparticles prepared through the d-film regime. In the film of the d-film regime, the connection between the particles was so weak that big holes appeared because of the evaporation of ethanol before the droplets were deposited on the substrate. In addition, more spraying cycles were needed to obtain the optimum thickness of the TiO_2 film through the d-film regime, which increased the roughness of film, while just one cycle was enough to achieve the thickness with the w-film regime. Thus, it is better to fabricate the TiO_2 compact layer using the w-film regime.

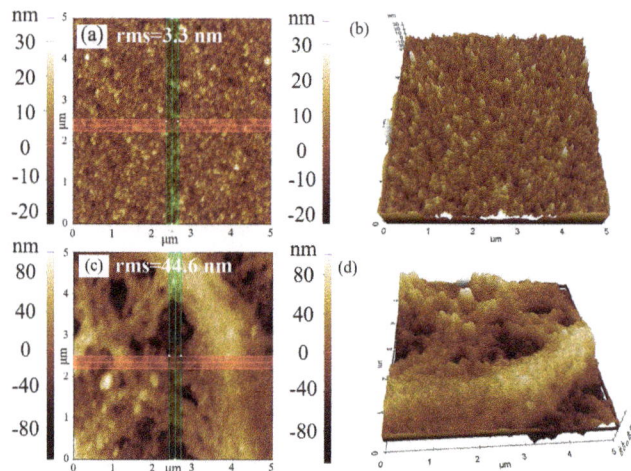

Figure 3. Atomic force microscope (AFM) images of the surface of the TiO_2 layers sprayed by (**a**,**b**) the w-film; (**c**,**d**) the d-film using the TiO_2 ethanol solution without TAA.

In order to discuss the influence of the thickness of the ultrasonically sprayed TiO_2 compact layer on the performance of PSC devices, ITO-glass substrates coated with TiO_2 films of different thicknesses (30–100 nm) were made into complete planar PSC devices. The thickness of the TiO_2 film was tuned by changing the concentration. Table 1 shows the photovoltaic parameters obtained from the current-voltage measurements of devices with different compact layers. It is obvious that the optimal concentration for the performance was 1.0 mg/mL, with a thickness around 60 nm. Figure 4 shows the surface image of different functional layers, and a dense TiO_2 compact layer was obtained (Figure 4b). Each functional layer can be clearly distinguished in the cross-section scanning electron microscopy (SEM) image (Figure 4d). The clear interface between the TiO_2 layer and perovskite layer confirmed that the TiO_2 layer was not utilized as a mesoscopic scaffold. It can be concluded that the uniform TiO_2 compact layer with low roughness is suitable for the preparation of perovskite (Figure 4c), and the optimal thickness ranges around 60 nm. Figure 5a shows the current-voltage curves of PSCs prepared with a TiO_2 solution of 1.0 mg/mL using the w-film regime, with an average PCE of 10.79%, and the highest PCE of 11.23%.

Table 1. The performance of devices fabricated based on the w-film ultrasonically sprayed TiO$_2$ with different thicknesses tuned by the concentration of TiO$_2$ dispersion.

Concentration of TiO$_2$ (mg/mL)	Average Thickness of TiO$_2$ Layer (nm)	V_{oc} (mV)	J_{sc} (mA/cm^2)	FF (%)	PCE (%)
0.5	30	654	18.74	0.38	4.61
1.0	60	896	17.57	0.69	10.82
2.0	100	838	15.72	0.59	7.81

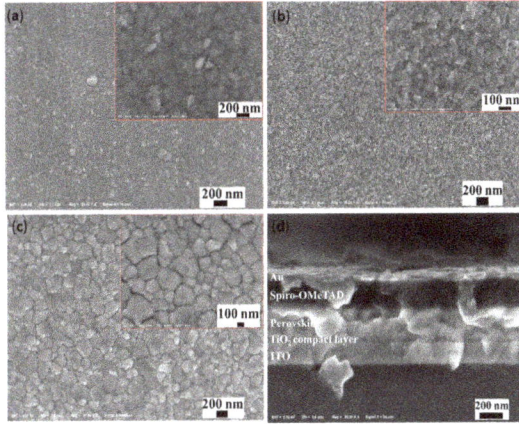

Figure 4. Scanning electron microscopy (SEM) image of (**a**) PEN-ITO substrate; (**b**) TiO$_2$ compact layer; (**c**) perovskite prepared by the gas-assisted method; and (**d**) cross-section SEM image of the PSCs.

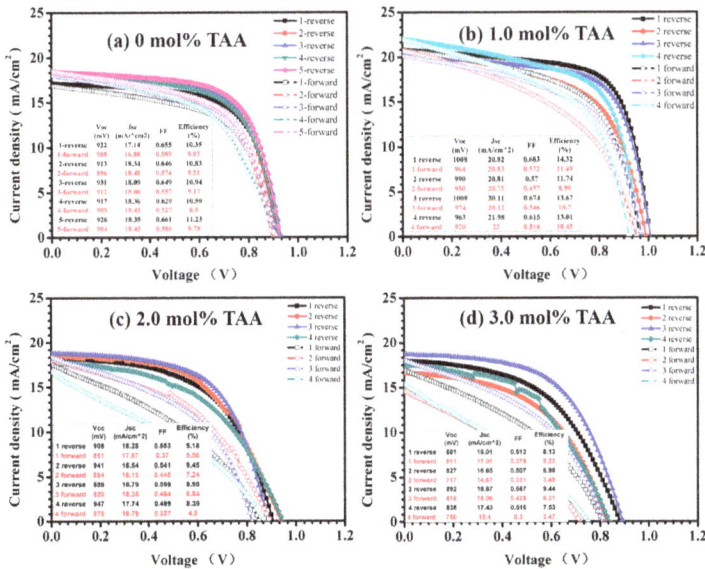

Figure 5. J-V characteristics of the gas-assisted processed PSCs based on TiO$_2$ compact layer deposited by the ultrasonic spray-coating of TiO$_2$ solution with the addition of TAA (**a**) 0 mol %; (**b**) 1.0 mol %; (**c**) 2.0 mol %; (**d**) 3.0 mol % with respect to the TiO$_2$ content.

In addition, more works have been done to improve the performance of TiO_2 by adding a small amount of TAA (1.0 mol %, 2.0 mol %, 3.0 mol % with respect to the TiO_2 content), with almost a negligible influence on the thickness of the TiO_2 film. The solutions with different amounts of TAA were ultrasonically sprayed on the substrates but with a similar thickness of the TiO_2 compact layers, around 60 nm. The RSM increased to 7.6 nm from 3.3 nm without a significant change of the roughness of the bare ITO after 1.0 mol % TAA was added to a TiO_2 solution of 1.0 mg/mL (Figure 6). It was found that the RMS increased with the additional increase of TAA. Furthermore, the increasing size of the large aggregates on the surface of the TiO_2 film was observed in the AFM images (Figure 6).

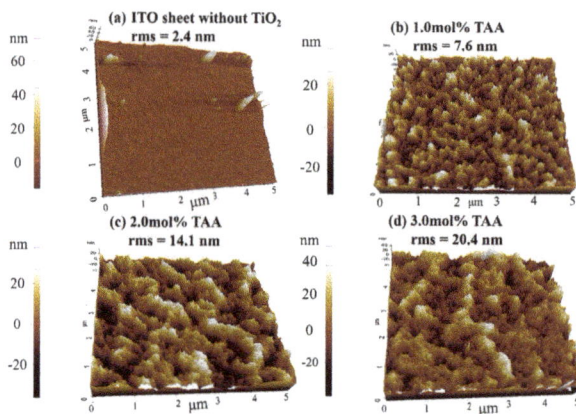

Figure 6. AFM images of the surface of the TiO_2 layers sprayed by the TiO_2 ethanol solution with the addition of TAA, (**a**) ITO sheet without TiO_2 film; (**b**) 1.0 mol % TAA; (**c**) 2.0 mol % TAA; and (**d**) 3.0 mol % TAA with respect to the TiO_2 content.

The devices with the highest performance were prepared based on the TiO_2 compact layer sprayed using the TiO_2 solution with 1.0 mol % of TAA. The highest PCE of 14.32% was achieved, with a V_{oc} of 1.008 mV, a J_{sc} of 20.82 mA/cm^2, and a FF of 68.3%. The improvement of the performance was attributed to the addition of TAA, which enhanced the connection of the TiO_2 nanoparticles. However, the addition of TAA increased the roughness of the film, which had a negative influence on the preparation of the perovskite. It was also observed that the hysteresis of the PSCs increased with the addition of TAA (Figure 5). To study the stability of PSCs based on ultrasonically sprayed TiO_2, a typical PSC on ITO-glass was placed in a N_2-filled glove box for long-duration stability testing. The PSC showed a good stability, as shown in Figure 7. The PCE of the device increased after two days, which can be attributed to the increase of the FF, caused by the enhancement of the connection between the TiO_2 particles and the oxidation of the Spiro-OMeTAD.

The optimized process of the ultrasonic spraying of 1.0 mg/mL TiO_2 solution with an addition of 1.0 mol % TAA was employed to fabricate PSCs on the flexible ITO-PEN substrate. A photographic image of the flexible device with a large area of 5.0 cm × 5.0 cm is shown in Figure 8a. The best-performing cell exhibited an outstanding photovoltaic performance with a V_{oc} of 945 mV, a J_{sc} of 20.68 mA/cm^2, a FF of 55.6% and a PCE of 10.87%, tested under AM 1.5 G sunlight (100 mW/cm^2) using a metal mask of 0.16 cm^2 (Figure 8b), which offers great potential for flexible PSCs with a large area. Figure 9 shows the current-voltage curve of a flexible large-area PSC with an active area of 13.5 cm^2, tested without a mask under standard sunlight. A PCE of 4.33% was achieved for the large-area flexible PSC.

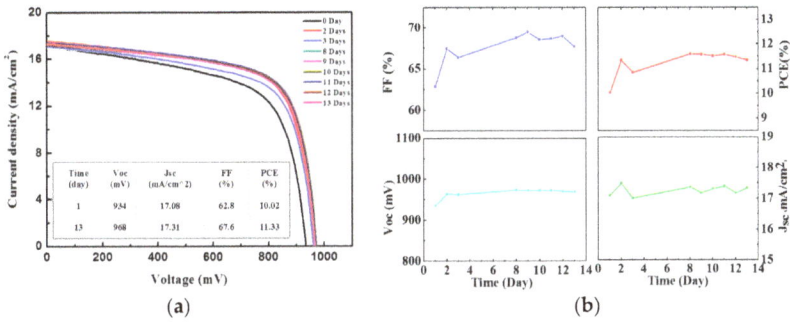

Figure 7. Long-duration stability test of a typical PSC based on ITO-glass, measured under AM 1.5 G sunlight (100 mW/cm^2) using a metal mask of 0.16 cm^2. (**a**) J-V curve of the PSC; and (**b**) FF, PCE, J_{sc}, V_{oc}.

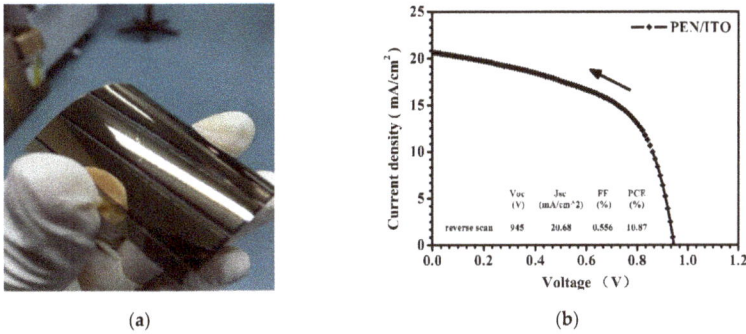

Figure 8. (**a**) Photographic image of the flexible device with a large area of 5.0 cm × 5.0 cm and (**b**) J-V curve of the flexible device fabricated by using the optimized parameters of ultrasonic spraying on polymer ITO-PEN substrate, tested under AM 1.5 G sunlight (100 mW/cm^2) using a metal mask of 0.16 cm^2.

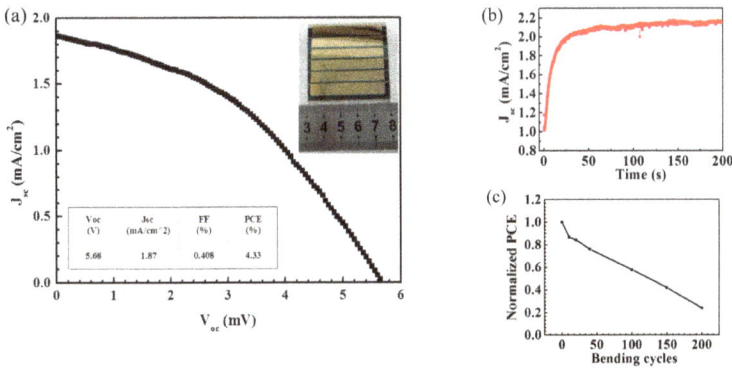

Figure 9. (**a**) J-V characteristics of a typical flexible PSC of 5 cm × 5 cm with an active area of 13.5 cm^2, tested without a mask. Inset photograph is the typical flexible PSC. (**b**) Photocurrent density measured as a function of time for the same flexible PSC held at 2.5 V forward bias. (**c**) Normalized PCE of the large-area flexible PSC after bending tests using 50 mm radii of curvature.

4. Conclusions

In summary, using an ultrasonic spraying technology, we prepared a low-temperature processed TiO_2 compact layer with low roughness compared to the thickness of the TiO_2 film on both rigid and flexible substrates with a large area. Two different drying process-related regimes were studied. The spraying method of the w-film regime showed advantages in preparing low-temperature processed TiO_2, resulting in a film with a RSM of 3.3 nm. A small amount of TAA was added into the TiO_2 dispersion to enhance the connection of the particles, thus improving the performance of the PSCs. The best-performing device was achieved with a PCE of 14.32% based on ITO-glass, and a PCE of 10.87% based on ITO-PEN. This method provides a potential route for the fabrication of efficient large-area flexible PSCs.

Acknowledgments: This work is financially supported by the National Natural Science Foundation of China (NSFC 51672202) and the Fundamental Research Funds for the Central Universities (WUT: 2016IVA085). Hubei Key Laboratory of Low Dimensional Optoelectronic Materials and Devices at Hubei University of Arts and Science was acknowledged for the measurement of AFM. The Analytical and Testing Center of Wuhan University of Technology and the Wuhan National Laboratory for Optoelectronics are also acknowledged for the SEM and TEM characterizations.

Author Contributions: Fuzhi Huang, Juan Zhao, and Peng Zhou conceived and designed the experiments; Peng Zhou performed the experiments; Fuzhi Huang and Peng Zhou analyzed the data; Wangnan Li, Tianhui Li, Tongle Bu, Xueping Liu, Jing Li, Jiang He, Rui Chen and Kunpeng Li contributed characterizations; Peng Zhou wrote the paper.

Conflicts of Interest: The authors declare no conflict of interest.

References

1. Park, N.G. Perovskite solar cells: An emerging photovoltaic technology. *Mater. Today* **2015**, *18*, 65–72. [CrossRef]
2. Kim, H.S.; Im, S.H.; Park, N.G. Organolead halide perovskite: New horizons in solar cell research. *J. Phys. Chem. C* **2014**, *118*, 5615–5625. [CrossRef]
3. Hodes, G. Perovskite-Based Solar Cells. *Science* **2013**, *342*, 317–318. [CrossRef] [PubMed]
4. Bai, S.; Wu, Z.; Wu, X.; Jin, Y.; Zhao, N.; Chen, Z.; Mei, Q.; Wang, X.; Ye, Z.; Song, T.; et al. High-performance planar heterojunction perovskite solar cells: Preserving long charge carrier diffusion lengths and interfacial engineering. *Nano Res.* **2014**, *7*, 1749–1758. [CrossRef]
5. Dong, Q.; Fang, Y.; Shao, Y.; Mulligan, P.; Qiu, J.; Cao, L.; Huang, J. Electron-hole diffusion lengths >175 µm in solution-grown $CH_3NH_3PbI_3$ single crystals. *Science* **2015**, *347*, 967–970. [CrossRef] [PubMed]
6. Stranks, S.D.; Eperon, G.E.; Grancini, G.; Menelaou, C.; Alcocer, M.J.P.; Leijtens, T.; Herz, L.M.; Petrozza, A.; Snaith, H.J. Electron-Hole Diffusion Lengths Exceeding 1 Micrometer in an Organometal Trihalide Perovskite Absorber. *Science* **2013**, *342*, 341–344. [CrossRef] [PubMed]
7. Bu, T.; Wen, M.; Zou, H.; Wu, J.; Zhou, P.; Li, W.; Ku, Z.; Peng, Y.; Li, Q.; Huang, F.; et al. Humidity controlled sol-gel Zr/TiO_2 with optimized band alignment for efficient planar perovskite solar cells. *Sol. Energy* **2016**, *139*, 290–296. [CrossRef]
8. Kojima, A.; Teshima, K.; Shirai, Y.; Miyasaka, T. Organometal halide perovskites as visible-light sensitizers for photovoltaic cells. *J. Am. Chem. Soc.* **2009**, *131*, 6050–6051. [CrossRef] [PubMed]
9. Kim, H.S.; Lee, C.R.; Im, J.H.; Lee, K.B.; Moehl, T.; Marchioro, A.; Moon, S.J.; Humphry-Baker, R.; Yum, J.H.; Moser, J.E.; et al. Lead iodide perovskite sensitized all-solid-state submicron thin film mesoscopic solar cell with efficiency exceeding 9%. *Sci. Rep.* **2012**, *2*, 591. [CrossRef] [PubMed]
10. Mahmood, K.; Swain, B.S.; Jung, H.S. Controlling the surface nanostructure of ZnO and Al-doped ZnO thin films using electrostatic spraying for their application in 12% efficient perovskite solar cells. *Nanoscale* **2014**, *6*, 9127–9138. [CrossRef] [PubMed]
11. Heo, J.H.; Song, D.H.; Han, H.J.; Kim, S.Y.; Kim, J.H.; Kim, D.; Shin, H.W.; Ahn, T.K.; Wolf, C.; Lee, T.W.; et al. Planar $CH_3NH_3PbI_3$ perovskite solar cells with constant 17.2% average power conversion efficiency irrespective of the scan rate. *Adv. Mater.* **2015**, *27*, 3424–3430. [CrossRef] [PubMed]

12. Yang, G.; Tao, H.; Qin, P.; Kw, W.; Fang, G. Recent progress in electron transport layers for efficient perovskite solar cells. *J. Mater. Chem. A* **2016**, *4*, 3970–3990. [CrossRef]

13. Peng, B.; Jungmann, G.; Jager, C.; Haarer, D.; Schmidt, H.W.; Thelakkat, M. Systematic investigation of the role of compact TiO$_2$ layer in solid state dye-sensitized TiO$_2$ solar cells. *Coord. Chem. Rev.* **2004**, *248*, 1479–1489. [CrossRef]

14. Kumar, M.H.; Yantara, N.; Dharahi, S.; Graetzel, M.; Mhaisalkar, S.; Boix, P.P.; Mathews, N. Flexible, low-temperature, solution processed ZnO-based perovskite solid state solar cells. *Chem. Commun.* **2013**, *49*, 11089–11091. [CrossRef] [PubMed]

15. Liu, D.; Yang, J.; Kelly, T.L. Compact layer free perovskite solar cells with 13.5% efficiency. *J. Am. Chem. Soc.* **2014**, *136*, 17116–17122. [CrossRef] [PubMed]

16. Kim, B.J.; Kim, D.H.; Lee, Y.Y.; Shin, H.W.; Han, G.S.; Hong, J.S.; Mahmood, K.; Ahn, T.K.; Joo, Y.C.; Hong, K.S.; et al. Highly efficient and bending durable perovskite solar cells: Toward a wearable power source. *Energy Environ. Sci.* **2015**, *8*, 916–921. [CrossRef]

17. Jiang, C.Y.; Koh, W.L.; Leung, M.Y.; Chiam, S.Y.; Wu, J.S.; Zhang, J. Low temperature processing solid-state dye sensitized solar cells. *Appl. Phys. Lett.* **2012**, *100*, 113901. [CrossRef]

18. Munoz-Rojas, D.; Sun, H.; Iza, D.C.; Weickert, J.; Chen, L.; Wang, H.; Schmidt-Mende, L.; MacManus-Driscoll, J.L. High-speed atmospheric atomic layer deposition of ultra thin amorphous TiO$_2$ blocking layers at 100 °C for inverted bulk heterojunction solar cells. *Prog. Photovolt. Res. Appl.* **2013**, *21*, 393–400.

19. Hart, J.N.; Menzies, D.; Cheng, Y.B.; Simon, G.P.; Spiccia, L. Microwave processing of TiO$_2$ blocking layers for dye-sensitized solar cells. *J. Sol-Gel Sci. Technol.* **2006**, *40*, 45. [CrossRef]

20. Kim, H.-J.; Jeon, J.-D.; Kim, D. Y.; Lee, J.-J.; Kwak, S.-Y. Improved performance of dye-sensitized solar cells with compact TiO$_2$ blocking layer prepared using low-temperature reactive ICP-assisted DC magnetron sputtering. *J. Ind. Eng. Chem.* **2012**, *18*, 1807–1812. [CrossRef]

21. Das, S.; Gu, G.; Joshi, P.C.; Yang, B.; Aytug, T.; Rouleau, C.M.; Geohegan, D.B.; Xiao, K. Low thermal budget, photonic-cured compact TiO$_2$ layers for high-efficiency perovskite solar cells. *J. Mater. Chem. A* **2016**, *4*, 9685. [CrossRef]

22. Wojciechowski, K.; Saliba, M.; Leijtens, T.; Abate, A.; Snaith, H.J. Sub-150 degrees C processed meso-superstructured perovskite solar cells with enhanced efficiency. *Energy Environ. Sci.* **2014**, *7*, 1142–1147. [CrossRef]

23. Yang, D.; Yang, R.; Zhang, J.; Yang, Z.; Liu, S.; Li, C. High efficiency flexible perovskite solar cells using superior low temperature TiO$_2$. *Energy Environ. Sci.* **2015**, *8*, 3208. [CrossRef]

24. Sanjib, D.; Yang, B.; Gu, G.; Joshi, P.C.; Lvanov, L.N.; Rouleau, C.M.; Aytug, T.; Grohegan, D.B.; Xiao, K. High-performance flexible perovskite solar cells by using a combination of ultrasonic spray-coating and low thermal budget photonic curing. *ACS Photonics* **2015**, *2*, 680–686.

25. Supasai, T.; Henjongchom, N.; Tang, I.-M.; Deng, F.; Rujisamphan, N. Compact nanostructured TiO$_2$ deposited by aerosol spray pyrolysis for the hole-blocking layer in a CH$_3$NH$_3$PbI$_3$ perovskite solar cell. *Solar Energy* **2016**, *136*, 515–524. [CrossRef]

26. Kim, M.-C.; Kim, B.J.; Yoon, J.; Lee, J.-W.; Suh, D.; Park, N.-G.; Choi, M.; Jung, H.S. Electro-spray deposition of a mesoporous TiO$_2$ charge collection layer: Toward large scale and continuous production of high efficiency perovskite solar cells. *Nanoscale* **2015**, *7*, 20725. [CrossRef] [PubMed]

27. Bose, S.; Keller, S.S.; Alstrom, T.S.; Boisen, A.; Almdal, K. Process optimization of ultrasonic spray coating of polymer films. *Langmuir* **2013**, *29*, 6911–6919. [CrossRef] [PubMed]

28. Tait, J.G.; Rand, B.P.; Heremans, P. Concurrently pumped ultrasonic spray coating for donor: Acceptor and thickness optimization of organic solar cells. *Org. Electron.* **2013**, *14*, 1002–1008. [CrossRef]

29. Wang, J.; Polleux, J.; Lim, J.; Dunn, B. Pseudocapacitive contributions to electrochemical energy storage in TiO$_2$ (Anatase) nanoparticles. *J. Phys. Chem. C* **2007**, *2*, 14925–14931. [CrossRef]

30. Huang, F.Z.; Dkhissi, Y.; Huang, W.; Xiao, M.; Benesperi, I.; Rubanov, S.; Zhu, Y.; Lin, X.; Jiang, L.; Zhou, Y. Gas-assisted preparation of lead iodide perovskite films consisting of a monolayer of single crystalline grains for high efficiency planar solar cells. *Nano Energy* **2014**, *10*, 10–18. [CrossRef]

Micromachines **2017**, *8*, 55

31. Li, X.; Bi, D.; Yi, C.; Décoppet, J.D.; Luo, J.; Zakeeruddin, S.M.; Hagfeldt, A.; Grätzel, M. A vacuum flash-assisted solution process for high-efficiency large-area perovskite solar cells. *Science* **2016**, *353*, 58–62. [CrossRef] [PubMed]
32. Deegan, R.; Bakajin, O.; Dupont, T. Capillary flow as the cause of ring stains from dried liquid drops. *Nature* **1997**, *389*, 827–829. [CrossRef]

micromachines

MDPI

Article

Direct Silver Micro Circuit Patterning on Transparent Polyethylene Terephthalate Film Using Laser-Induced Photothermochemical Synthesis

Chen-Jui Lan, Song-Ling Tsai and Ming-Tsang Lee *

Department of Mechanical Engineering, National Chung Hsing University, Taichung 402, Taiwan; lan.andy@gmail.com (C.-J.L.); s3677265@gmail.com (S.-L.T.)
* Correspondence: mtlee@nchu.edu.tw; Tel.: +886-4-2284-0433

Academic Editors: Seung Hwan Ko, Daeho Lee and Zhigang Wu
Received: 16 January 2017; Accepted: 6 February 2017; Published: 13 February 2017

Abstract: This study presents a new and improved approach to the rapid and green fabrication of highly conductive microscale silver structures on low-cost transparent polyethylene terephthalate (PET) flexible substrate. In this new laser direct synthesis and pattering (LDSP) process, silver microstructures are simultaneously synthesized and laid down in a predetermined pattern using a low power continuous wave (CW) laser. The silver ion processing solution, which is transparent and reactive, contains a red azo dye as the absorbing material. The silver pattern is formed by photothermochemical reduction of the silver ions induced by the focused CW laser beam. In this improved LDSP process, the non-toxic additive in the transparent ionic solution absorbs energy from a low cost CW visible laser without the need for the introduction of any hazardous chemical process. Tests were carried out to determine the durability of the conductive patterns, and numerical analyses of the thermal and fluid transport were performed to investigate the morphology of the deposited patterns. This technology is an advanced method for preparing micro-scale circuitry on an inexpensive, flexible, and transparent polymer substrate that is fast, environmentally benign, and shows potential for Roll-to-Roll manufacture.

Keywords: laser direct synthesis and patterning; flexible electronics; laser direct write; transparent conductors

1. Introduction

The market and applications of flexible, stretchable, and wearable electronics have increased remarkably in recent years. The search for rapid and cost-effective ways to combine electronics with flexible and stretchable substrates has also grown enormously. One central issue that challenges the fabrication of flexible electronics is that most flexible substrates, such as polymers, are unsuitable for the conventional photolithography used in microelectronic fabrication because they have poor chemical and thermal resistance. The thermal issue also leads to poor adhesion between the metallic materials and the polymer substrate caused by significantly large differences in thermal and mechanical properties [1]. To overcome these problems, especially the thermally related limitations, several advanced low-temperature manufacturing techniques have been developed that allow high resolution patterning [2]. These include nano-imprinting [3], transfer printing [4], precision deposition and low temperature process using nanoparticle ink [5], and laser-assisted direct sintering and patterning of nanomaterials [6–8]. These techniques take advantage of the melting point depression effect of nanomaterials, which allows the nanoparticles of metal to be easily sintered at moderate temperature [9]. The material is deposited on the substrate from a solution followed by low temperature sintering to form highly conductive nano-/microscale structures. The pattern can be produced by selective deposition or selectively sintering. The nanomaterial-based fabrication technique is fast, maskless, and gives reasonably good resolution mainly depending on the laser parameters and optics. Large area processing for Roll-to-Roll manufacturing is also possible by the laser-nanomaterial-based microfabrication technique.

However, all these nanomaterial-based processes require the synthesis of metal nanomaterials, which is usually a complex, time-consuming, and costly procedure. The coating and recycling of nanomaterial on a polymer substrate is also very challenging. The novel laser direct synthesis and pattering (LDSP) technology presented here provides a low temperature, green, nanoparticle-free process that does not require masks, or vacuum, and is ideal for the rapid fabrication of flexible electronics [10,11]. In contrast to the nanomaterial-based laser-assisted manufacturing processes such as laser direct write that utilizes a pre-synthesized metal nanomaterial solution to deposit the material followed by laser sintering, a transparent and particle-free metallic ion reaction solution is applied to the polymer substrate in the LDSP process. A focused continuous wave (CW) laser beam is scanned across the polymer surface to create a pattern. The laser energy absorbed by the tinted polymer substrate heats the reaction solution near the interface. Metallic ions are reacted and precipitate to form the predetermined pattern on the substrate surface. LDSP uses a nanoparticle-free metal ion precursor—not metal nanoparticle ink—which makes the process significantly faster. The process is simpler and green when compared to conventional laser direct sintering of metal nanoparticles. However, one limitation of the LDSP process is that the polymer substrate needs to absorb the scanned laser light. In other words, if an economic CW visible laser is to be used as a heat source, then the polymer substrate has to be tinted. However, for applications needing highly transparent and flexible electrodes, such as touch panels and flexible solar cells, a transparent polymer substrate has to be used. In such a case, the reactive ionic solution itself must absorb the incident laser light. A possible solution is to pre-deposit an absorbing layer on the substrate [12]. However, this could induce the complexity of the fabrication process and could also affect the transparency of the substrate. Another straightforward solution involves mixing the transparent ionic solution with a highly absorptive dye. The laser beam can then be directed through the transparent substrate to the top surface from below where it will be absorbed by the adjacent solution to initiate a thermochemical reaction and cause silver synthesis and patterning.

Although conceptually promising, the feasibility and the consequences of this approach remain to be proved. In addition, for realization of economic flexible electronics, low-cost transparent substrates such as polyethylene terephthalate (PET) film should be used. However, this kind of substrate is usually vulnerable to elevated temperature during the fabrication process. The thermal impact on the PET substrate subjected to the modified LDSP process also needs to be investigated. In this study, an improved LDSP process has been modified as described. Specifically, a red dye was added to the transparent nanoparticle-free reactive ion solution. The solution was applied to the transparent PET substrate, after which the LDSP process, where a focused laser beam was directed through the underside of the transparent PET substrate, was carried out. The red-colored reactive solution serves as the absorber of photon energy and converts it to thermal energy for the chemical reaction. The effects of the additive materials, concentration and the number of laser scans were all investigated. Electrical conductivities and the mechanical durability of the resulting nano-/microscale conductive patterns on the transparent PET substrate were measured and discussed. It is emphasized that, and shown in the results, using the red dye to make the reactive solution absorptive is a simple and cost-effective way to expand the application of the LDSP technology to include transparent polymer substrates. Unlike the other possible types of additives where nano-powders are used to form adsorptive colloids, the reactive solution mixed with red dye used in this study is free from precipitation and unwanted chemical reactions caused by the use of additives. Numerical simulation was also carried out on the heat and flow transport near the laser focal spot. Discussions of the impact on the resulting silver line structures from the temperature and velocity fields are also provided.

2. Materials and Methods

2.1. Processing Solution Preparation

The transparent reactive silver ion solution was prepared following a reported procedure [13] with modification. In brief, 1.0 g of silver acetate (anhydrous 99%, Alfa Aesar, Ward Hill, MA, USA) was dissolved in 2.5 mL of aqueous ammonium hydroxide (28%~30%, ACS Reagent, J.T. Baker, Phillipsburg, NJ, USA) at room temperature with vigorous stirring, followed by the drop-wise addition of 0.2 mL of formic acid (88%, ACS Reagent, J.T. Baker). The solution was filtered through a 220 nm poly(vinylidene fluoride) (PVDF) filter. The resulting particle-free clear ionic solution contains approximately 22% silver. Separately, a red dye, Allura red AC, was dissolved in ethylene glycol (EG, 99%, Alfa Aesar) under sonication to give solutions with molarities of 8 mM, 16 mM, and 32 mM. The red-EG was then mixed with the ionic solution at a volumetric ratio of 1:1 to complete the preparation of the EG–silver ion reaction solutions. Consequently, process solutions with red-dye concentration of 4 mM, 8 mM, and 16 mM were utilized in the current study. It should be pointed out that the red dye was chosen in the current study owing to its effective absorption to the green light laser energy, as also to be shown and discussed in the following section regarding the optical property of the red dye.

2.2. The LDSP Process

The diagram in Figure 1 illustrates the setup of the experimental apparatus used in the LDSP process. The 50-μm-thick PET film with an area of 5 cm × 5 cm used for this experiment was cleaned by rinsing with ethanol and DI water consecutively. The film was then treated with oxygen plasma in an Atmospheric Pressure Plasma Cleaner (Output voltage: 10,000 V to 50,000 V, output frequency: 4 MHz to 5 MHz) for 10 s to make the surface hydrophilic and improve adhesion of the silver to the polymer surface. The PET substrate was then placed carefully on a glass slide sample holder. The process solution was pipetted onto the substrate surface to form a liquid film typically 2~3 mm thick. The thickness of the process solution was controlled by controlling the amount of the solution being dispensed. It should be noted that the LDSP process may be sensitive to liquid film thickness as discussed in our previous studies [10,14]. As also to be shown from the numerical simulation results, the range of temperature and fluid flow in the process solution affected by the focused laser is confined within 1 mm surround the focal point. Therefore, 2~3-mm-thick process solution in the current study is sufficient to prevent the impact from the liquid-air boundary. The laser beam from a continuous wave (CW) diode laser ($\lambda = 532$ nm) was aligned and focused to 50 μm in diameter on the top surface of the PET substrate from underneath with the assistance of a beam expander and beam profile analyzer (BeamGage® Beam Profiler SP620U, Ophir Optronics®, Jerusalem, Israel). The focused laser beam was then directed by a programmable galvanometer scanner system (SCANLAB hurryScan® II-7, SCANLAB GmbH, Munich, Germany) to selectively scan a predetermined light pattern through the transparent PET substrate, the energy of which was absorbed by the process solution. Since the PET substrate is transparent and thin, there was no significant extinction of the laser light and the process solution absorbed the laser energy very effectively. Figure 2 shows the absorption coefficient of the process solutions with respect to the concentration of the red dye. Measurements were carried out using a UV-Vis spectrophotometer (Genesys 10S, Thermo Scientific, Waltham, MA, USA). Due to the limited measurement range of the spectrophotometer, three diluted solution concentrations were used. Assuming negligible scattering and from extrapolation of the measured absorption coefficient using the fitted curve as shown in Figure 2, the penetration distance attributed to absorption by the 16 mM process solution was approximately 65 μm [15]. This estimated result suggests that the absorption of the incident laser energy, and consequent heating of the process solution, would be confined to a vicinity of ~$O(10^2$ μm) to the PET surface. A small and well controlled thermal impact area is essential for the LDSP process to ensure good resolution as well as the recovery of unreacted process solution. After the LDSP process was completed, the PET surface was washed carefully with deionized water and ethanol and air-dried to remove any remaining process solution.

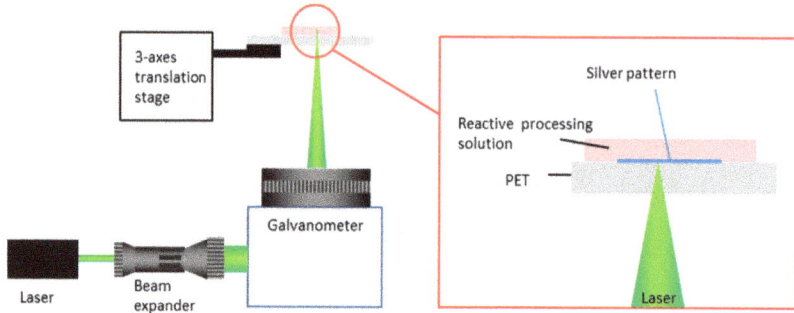

Figure 1. Experimental apparatus for the modified laser direct synthesis and pattering (LDSP) process.

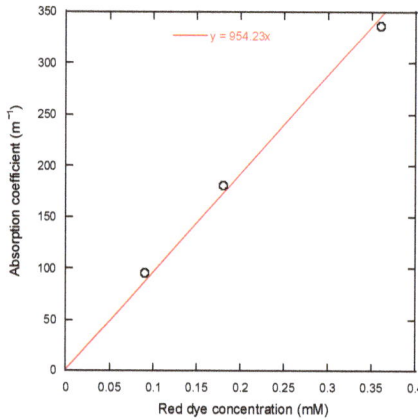

Figure 2. The absorption coefficient with respect to the red dye concentrations in the process (Ag–EG) solution.

3. Results

In this study, the laser power was fixed at 100 mW, and the laser scan speed was 10 mm/s. The number of laser scans and the red-dye concentrations were varied to conduct the parametric studies. Figure 3 shows the silver lines fabricated by the LDSP process on PET substrate with a range of different laser scans (10×, 20×, 30×) and red-dye concentrations (4 mM, 8 mM, 16 mM, based on the mixed process solution). It can be clearly seen that the 4 mM red-dye process solution did not yield a continuous silver line even when the maximum number of scans (30×) was applied. The formation of such broken lines also indicates that the thermochemical reduction of silver ions on the PET surface is not uniform and steady. The temperature and the reactant concentration are two major factors affecting the chemical reaction rate and it is possible that the concentration of reactants and the red dye was not uniform along the laser scan path. In these experiments, the red dye in the reactive solution absorbs laser energy and participates in radiative heat transfer. The reaction temperature and thermally induced convection is affected by the absorption of the red dye in the solution and although the process solution appears to be homogenous, there could still be microscale fluctuations in concentration. Thermally induced fluid motion such as buoyancy-driven flow, the Soret effect, and Brownian motion will also affect the concentration of species in the solution [16]. These fluctuations are highly random, especially for solutions of low red-dye concentration, and increasing the number of laser scans can build the line incrementally. Nevertheless, for a real application, the number of scans should be minimal and the red-dye concentration was increased to improve uniformity and the absorption of laser light. As can be seen in Figure 3, the silver lines fabricated using a red-dye concentration of 8 mM yields continuous conductive lines after 20 laser scans and 10 laser scans were enough to give continuous lines with a 16 mM solution. The width of the deposited line is approximately 50 μm. The size and the resolution of the conductive patterns being fabricated by the LDSP process is mainly controlled by the size of the heat affect area [14], which are determined by the process parameters such as the laser intensity, scanning speed, and heat transfer properties of the process solution. In addition, the size and morphology of the LDSP deposited patterns could also be controlled with the assistance of microfluidic dispensers [17] in conjunction with adjusted laser parameters. Figure 4 shows the electrical resistivity, where measurable, of these silver lines. The resistivity decreased to become stable with an increase in the number of laser scans or red-dye concentration. It should be noted that, for the 16 mM solution, silver lines with 20 laser scans and 30 laser scans yield compatible resistivity, which suggests that the effective thickness of the line did not increase significantly with more than 20 laser scans. A similar trend had also been observed earlier [10,14]. This effect is attributed to the fact that, as the thickness of the silver line increases, it begins to reflect the incident laser light, the energy is no longer absorbed by the process liquid, and the chemical reaction therefore stops.

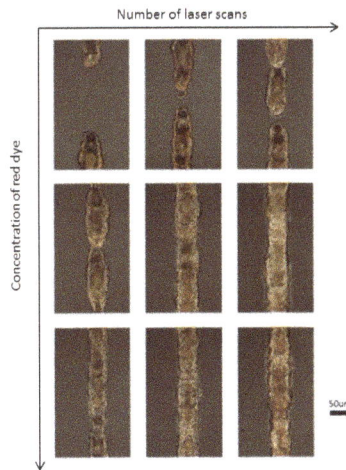

Figure 3. Silver lines fabricated by the LDSP process on the polyethylene terephthalate (PET) substrate with different numbers of laser scans: 10×, 20×, 30× and red-dye concentrations in the process solution of 4 mM, 8 mM, and 16 mM.

Figure 5 shows the silver line fabricated with the 16 mM process solution and 30 laser scans. The rough surface mainly results from solution flow and the possible sintering of silver nanoparticles that were produced during the laser scan [18], which could be improved by increasing the viscosity of the fluid [10]. The cross section of the silver line exhibits a concave shape. This surface morphology is one of the typical shapes found for patterns fabricated by laser chemical vapor deposition (LCVD), which is similar to the mechanism of LDSP. The formation of the morphology is closely related to the temperature distribution, the transport of reactants and products in the liquid (in this case, possibly silver nanoparticles being produced during the laser scan) and the surface diffusion of species [19]. The concave and round shape indicates strong impact from capillary-driven flow and thermophoresis, which are significant when the local temperature gradient in the fluid near the laser focal spot is large.

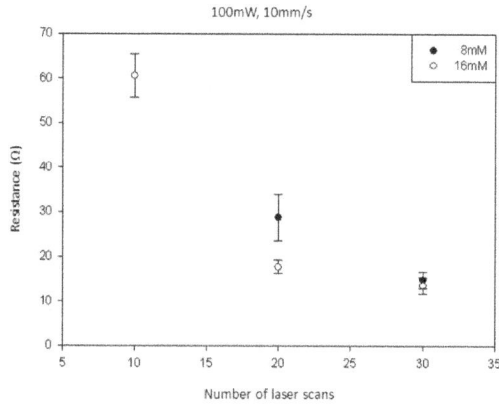

Figure 4. Electrical resistivity of silver lines with respect to the red-dye concentration and number of laser scans.

Figure 5. Scanning electron microscope (SEM) image of a silver line fabricated using a 16 mM process solution and 30 laser scans.

Figure 6 shows the results of energy dispersive X-ray analysis (EDX) analysis. It can be seen that the PET substrate after the LDSP in the process solution without silver acetate (but with red dye) contains no silver element, and the ratio of carbon to oxygen slightly increased compared with the original PET substrate. It can be assumed that very little red dye would be sintered and deposited on the PET substrate during the LDSP process. Silver lines fabricated with the process solution containing silver acetate shows dominant silver content and a small amount of carbon and oxygen. The carbon and oxygen could be from the red dye embedded in the silver line. These carbon and oxygen components in the conductive pattern are thought to be the main reason for its higher resistivity compared to pure silver.

Figure 7 shows the morphology of the cross section of the silver lines fabricated with LDSP (16 mM process solution, 30 laser scans). The measurements were carried out using a white light interferometer. The concave morphology can be clearly seen. The negative value at the center of the silver line could be attributed to slight melting of the PET substrate during the LDSP process. Using the average altitude profile shown in Figure 7 to calculate the cross-sectional area of the line, and assuming the canyon at the center under the plain substrate surface does not contribute to significant electrical conductivity, the electrical resistivity of the silver line was estimated to be 2.25×10^{-7} Ω·m, which is approximately 10 times that of pure silver. This can again be attributed to the fact that the composition of the conductive line contains a fraction of carbon and oxygen. Nevertheless, the conductivity of the line is sufficient to conduct electricity effectively as demonstrated in Figure 8. The durability test apparatus and the results are shown in Figure 9. After 10,000 repetitive bends, the drift in electrical resistance of the silver line was approximately 3 times higher than the initial value. Referring to the microstructures of the silver line showing in Figure 9b, there are silver micro-flakes that tend to be easily deformed and deteriorated. The variation in the electrical property is mainly attributed to the change in mechanical structure of the line. Further effort should be made to improve the mechanical robustness and the durability of the conductive patterns fabricated by this technique. It should also be noted that the processing speed could be increased by increasing the laser intensity, the absorptivity, and/or the reactivity of the process solution. The productivity of the LDSP process can thus be improved.

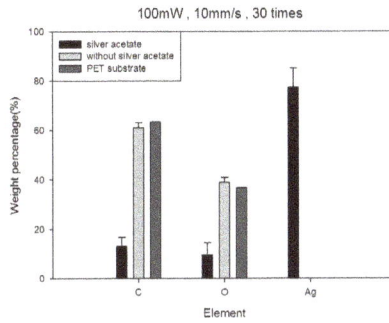

Figure 6. Energy dispersive X-Ray analysis (EDX) elemental analyses.

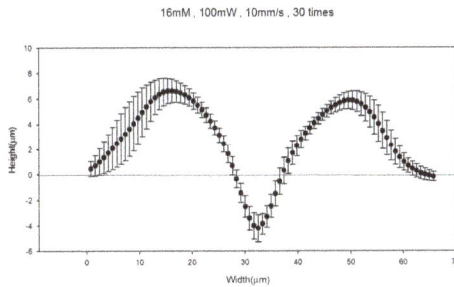

Figure 7. The morphology of the cross section of a silver line fabricated with LDSP (16 mM process solution, 30 laser scans).

Figure 8. Demonstration of conductive silver line fabricated by LDSP on the transparent PET substrate.

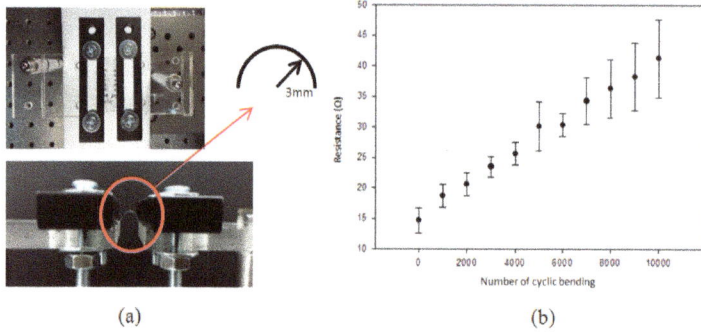

(a)   (b)

Figure 9. (**a**) Bending test apparatus; (**b**) resistance drift.

Numerical simulations were carried out to further investigate the transport phenomena during the fabrication process. COMSOL Multiphysics® software (COMSOL Inc., Burlington, MA, USA) was applied in the current study. Details of the numerical model can be found in an earlier report [14] and in [19,20]. In comparison to our previous numerical study of transport phenomena in the LDSP process, the main difference in the current numerical model is the absorbing material. In earlier studies, the substrate (polyimide film) served as a laser light absorbing material, while in the current study the laser energy was absorbed by the process solution. The heat source Q_{abs} from the absorption of a Gaussian laser beam moving in the liquid can be determined by [3,20,21]:

$$Q_{abs} = (1 - R)\gamma I_{pk} \exp\left[-\left(\frac{x - Ut}{\omega}\right)^2 - \left(\frac{y}{\omega}\right)^2 - \gamma z \right] \tag{1}$$

where R is the reflectivity, γ is the absorption coefficient of the process solution estimated from Figure 2, I_{pk} is the peak laser intensity, and ω is determined from the size of the laser beam on the x-y plane of the PET substrate. In the current simulation, the surface reflectivity from the PET substrate and the interfaces of the PET and process solution were assumed to be negligible. Chemical reactions and liquid-vapor phase change were also neglected in this analysis considering that there was no significant boiling observed during the process.

Figure 10a shows the temperature profile of the cross section at the center of the laser beam in the scan direction. It can be seen that the highest temperature spot is inside the reaction fluid and is some tens of microns above the PET surface. In contrast to the previous studies where the laser beam energy was absorbed by the polymer substrate itself, so that the highest temperature in the reaction fluid was on the interface between the solid substrate and the fluid, the process solution absorbs the incident laser energy gradually along the laser path within the fluid, and the hottest spot is in the interior of the process fluid region. This, in turn, results in a strong in-plane flow right on the PET surface below the hottest spot. Strong in-plane flow, especially near the centerline (the symmetric line), could adversely affect the deposition and adhesion of the reduced silver (from the hottest spot) on the PET surface. Additionally, the symmetric swirling flow as shown in Figure 10b indicates that the reduced silver could precipitate on the shoulders along the laser scanning path, which also corresponds to the morphology of the silver line patterns shown in Figure 5. It should also be noted that the upward flow at the centerline due to buoyancy exists with an absorptive substrate (polyimide) or an absorptive fluid as in the current case. The morphology of the deposited pattern, however, is closely related to the intensity of the buoyancy flow, which generally increases with laser intensity, which determines the temperature of the focal point [19]. Other forces such as surface tension would be important for improving the surface morphology [22]. Furthermore, the roughness of the surface of the deposited patterns could yield a random scattering of the incident laser irradiation [23]. Thus, the reflectance to the laser light by the deposited patterns should be carefully considered. Continuous study in this regard is being carried out.

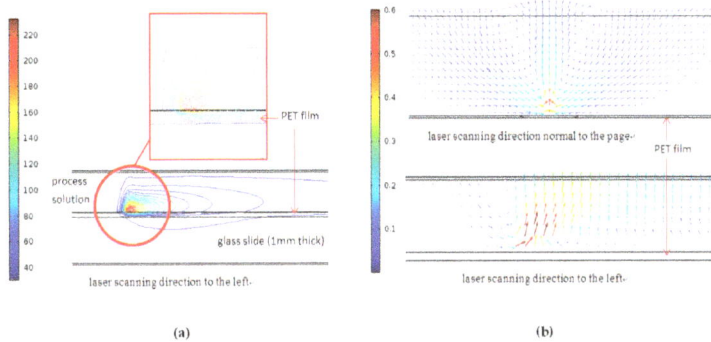

(a)   (b)

Figure 10. (**a**) Temperature (unit: °C) profile; and (**b**) velocity (unit: mm/s) profile near the laser focal point.

4. Conclusions

In the current study, a new and improved approach to the rapid and green fabrication of highly conductive nano-/micro-scale silver structures on low-cost transparent polyethylene terephthalate (PET) flexible substrate was successfully demonstrated. Predetermined silver pattern was formed by photothermochemical reduction of the silver ions induced by the focused CW laser beam in a processing solution containing a red azo dye as the absorbing material. The effects of the additive materials, concentration, and the number of laser scans were all investigated. The electrical resistivity of the silver line was estimated to be 2.25×10^{-7} $\Omega \cdot$m, which is approximately 10 times that of pure silver. Numerical simulation was also carried out on the heat and flow transport near the laser focal spot to provide insights of the impact on the resulting silver line structures from the temperature and velocity fields. The concave and round shape of the conductive lines indicates strong impact from capillary-driven flow, thermophoresis, and buoyancy flow in the current LDSP configuration. This study and the revised LDSP technology provides a potential approach for preparing nano/micro-scale circuitry on inexpensive, flexible, and transparent polymer substrates that are fast, environmentally benign, and cost-effective.

Acknowledgments: The authors would like to acknowledge the financial support for this study from the Ministry of Science and Technology, Taiwan R.O.C. Project Nos. 105-2221-E-005-041 and 105-2218-E-005-005.

Author Contributions: Chen-Jui Lan, Song-Ling Tsai, and Ming-Tsang Lee conceived and designed the experiments; Chen-Jui Lan and Song-Ling Tsai performed the experiments; Song-Ling Tsai performed the numerical simulations; Chen-Jui Lan and Ming-Tsang Lee analyzed the data; Ming-Tsang Lee wrote the paper.

Conflicts of Interest: The authors declare no conflict of interest.

References

1. Perelaer, J.; Smith, P.J.; Mager, D.; Soltman, D.; Volkman, S.K.; Subramanian, V.; Korvink, J.G.; Schubert, U.S. Printed electronics: The challenges involved in printing devices, interconnects, and contacts based on inorganic materials. *J. Mater. Chem.* **2010**, *20*, 8446–8453. [CrossRef]
2. Subramanian, V.; Lee, T. Nanotechnology-based flexible electronics. *Nanotechnology* **2012**, *23*, 340201. [CrossRef] [PubMed]
3. Lee, D.; Pan, H.; Sherry, A.; Ko, S.H.; Lee, M.T.; Kim, E.; Grigoropoulos, C.P. Large-area nanoimprinting on various substrates by reconfigurable maskless laser direct writing. *Nanotechnology* **2012**, *23*, 344012. [CrossRef] [PubMed]

4. Yu, X.; Mahajan, B.; Shou, W.; Pan, H. Materials, mechanics, and patterning techniques for elastomer-based stretchable conductors. *Micromachines* **2016**, *8*, 7. [CrossRef]

5. Kullmann, C.; Schirmer, N.C.; Lee, M.-T.; Ko, S.H.; Hotz, N.; Grigoropoulos, C.P.; Poulikakos, D. 3D micro-structures by piezoelectric inkjet printing of gold nanofluids. *J. Micromech. Microeng.* **2012**, *22*, 055022. [CrossRef]

6. Yeo, J.; Hong, S.; Lee, D.; Hotz, N.; Lee, M.T.; Grigoropoulos, C.P.; Ko, S.H. Next generation non-vacuum, maskless, low temperature nanoparticle ink laser digital direct metal patterning for a large area flexible electronics. *PLoS ONE* **2012**, *7*, e42315. [CrossRef] [PubMed]

7. Lee, M.-T.; Lee, D.; Sherry, A.; Grigoropoulos, C.P. Rapid selective metal patterning on polydimethylsiloxane (PDMS) fabricated by capillarity-assisted laser direct write. *J. Micromech. Microeng.* **2011**, *21*, 095018. [CrossRef]

8. Hong, S.; Lee, H.; Lee, J.; Kwon, J.; Han, S.; Suh, Y.D.; Cho, H.; Shin, J.; Yeo, J.; Ko, S.H. Highly stretchable and transparent metal nanowire heater for wearable electronics applications. *Adv. Mater.* **2015**, *27*, 4744–4751. [CrossRef] [PubMed]

9. Buffat, P.; Borel, J.P. Size effect on the melting temperature of gold particles. *Phys. Rev. A* **1976**, *13*, 2287–2298. [CrossRef]

10. Liu, Y.K.; Lee, M.T. Laser direct synthesis and patterning of silver nano/microstructures on a polymer substrate. *ACS Appl. Mater. Interfaces* **2014**, *6*, 14576–14582. [CrossRef] [PubMed]

11. Liu, Y.-K.; Sie, Y.-Y.; Liu, C.-A.; Lee, M.-T. A novel laser direct writing system integrated with A&F XXY alignment platform for rapid fabrication of flexible electronics. *Smart Sci.* **2015**, *3*, 87–91.

12. Yeo, J.; Hong, S.; Kim, G.; Lee, H.; Suh, Y.D.; Park, I.; Grigoropoulos, C.P.; Ko, S.H. Laser-induced hydrothermal growth of heterogeneous metal-oxide nanowire on flexible substrate by laser absorption layer design. *ACS Nano* **2015**, *9*, 6059–6068. [CrossRef] [PubMed]

13. Walker, S.B.; Lewis, J.A. Reactive silver inks for patterning high-conductivity features at mild temperatures. *J. Am. Chem. Soc.* **2012**, *134*, 1419–1421. [CrossRef] [PubMed]

14. Tsai, S.-L.; Liu, Y.-K.; Pan, H.; Liu, C.-H.; Lee, M.-T. The coupled photothermal reaction and transport in a laser additive metal nanolayer simultaneous synthesis and pattering for flexible electronics. *Nanomaterials* **2016**, *6*, 12. [CrossRef]

15. Siegel, R.; Howell, J. *Thermal Radiation Heat Transfer*, 4th ed.; Taylor & Francis: New York, NY, USA, 2002.

16. Xu, L.; Lee, B.J.; Hanson, W.L.; Han, B. Brownian motion induced dynamic near-field interaction between quantum dots and plasmonic nanoparticles in aqueous medium. *Appl. Phys. Lett.* **2010**, *96*, 174101. [CrossRef]

17. Pan, P.-T.; Hsieh, M.-C.; Liu, C.-Y.; Pan, C.-W.; Zhao, K.-W.; Hsieh, Y.-W.; Wang, A.-B. A precision capillary coating system and applications. *Smart Sci.* **2016**, *4*, 151–159. [CrossRef]

18. Zhang, R.; Lin, W.; Moon, K.S.; Wong, C.P. Fast preparation of printable highly conductive polymer nanocomposites by thermal decomposition of silver carboxylate and sintering of silver nanoparticles. *ACS Appl. Mater. Interfaces* **2010**, *2*, 2637–2645. [CrossRef] [PubMed]

19. Grigoropoulos, C.P. *Transport in Laser Microfabrication*; Cambridge University Express: Cambridge, UK, 2009.

20. Lee, D.G.; Kim, D.K.; Moon, Y.J.; Moon, S.J. Effect of laser-induced temperature field on the characteristics of laser-sintered silver nanoparticle ink. *Nanotechnology* **2013**, *24*, 265702. [CrossRef] [PubMed]

21. Paeng, D.; Yeo, J.; Lee, D.; Moon, S.-J.; Grigoropoulos, C.P. Laser wavelength effect on laser-induced photo-thermal sintering of silver nanoparticles. *Appl. Phys. A* **2015**, *120*, 1229–1240. [CrossRef]

22. Fukuda, K.; Sekine, T.; Kumaki, D.; Tokito, S. Profile control of inkjet printed silver electrodes and their application to organic transistors. *ACS Appl. Mater. Interfaces* **2013**, *5*, 3916–3920. [CrossRef] [PubMed]

23. Gu, M.-J.; Chen, Y.-B. Modeling bidirectional reflectance distribution function of one-dimensional random rough surfaces with the finite difference time domain method. *Smart Sci.* **2014**, *1*, 101–106. [CrossRef]

micromachines

MDPI

Article

Wrinkled Graphene–AgNWs Hybrid Electrodes for Smart Window

Ki-Woo Jun, Jong-Nam Kim, Jin-Young Jung and Il-Kwon Oh *

Creative Research Initiative Centre for Functionally Antagonistic Nano-Engineering, Department of Mechanical Engineering, Korea Advanced Institute of Science and Technology (KAIST), 291 Daehak-ro, Yuseong-gu, Daejeon 34141, Korea; jkw82@kaist.ac.kr (K.-W.J.); jnkim4388@kaist.ac.kr (J.-N.K.); jinyjung@kaist.ac.kr (J.-Y.J.)
* Correspondence: ikoh@kaist.ac.kr; Tel.: +82-42-350-1520

Academic Editors: Seung Hwan Ko, Daeho Lee and Zhigang Wu
Received: 31 December 2016; Accepted: 26 January 2017; Published: 1 February 2017

Abstract: Over the past few years, there has been an increasing demand for stretchable electrodes for flexible and soft electronic devices. An electrode in such devices requires special functionalities to be twisted, bent, stretched, and deformed into variable shapes and also will need to have the capacity to be restored to the original state. In this study, we report uni- or bi-axially wrinkled graphene–silver nanowire hybrid electrodes comprised of chemical vapor deposition (CVD)-grown graphene and silver nanowires. A CVD-grown graphene on a Cu-foil was transferred onto a bi-axially pre-strained elastomer substrate and silver nanowires were sprayed on the transferred graphene surface. The pre-strained film was relaxed uni-(or bi-)axially to produce a wrinkled structure. The bi-axially wrinkled graphene and silver nanowires hybrid electrodes were very suitable for high actuating performance of electro-active dielectric elastomers compared with the wrinkle-free case. Present results show that the optical transparency of the highly stretchable electrode can be successfully tuned by modulating input voltages.

Keywords: graphene; silver-nanowires; wrinkles; elastomer actuators; tunable transparency

---

1. Introduction

Smart windows, which can control the level of light transmission, have recently been attracting great interest, because they can be applied to a variety of applications, including vehicle windows, exterior wall windows and skylight windows. Until now, most smart window technology has been based on the redox reaction of a molecular element and a chromic material in response to external stimuli such as light, electricity, or temperature [1–4]. However, such materials are chemically unstable during the conversion processes, and are further disadvantageous in that they are difficult to control [4]. As a result, there is a need for a new smart window that is efficient in manufacturing, simple to implement, durable to operate and has a fast response time.

Among various approaches to smart window technology, there has been recent interest in techniques for modulating transmittance by changing the surface morphology of an elastomer actuator. In order to generate such a morphological change of soft materials, a flexible and stretchable conducting material should be used as compliant electrodes, which can be stretched or shrunk according to the movement of the soft material, such as nanowires, graphene, and hybrid materials [5–12]. In particular, techniques using a soft dielectric elastomer and stretchable electrodes are being studied for controlling surface morphology by applying electric input voltages. Smart window technology based on such soft materials can be applied to micro-lens arrays [13,14], flexible electronic devices [15,16], and variable diffraction gratings [17].

However, since most devices were operated with flat electrodes, the changes in transmittance were not so large, resulting in a limited application. In order to resolve this performance problem, research on smart windows using graphene and silver nanowire hybrid electrodes has been carried out, but a wrinkled structure implemented with stretchable hybrid materials has not been reported yet.

In this study, we developed a wrinkled hybrid electrode, comprised of CVD-grown graphene and silver nanowires, and reported a tunable transmittance by applying electrical fields to the dielectric elastomer actuators sandwiched with the wrinkled graphene–silver nanowire hybrid electrodes. A monolayer of graphene grown by CVD on a Cu-foil was transferred onto a bi-axially pre-strained elastomer substrate and silver nanowires were sprayed on the transferred graphene surface. The pre-strained film was released bi-(or uni-)axially to develop a wrinkled structure. The morphological and electrical characteristics of the wrinkled structures with bi-(or uni-)axially compressive strains were investigated using scanning electron microscopy (SEM). Also, the bi-axially wrinkled graphene and silver nanowire hybrid nanostructures were successfully applied to produce high actuating performance compared with the wrinkle-free case. Using the fabricated electrode, we demonstrated a device whose optical transparency was tunable with input voltages.

2. Preparation and Experiment

2.1. Materials

VHB 4905 film (3M Corp., Maplewood, MN, USA) is an appropriate material for dielectric elastomer actuators because the film is easy to handle, and both sides of the film are sticky enough for coating with electrode materials such as graphene and silver nanowires. In addition, the VHB 4905 is almost transparent, with a thickness of around 5 mm and a density of 960 kg/m^3. The dielectric constant of the VHB 4905 film is 3.21 at 1 kHz. The dissipation factor and dielectric breakdown strength are 0.0214 and 630 V/mm, respectively [18,19].

Silver nanowires (AgNWs) are nanostructures, generally with a diameter of ten to several hundred nanometers and lengths of tens of microns. Silver nanowire material is a kind of grayish powder and is often dispersed in solvents, such as water, ethanol, and isopropanol [20]. The silver nanowires used in this study were synthesized by the polyol process [21–23]. A cleaned two-neck flask was placed in a heating mantle. Polyvinylpyrrolidone (PVP, Mw = 55,000) 5.86 g in 190 mL of glycerol was heated at 150 °C for 2 h to remove moisture in the PVP. The solution was injected into the cleaned two-neck flask and cooled down to 55 °C. A mixture of 0.059 g of sodium chloride, 0.5 mL of deionized water, and 10 mL of glycerol was added into the prepared solution with 1.58 g silver nitrate. The suspension was heated up to 155 °C and left for 20 min for the additional reaction. The final product was washed using glass filters and stored in methanol.

2.2. Fabrication

The fabrication process of the graphene and silver nanowire hybrid electrode with a wrinkled structure is illustrated in Figure 1. In this process, the wrinkled structure of the elastomer surface was formed mainly due to the difference in modulus between the elastomer film and the graphene and silver nanowire electrode.

The first step is the transfer of the CVD-grown graphene onto the surface of the pre-stretched elastomer film. In this step, the copper layer is removed using a wet-etching process to allow the graphene transfer. In order to maintain electrical conductivity between the graphene islands on the elastomer surface, silver nanowires were sprayed on the wrinkled graphene surface. Then, the pre-stretched elastomer film was released, to prepare a hybrid electrode with a wrinkled structure. In order to fabricate a dielectric elastomer actuator with a wrinkled structure, the electrode formation method described above was also applied to the opposite side of the elastomer. Using this method, a wrinkled elastomer actuator whose light transmittance could be controlled was manufactured.

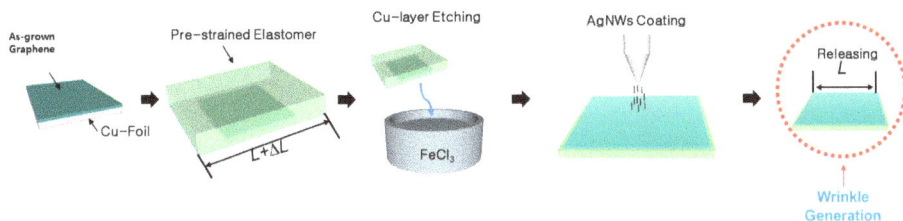

Figure 1. Fabrication process of the wrinkled graphene–AgNWs hybrid electrode.

2.3. Experiment Method

To induce the electro-mechanical deformation of the dielectric elastomer actuator, a high input voltage over 1 kV needs to be applied to both electrodes. For this purpose, a high voltage power supply (Trek Model 610E, Lockport, NY, USA) was connected to both the top and bottom electrodes of the actuator. To connect each compliant electrode to a cable from the power supply, conductive copper tape was used instead of the conventional cable-wires. Actuation testing was performed on an experiment table, and video data was acquired using a complementary metal-oxide-semiconductor (CMOS) image sensor to show the areal expansion of the actuator. To observe the variation in optical transmittance under actuating conditions, we placed the fabricated actuator on the sample holder of a UV-VIS spectroscopy. The optical transmittance was then examined, as input voltages were applied to the actuator.

3. Results and Discussions

3.1. Uni- and Bi-axially Wrinkled Electrode

Figure 2 shows SEM images of the hybrid electrode, which forms a wrinkled structure in the uni- and bi-axial directions on the elastomer film. Because the as-transferred graphene layer is fragmented, it has low electrical conductivity, and so by itself cannot function as an actuator electrode. In order to solve this problem, we additionally used silver nanowires to act as conductive bridges between the graphene islands. Silver nanowires with a length of about 10 to 20 μm were used, and the synthesized silver nanowires were thermally annealed at a temperature of 120 °C.

In Figure 2b,c, the surface morphology of the wrinkled structure formed along the uni- and bi-directions can be observed. A stretching–releasing method was used to create the wrinkled structure of the hybrid electrode. Pre-stretching was used to elongate the state of the elastomer film, which was used as the elastic body of the actuator. Different wrinkled structures can be formed depending on the pre-stretching direction. In this experiment, we formed wrinkled structures with uni- and bi-directions. The wavelength of the wrinkles is about 5 to 15 μm, as shown in Figure 2b,c.

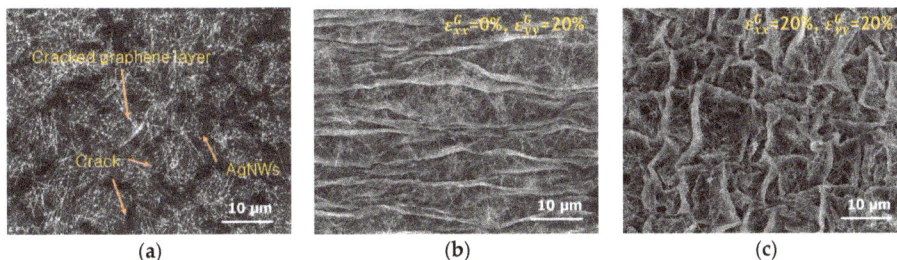

Figure 2. SEM images of the graphene–AgNWs hybrid electrode: (**a**) flat surface morphology, (**b**) uniaxially compressive strain of 20% and (**c**) biaxially compressive strain of 20%.

3.2. Morphology Variation with Tensile Strength

The working principle of the elastomer actuator is based on an attraction force which is generated between the two electrodes. This force causes an expansion of the electrode area in the in-plane direction. In order to examine the change of the surface in the multilayered elastomer-electrodes according to the pulling forces in the in-plane direction, the surface morphology was examined under tensile conditions, as shown in Figure 3. The uni- or bi-axial mechanical forces were applied to make uni- or bi-axially wrinkled patterns. The upper figure shows a wrinkled structure formed in one axis direction, and the lower figure shows a wrinkle structure formed in two axis directions. The surface patterns were observed as 0 to 50% tensile strain applied along the wrinkle direction. As the strain value increases, it can be seen that the wrinkles spread out. In the strained condition, it was confirmed that the interval between the pitches is increased, the height of the wrinkles and the density of electrical networks are lowered.

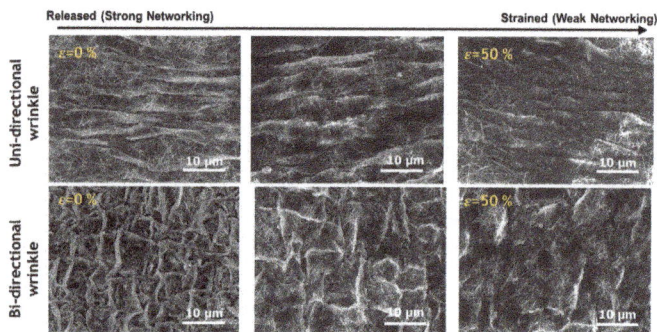

Figure 3. Morphology changes in the electrode surfaces with mechanical tensile strains.

3.3. Electrical Measurement

The relative resistance of the electrodes was examined in order to investigate the influence of the wrinkling morphology. Three types of specimens with different compressive strain values were prepared for this experiment. The compressive strain values applied to the specimens were 20%, 33% and 42%, respectively. As shown in Figure 4, the relative resistance ($\Delta R/R_0$) remains stable in the initial section, and as the strain value is increased, the magnitude of the relative resistance value is changed sharply. These results can be explained based on the wrinkled structure of the electrode. As shown in Figure 3, when the value of the tensile strain is increased, the hybrid electrode material, that is, the connectivity between the graphene and the silver nanowires, is weakened. For this reason, it can be understood that the resistance increases sharply in the section where there are fewer wrinkled structures.

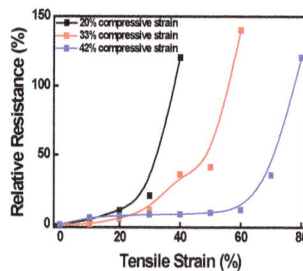

Figure 4. Resistance changes of wrinkled hybrid electrodes according to mechanical tensile strains.

3.4. Electromechanical Transducer with Elastomer Films

The dielectric elastomer actuator has a sandwich structure consisting of an elastomer film with two electrodes on both sides. For application, the elastomer film was fixed in a rigid frame made of polyacetal. Figure 5a illustrates the working principle of the elastomer actuator. When an input voltage is applied to the electrodes, electrons are accumulated on both sides of the wrinkled electrode. At this time, Maxwell stress is generated in the thickness direction of the elastomer film. Deformation occurs in the in-plane direction from the active area where the electrodes are formed.

In order to verify the operating performance of the actuator, we connected the electrode of the actuator to a DC high voltage supply capable of generating 0–10 kV. Then, the applied voltage was gradually increased from 0 to 6 kV, and the change in the area of the actuator was examined. For more precise performance analysis, it is essential to determine the area of the increased actuator under electrical stimuli. For this purpose, the moving images of the actuator were captured frame-by-frame, and analyzed using an image processing program. The actuation of conventional dielectric elastomer actuators using a high input-voltage caused the electrodes to expand with an isotropic areal strain as shown in Figure 5a. Isotropic behaviors of the dielectric elastomer actuator can be differently realized by the morphological configuration of hybrid electrodes. Figure 5b shows the experimental results to confirm the performance difference of the wrinkled and wrinkle-free electrodes when electrical input voltages are applied to two dielectric elastomer actuators. The actuator sample with wrinkled electrodes shows an areal change ($\Delta A / A_0$) of 33%. Within 4 kV, the areal change of the actuator is almost similar in both cases, but there is a remarkable maximum difference of about 6% at 6 kV.

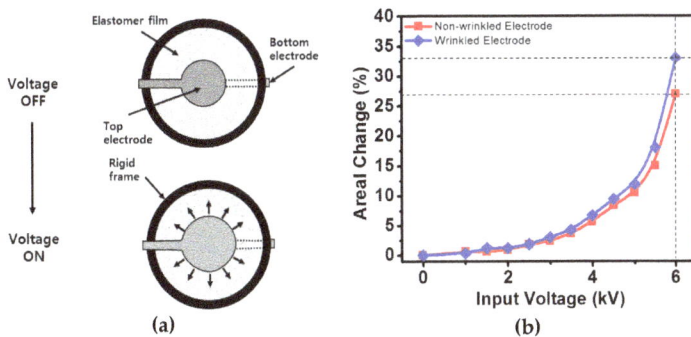

Figure 5. Actuation behaviors: (**a**) concept of isotropic elastomer actuation, (**b**) comparison of areal changes of wrinkle-free and wrinkled elastomer actuator.

As the applied voltage increases, the strength of the interaction force in the thickness direction increases in the elastomer actuator. This force has the effect of reducing the thickness of the film and increasing the area of the electrode. At this time, the transmittance of the active area is increased due to the reduced thickness of the film and the change in surface morphology on the surface of the elastomer film. Figure 6 is a graphical representation of the relationship between the rate of areal change and transmittance obtained when a voltage is applied to the actuator. Using a sample without the wrinkled electrodes, experimental results confirmed that the areal change rate was about 27% and the transmittance at 550 nm was changed by about 8%.

Using the same method as that used for the wrinkle-free sample, the change in active area of the dielectric elastomer actuator with wrinkled graphene–AgNWs hybrid electrodes was measured with respect to the applied voltages. Then, the transmittance change according to area changes was also examined. The wrinkled electrode structure exhibited about a 33% change in area ratio, and a transmittance change of about 15%. That is, although the wrinkled structure showed the same area change rate, it can be confirmed that the transmittance was about twice as large as that in the case of the wrinkle-free sample. When compared with the structure without wrinkles, the cause of this phenomenon is understood to be the difference in transmittance due to the scattering of light.

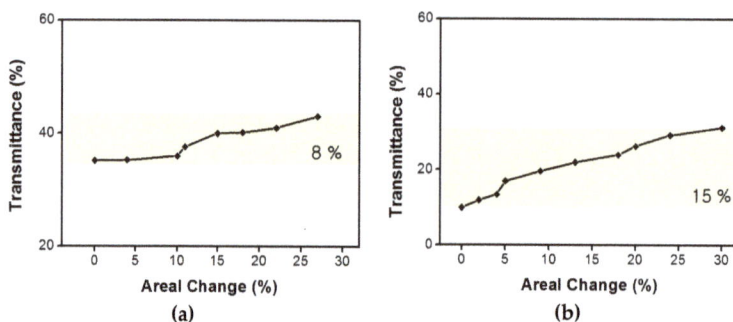

Figure 6. Optical transmittance control of (**a**) a wrinkle-free and (**b**) wrinkled elastomer actuator with electrical input.

4. Conclusions

A novel dielectric elastomer actuator with wrinkled graphene and silver nanowire hybrid electrodes was developed in this study. Graphene and silver nanowires were used to realize the highly conductive and compliant electrodes. The developed dielectric elastomer actuator with the wrinkled surfaces exhibited a large in-plane deformation of up to 33%. As compared with wrinkle-free electrodes, the proposed actuator with wrinkled electrodes exhibited a 10% higher areal change rate because of the synergistic effect of the ultrahigh conductivity of the silver nanowires and the remaining capacitance of the graphene under large deformation. This novel elastomer actuator with wrinkled hybrid electrodes may be a candidate to tune the transmittance of a smart window system.

Acknowledgments: This work was partially supported by the Ministry of Trade, Industry and Energy through Technology Innovation Program (Grant 10044410) and by the Creative Research Initiative program funded by the National Research Foundation of Korea (No. 2015R1A3A2028975). This work was supported by the Center for Advanced Soft Electronics funded by the Ministry of Science, ICT and Future Planning as Global Frontier Project (CASE-2011-0031640).

Author Contributions: K.W.J. performed the experiments and analyzed the data. J.N.K. and J.Y.J. supported the experiments and revised the paper. I.K.O. supervised the research work at the whole stage. All authors read and approved the final manuscript.

Conflicts of Interest: The authors declare no conflict of interest.

References

1. Beaujuge, P.M.; Reynolds, J.R. Color control in π-conjugated organic polymers for use in electrochromic devices. *Chem. Rev.* **2010**, *110*, 268–320. [CrossRef] [PubMed]
2. Bechinger, C.; Ferrer, S.; Zaban, A.; Sprague, J.; Gregg, B.A. Photoelectrochromic windows and displays. *Nature* **1996**, *383*, 608–610. [CrossRef]
3. Granqvist, C.G. Electrochromic tungsten oxide films: Review of progress 1993–1998. *Sol. Energy Mater. Sol. Cells* **2000**, *60*, 201–262. [CrossRef]
4. Lampert, C.M. Chromogenic smart materials. *Mater. Today* **2004**, *7*, 28–35. [CrossRef]
5. An, B.W.; Gwak, E.-J.; Kim, K.; Kim, Y.-C.; Jang, J.; Kim, J.-Y.; Park, J.-U. Stretchable, transparent electrodes as wearable heaters using nanotrough networks of metallic glasses with superior mechanical properties and thermal stability. *Nano Lett.* **2016**, *16*, 471–478. [CrossRef] [PubMed]
6. An, B.W.; Hyun, B.G.; Kim, S.-Y.; Kim, M.; Lee, M.-S.; Lee, K.; Koo, J.B.; Chu, H.Y.; Bae, B.-S.; Park, J.-U. Stretchable and transparent electrodes using hybrid structures of graphene–metal nanotrough networks with high performances and ultimate uniformity. *Nano Lett.* **2014**, *14*, 6322–6328. [CrossRef] [PubMed]

7. He, T.; Xie, A.; Reneker, D.H.; Zhu, Y. A tough and high-performance transparent electrode from a scalable and transfer-free method. *ACS Nano* **2014**, *8*, 4782–4789. [CrossRef] [PubMed]
8. Hsu, P.-C.; Wang, S.; Wu, H.; Narasimhan, V.K.; Kong, D.; Ryoung Lee, H.; Cui, Y. Performance enhancement of metal nanowire transparent conducting electrodes by mesoscale metal wires. *Nat. Commun.* **2013**, *4*, 2522. [CrossRef] [PubMed]
9. Hsu, P.-C.; Wu, H.; Carney, T.J.; McDowell, M.T.; Yang, Y.; Garnett, E.C.; Li, M.; Hu, L.; Cui, Y. Passivation coating on electrospun copper nanofibers for stable transparent electrodes. *ACS Nano* **2012**, *6*, 5150–5156. [CrossRef] [PubMed]
10. Kim, J.; Lee, M.-S.; Jeon, S.; Kim, M.; Kim, S.; Kim, K.; Bien, F.; Hong, S.Y.; Park, J.-U. Highly transparent and stretchable field-effect transistor sensors using graphene–nanowire hybrid nanostructures. *Adv. Mater.* **2015**, *27*, 3292–3297. [CrossRef] [PubMed]
11. Kim, K.; Kim, J.; Hyun, B.G.; Ji, S.; Kim, S.-Y.; Kim, S.; An, B.W.; Park, J.-U. Stretchable and transparent electrodes based on in-plane structures. *Nanoscale* **2015**, *7*, 14577–14594. [CrossRef] [PubMed]
12. Wu, H.; Kong, D.; Ruan, Z.; Hsu, P.-C.; Wang, S.; Yu, Z.; Carney, T.J.; Hu, L.; Fan, S.; Cui, Y. A transparent electrode based on a metal nanotrough network. *Nat. Nanotechnol.* **2013**, *8*, 421–425. [CrossRef] [PubMed]
13. Chan, E.P.; Crosby, A.J. Fabricating microlens arrays by surface wrinkling. *Adv. Mater.* **2006**, *18*, 3238–3242. [CrossRef]
14. Chandra, D.; Yang, S.; Lin, P.-C. Strain responsive concave and convex microlens arrays. *Appl. Phys. Lett.* **2007**, *91*, 251912. [CrossRef]
15. Khang, D.-Y.; Jiang, H.; Huang, Y.; Rogers, J.A. A stretchable form of single-crystal silicon for high-performance electronics on rubber substrates. *Science* **2006**, *311*, 208–212. [CrossRef] [PubMed]
16. Scalisi, R.G.; Paleari, M.; Favetto, A.; Stoppa, M.; Ariano, P.; Pandolfi, P.; Chiolerio, A. Inkjet printed flexible electrodes for surface electromyography. *Org. Electron.* **2015**, *18*, 89–94. [CrossRef]
17. Harrison, C.; Stafford, C.M.; Zhang, W.; Karim, A. Sinusoidal phase grating created by a tunably buckled surface. *Appl. Phys. Lett.* **2004**, *85*, 4016–4018. [CrossRef]
18. Opris, D.M.; Molberg, M.; Walder, C.; Ko, Y.S.; Fischer, B.; Nüesch, F.A. New silicone composites for dielectric elastomer actuator applications in competition with acrylic foil. *Adv. Funct. Mater.* **2011**, *21*, 3531–3539. [CrossRef]
19. Jun, K.W.; Lee, J.M.; Lee, J.Y.; Oh, I.K. Bio-inspired dielectric elastomer actuator with agnws coated on carbon black electrode. *J. Nanosci. Nanotechnol.* **2014**, *14*, 7483–7487. [CrossRef] [PubMed]
20. Tien, H.-W.; Hsiao, S.-T.; Liao, W.-H.; Yu, Y.-H.; Lin, F.-C.; Wang, Y.-S.; Li, S.-M.; Ma, C.-C.M. Using self-assembly to prepare a graphene-silver nanowire hybrid film that is transparent and electrically conductive. *Carbon* **2013**, *58*, 198–207. [CrossRef]
21. Jin, J.; Lee, J.; Jeong, S.; Yang, S.; Ko, J.-H.; Im, H.-G.; Baek, S.-W.; Lee, J.-Y.; Bae, B.-S. High-performance hybrid plastic films: A robust electrode platform for thin-film optoelectronics. *Energy Environ. Sci.* **2013**, *6*, 1811–1817. [CrossRef]
22. Lee, J.; Lee, I.; Kim, T.-S.; Lee, J.-Y. Efficient welding of silver nanowire networks without post-processing. *Small* **2013**, *9*, 2887–2894. [CrossRef] [PubMed]
23. Yang, C.; Gu, H.; Lin, W.; Yuen, M.M.; Wong, C.P.; Xiong, M.; Gao, B. Silver nanowires: From scalable synthesis to recyclable foldable electronics. *Adv. Mater.* **2011**, *23*, 3052–3056. [CrossRef] [PubMed]

micromachines

MDPI

Article

Suspended Graphene-Based Gas Sensor with 1-mW Energy Consumption

Jong-Hyun Kim [1], Qin Zhou [2] and Jiyoung Chang [1,*]

[1] Department of Mechanical Engineering; University of Utah, Salt Lake City, UT 84112, USA;
    jonghyun.kim@utah.edu
[2] Department of Mechanical Engineering; University of Nebraska–Lincoln, Lincoln, NE 68588, USA;
    zhou@unl.edu
*   Correspondence: jy.chang@utah.edu; Tel.: +1-801-581-7400

Academic Editors: Seung Hwan Ko, Daeho Lee and Zhigang Wu
Received: 28 December 2016; Accepted: 24 January 2017; Published: 1 February 2017

Abstract: This paper presents NH_3 sensing with ultra-low energy consumption for fast recovery and a graphene sheet based on a suspended microheater. Sensitivity and repeatability are important characteristics of functional gas sensors embedded in mobile devices. Moreover, low energy consumption is an essential requirement in flexible and stretchable mobile electronics due to their small dimension and fluctuating resistivity during mechanical behavior. In this paper, we introduce a graphene-based ultra-low power gas detection device with integration of a suspended silicon heater. Dramatic power reduction is enabled by a duty cycle while not sacrificing sensitivity. The new oscillation method of heating improves the sensitivity of 0.049 ($\Delta R/R_0$) measured at a flow rate of 18.8 sccm $NH_{3(g)}$ for 70 s. Our experimental tests show that a 60% duty cycle does not sacrifice sensitivity or recovery by dropping the total power consumption from 1.76 mW to 1.05 mW. The aforementioned low energy consuming gas sensor platform not only attracts environmentally-related industries, but also has the potential to be applied to flexible and stretchable mobile electronic devices.

Keywords: graphene; sensor; heater; NH_3; oscillation; pulse; sensitivity; recovery; energy; flexible

1. Introduction

Graphene's unique and excellent electrical and mechanical properties make it a tremendous contribution to flexible and stretchable electronics [1–3]. In particular, gas detection at low temperatures has been one of the main topics explored in applications based on 2D materials [4–6]. Among many gases, ammonia (NH_3) is a one of major target compounds of toxic gases in the field of safety monitoring [2] Conventional toxic gas detection relies on catalytic reactions based on metal oxide, however recent study suggests that the sensitivity can be greatly enhanced by integration of atomically thin layered materials. There are multiple known advantages of graphene. First of all, graphene sheets can be used electronically in single electronic detectors operating at room temperature and in ultra-high sensitivity sensors with mechanical strain or magnetic fields [3]. In terms of mechanical aspects, graphene strength is an essential characteristic for wearable and embedded gas sensors in mobile devices, and graphene-based gas sensors are known to tolerate extreme sensitivity [4]. Thus, many attempts have been made recently to increase the performance of gas detection using chemical doping, array structures and UV-light exposure. The large arrays of sensors can raise the sensitivity due to broadening of the active detecting area [4]. However, there is a limit in size because small size and high sensitivity are required in order to match modern electronic devices. Increased concentration of charge carrier in graphene induced by adsorbed gas molecules can be useful in highly sensitive sensors [5]. Metal oxide semiconductor and solid electrolyte sensors with a wide typical detection range at an operating temperature of a few hundred degrees Celsius are commercially available. Although these types of sensors are inexpensive and robust, they require high energy consumption and cannot be fabricated on flexible substrates [6]. Other research revealed that the UV light exposure shows much improved sensitivity [7], but there are critical limitations to the method, specifically with respect to its reliability, as the sensor eventually loses its performance due to UV exposure. This is because single-walled carbon nanotube, sensing probes, are gradually removed by the continuous UV light irradiation [8]. In addition, several researchers focused on reducing recovery time. According to the experiment by Schedin et al., adsorbates can be removed by annealing the device at 150 °C [3]. Additionally, another experiment showed that the low temperature in range of 50~150 °C also increasingly contributed to molecular desorption. Semantic et al. fabricated micro-scale suspended hot plate arrays and deposited metals onto the plate by post-chemical vapor deposition (CVD) method [9,10]. The post-CVD method for metal deposition can provide thermal shock effects on the device. In addition, the study also indicated that the manufacturing process of the suspended heater structure is a challenging task. As a separate study, Fowler et al. made a suspended nitride heater and deposit graphene dispersion using spin-coating [11]. However, those suspended heaters have large dimensions (100–500 μm^2), that are three-fold larger compared to the device in this paper. UV exposure for a short time offered an alternative to thermal annealing, however use of UV exposure is not recommended in general due to possible corrosion of carbon on the sensing sheet [8]. The temperature-programmed desorption curve using the Monte Carlo simulation methods and force-field parameters gives predictable data to grasp the fast recovery rate [12]. Therefore, there are clear needs for a device which can detect toxic gases at low power consumption without compromising sensitivity and recovery time. In this paper, a silicon-based microheater and single layered graphene are implemented to enable ultra-low energy consumption. The suspended structured heater is fabricated by surface micromachining followed by backside wet etching. Then, the graphene is deposited via poly-methyl-methacrylate (PMMA) transferring method. Then the sensing probe, covered with graphene, is formed by e-beam lithography. We have demonstrated ultra-low power consumption of up to 1.05 mW by implementing several factors including various input vibration parameters such as air-suspended microheaters, voltage, duty cycle and power input frequency. Multiphysics simulations are used to predict the temperature of a silicon heater and the results are used when designing a suspended heater.

2. Device and Methods

2.1. Structure

Figure 1a shows the working principle of low-power gas sensor. Gas molecules are absorbed on the graphene and sensing electrodes placed on top of silicon nitride layer. A structural side view of the sensor is shown in Figure 1b. The microheater layer is sandwiched between two low-stress silicon nitride (LSN) membranes and suspended in the air to minimize heat loss. Ti and Pt are patterned to be electrically connected to the graphene and the packaging. A single layer of graphene on the copper film synthesized by CVD is transferred onto the device surface and patterned by e-beam lithography as shown in Figure 1c. The graphene sheet, as an absorber, is located on the narrow bridge region of the microheater layer. Each device has four sensors which perform independently and they are positioned on top, left, right and bottom in Figure 1d. Each sensor has four yellow-colored Ti/Pt legs at the center of the device. Two square shaped electrical connecting pads in the middle are opened to the input power source. The other two pads on the outside are connected to the electrical measurement setup. In Figure 1e, the device is mounted on a packaging platform that can be connected by wire bonding.

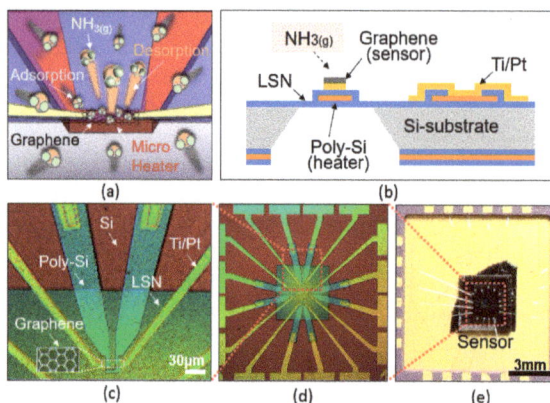

Figure 1. Schematic view of the sensor. (**a**) Concept of a microheater-based graphene gas sensor. Molecules are adsorbed on the graphene and fast recovery is achieved via heat-assisted desorbtion. (**b**) Cross-sectional view of the sensor. Heater is suspended in air to reduce heat loss. (**c**) The graphene is patterned on the bridge of the heater that generates high temperature due to narrow carrier pathway. (**d**) Top view of the device consists of four sensors on a single device. (**e**) A device is connected to a silicon-package by wirebonding. LSN: low-stress silicon-rich nitride.

2.2. Working Principle

A mechanism of adsorption and desorption of NH_3 molecules is shown in Figure 2. When the graphene surface is exposed to a stable environment composed of atmospheric pressure and room temperature, the graphene sheet exhibits an ohmic response before gas adsorption as shown in Figure 2a. When the graphene is exposed to NH_3 as shown in Figure 2b, molecules are adhered on the graphene surface. This $NH_{3(g)}$ sensing is based on changes in the resistivity due to molecular adsorption on the graphene sheet that act as donors [5]. Figure 2c shows desorption by air purification and annealing to accelerate the desorption rate. Molecular desorption reduces the electrical resistance by removing molecules from the graphene surface. The device is then purged with dry air to return graphene to its original resistance value [13]. In addition, the ultraviolet irradiation on graphene is known to increase the electrical resistance of the graphene [14]. Ultraviolet radiation catalyzes the sensitivity of graphene for sensing different types of gases, but graphene has weak light absorption and therefore light irradiation systems are difficult to apply to microelectronics [15]. The sensor's sensitivity has a strong relationship with the activation temperature, therefore, accurately knowing the temperature is important.

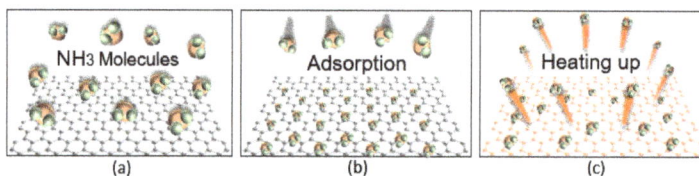

Figure 2. Conceptual process of interaction between molecules and graphene sheet. (**a**) The graphene has a steady condition without gas adsorption. (**b**) Molecules are adhered by interaction bonding. (**c**) Desorption occurs by air purging and annealing that accelerates desorption.

However, due to the size of the microheater, it is difficult to directly measure the temperature of the microheater. We have predicted the temperature of the heater to be about 200 °C by the 2 V input source using commercial finite element analysis (FEA) packaging. In addition, a silicon heater with the same dimensions was fabricated and tested with thermal couple on top of it to monitor the temperature variation along with voltage input. The heater's micro-scale bridge can be used over a wide temperature range of up to 800 °C, driven by input power. The emission of a 'U' shaped microheater (bridge structure) on the left side of Figure 3 shows a silicon heater that glows around an input voltage of 8 volts with an expected temperature of 900 °C.

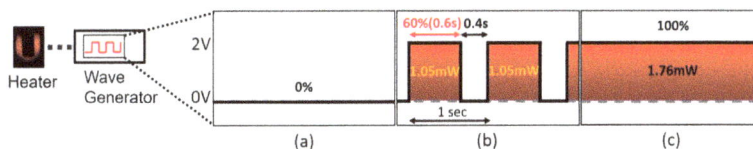

Figure 3. Schematic view of connection with microheater and power wave generator. (**a**) A 0% duty cycle means non-heating. (**b**) A 60% duty cycle of oscillation with 2 V shows 1.05-mW energy consumption. (**c**) Full duty cycle (continuous heating) generates 1.76 mW of energy consumption.

The zero duty cycle in Figure 3a has no energy consumption, and the 60% duty cycle with 2 V in Figure 3b means the generator releases 2 V power source for 0.6 s, then takes a rest for 0.4 s. Overall energy consumption during 1 s is equivalent to 1.05 mW of power consumption. In Figure 3c, the overall duty cycle shows a continuous power supply, reaching a higher energy consumption of 1.76 mW. Consolidating the duty cycle in the operation of the microheater greatly reduces power consumption.

2.3. Device Fabrication

The fabrication process of the sensor is shown in Figure 4. (a) First, a low-stress silicon-rich nitride (LSN) of 100 nm is deposited on 500-µm-thick silicon substrate using low-pressure chemical vapor deposition (LP-CVD). (b) A boron-doped poly-silicon is deposited by CVD. Then the substrate is annealed at 1050 °C. Micro heater structure is patterned by photolithography process followed by reactive ion etching (RIE) of silicon. (c) LSN on the top surface is removed by plasma etching. (d) The LSN is deposited to make a sandwich structure on both sides of the heater. (e) The LSN is patterned to open the electrical connection between the poly-silicon and Ti/Pt that is formed. (f) Ti/Pt, 10/90 nm, are deposited on the top surface and patterned for the electrical pathway using supplemental items including a graphene connector and wire bonding pads. The device is annealed at 350 °C for 1 h in a nitrogen environment. (g) The silicon substrate is etched from the bottom by KOH solution for the suspended microheater. (h) The PMMA-coated graphene sheet is transferred onto the top surface. (i) The PMMA is patterned to shape the graphene sheet as an absorber on the heater region using e-beam lithography. The graphene layer is etched to define the heater region. Single-layer graphene synthesis is as follows.

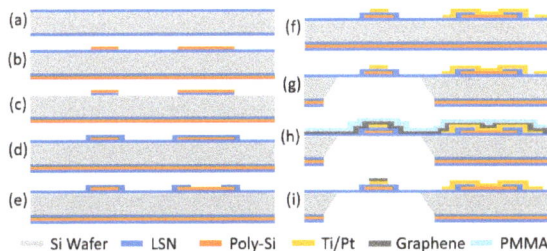

Figure 4. Fabrication process of suspended graphene gas sensor mounted on silicon based heater. Details of each process (**a**~**i**) are explained in above paragraph.

The single-layered graphene is grown using the CVD process. A copper film is used for the growth substrate. The film is exposed to H_2 for 100 sccm for 1 min and CH_4 for 30 sccm for 3 min in the 550 °C furnace. Then, 10 sccm of H_2 flows for 100 min in the 1050 °C furnace. After a pre-heating process, the film is exposed to both H_2 of 10 sccm and CH_4 22 sccm for 85 min simultaneously. The urnace is shut-off after 25 min in a growth process. All gas valves are closed after 85 min. The film is stored in the furnace for cooling to room temperature overnight. Based on the characterization by Raman spectroscopy, the grown film is identified by matching the characteristic wavelength of the single layer graphene [16].

2.4. Measurement

Figure 5 shows an experimental setup for the gas adsorption and desorption. The test is performed in ambient pressure and room temperature environment. Here, 18.8 sccm of air is supplied from the syringe on the left side of the schematic image. The liquid form of ammonium hydroxide $NH_{3(L)}$ evaporates and dissolves in the air. Then the air containing NH_3 molecules is pumped into the next flask containing drierites, which absorbs water vapor from a mixture of NH_3 and air. Then dry $NH_{3(g)}$ molecules penetrate a test chamber. For desorption of the molecules, a wave form generator (SDG1025, Siglent, Shenzhen, China) is used to generate power for the heating. Multiple sets of temperatures and input power are tested by varying input voltage (0~2.5V), duty cycle (45%~100%) and frequency (1100 Hz). As a reference, the same set of tests is performed using argon$_{(g)}$ gas. A multi-meter and a digital oscilloscope (TBS1052B, Tektronix, OR, USA) is used to measure the resistivity and wave form of the system. The detection capability is evaluated by the relative change in resistance per adsorption and desorption time $(R_1 - R_0)/R_0$. $NH_{3(g)}$ 70-s adsorption and air desorption 70 s was repeated four times. This is a cycle that repeats every 70 s.

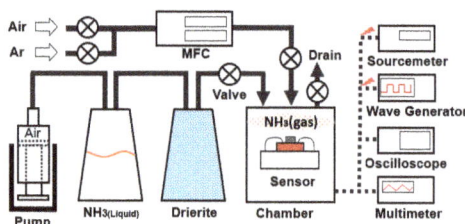

Figure 5. Experimental setup to flow target gas and measure the output from the gas sensor. MFC: mass flow controller.

3. Results

The graphene-based sensor is evaluated as a change in electrical resistance based on NH_3 gas and ultraviolet exposure. When the sensor is exposed to gas/UV, the resistance increases in both UV and NH_3. However, low temperature heating between room temperature and 100 °C does not significantly improve sensitivity/recovery of UV detection. On the other hand, the adsorption sensitivity of NH_3 increases by 40% when 0.5 V is applied compared to the same test conducted at room temperature. Argon (Ar) gas is supplied through a mass flow controller (MFC) in the gas tank. While heating from 100 to 200 °C, $\Delta R/R_0$ per 10 s recovery rate is monitored to be in the same range over all temperature ranges. The 200 sccm of Ar flow also provides 26% faster response than at 10 sccm. The air purging by uncovering with higher temperature gives a 6%~10% faster rate than 200 sccm Ar purging in detail. Using the air purging is the simplest method to refresh the sensor's surface. In addition, the air is able to be regularly utilized when the sensor is operating on the wearable device. Thus, NH_3 and the air worked in shifts for the evaluation of gas sensing and recovery. When the heater is heated to 100 °C or more by applying 1.5 V, the recovery speed is significantly reduced. In other words, the faster the recovery speed, the better the high voltage, but the constant power input of 2.0 V greatly reduces the sensitivity. Based on the tests, input power ranging between 1.5 and 2.0 V can serve as optimized refreshing input power, satisfying both reasonable recovery and sensitivity. As shown in Figure 6, increasing the temperature around 200 °C by 2 V input voltage with 60% duty cycle of oscillation improves the sensitivity and the recovery rate of $NH_{3(g)}$ sensing. It is noted that a test without heating (no input power source) clearly shows the poorest recovery for the gas. Consolidating the duty cycle to reduce power consumption significantly reduces power consumption over continuous heating without compromising sensitivity and recovery time. If the total input power is fixed at 0.6 mW through 40% of the duty cycle, the influence of the input frequency is not apparent. A 1-Hz input frequency improves performance if the total input power is fixed at 1.76 mW over 75% of duty cycle and 1.05 mW over 60%. For example, red (1 Hz) represents a higher absorption and desorption response than orange (100 Hz) with a total input power of 1 mW. However, the 45% and 75% of duty cycles with 1 Hz of frequency give low adsorption/desorption rates that are 0.041/0.041 and 0.038/0.047 ($\Delta R/R_0$) per 70 s, respectively, compared to the rates 0.049/0.052 ($\Delta R/R_0$) per 70 s at 60%. Next, the $NH_{3(g)}$ sensing device's dependence on input power frequency is tested and a small noticeable difference is observed as shown in Figure 7a,b. The 60% duty cycle at 1 and 100 Hz reaches the highest $\Delta R/R_0$ of sensitivity. A 1-Hz frequency shows a higher sensitivity/recovery rate than 100 Hz. Lastly, in an optimal test condition, ultra-low 1.05-mW energy consumption is achieved by a 2.0-V 1-Hz 60% duty cycle input power source that shows as high as 0.049 ($\Delta R/R_0$) sensitivity with 18.8 sccm of $NH_{3(g)}$ input flow.

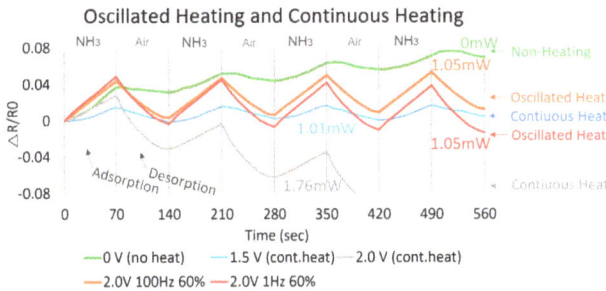

Figure 6. Comparison of sensing and recovery with varying duty cycles (0%, 60%, 100%). Adsorption cycle with $NH_{3(g)}$ 18.8 sccm for 70 s and desorption cycle with air for 70 s are repeated for four cycles each. Ultra-low 1.05-mW energy consumption by optimized heating mode (2.0 V, 1 Hz, 60% duty cycle, red color) shows significant adsorption rate 0.049 ($\Delta R/R_0$) per 70 s of sensitivity and desorption rate 0.052 ($\Delta R/R_0$) per 70s of recovery.

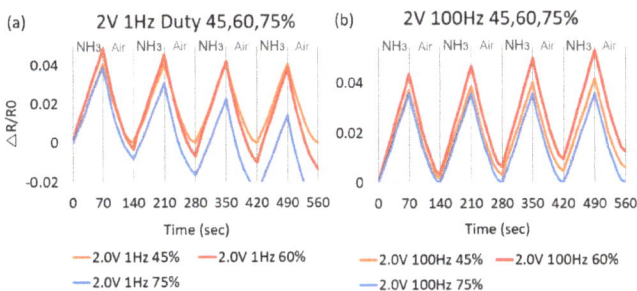

Figure 7. (**a**,**b**) shows a dependence of the frequency. A 60% Duty cycle achieves the highest 0.049 ($\Delta R/R_0$) of sensitivity. A 1-Hz frequency of all duty cycles shows higher sensitivity/recovery to gas. An optimized heating mode (2.0 V, 1 Hz, 60% duty cycle) shows the best performance.

4. Discussion

Various types of interaction, from weak van der Waals to strong covalent bonding, occur between adsorbed molecules and the graphene atoms during the experiment. This leads to a noticeable change in electrical conductivity of graphene [17]. Molecules adsorbed on the graphene sheet act as scattering centers, resulting in an increase in the resistivity of graphene [18]. Higher density of gas can decrease the greater conductivity. It can also be seen that, depending on the change in the emission of the microheater, the heater can generate various temperatures by changing the input power, as well as predicting the temperature in the low temperature range of 100 to 300 °C. The desorption is accelerated, especially with an input voltage of 1.5 V or more. This simulation shows that the 1.5 and 2.0 V input voltages produce approximately 100 and 200 °C respectively. Thus, the relationship between simulation and experimental data can accurately predict low heating temperatures. It is noted that when thermal energy is added, separation of molecules can be accelerated. This is because as the temperature increases, the hydrogen bond network is reorganized, and the NH_3 molecules are gradually desorbed from the surface by breaking of their intermolecular hydrogen bonds [12]. Moreover, there are various binding energies between molecules and graphene within one hexagon of graphene [19]. Thus, the bindings are increasingly broken as time goes on when the graphene is heated. Heating generally increases the sensitivity and at the same time increases the desorption rate. However, when the continuous heating is applied, the sensitivity of continuous heating does not show significant improvements compared to both non-heating and optimized oscillations. The optimized oscillation input is therefore suitable for overall performance and energy consumption. To further reduce energy consumption, the sensor can employ an oscillating power delivery scheme that uses a variety of duty cycles, frequencies, and voltages. According to various sources, high temperature and continuous power can gradually increase the supported recovery rate by weakening the interatomic coupling due to absorption of external atomic energy.

5. Conclusions

Ultra-low power consumption with graphene and micro-electro-mechanical-system (MEMS) based gas sensors have been successfully fabricated and demonstrated. A suspended microheater structure plays a key role in reducing power consumption. In addition, the duty cycle is integrated to further reduce power consumption to 1.05 mW. Although NH_3 gas is tested as one of the typical toxic gases, the platform is open to other types of gas detection through simple calibration. The sensor shows rapid recovery through heating of the silicon heater. As the temperature increases, the desorption gradually increases by reducing hydrogen bonding. With optimized configuration parameters, a 60% duty cycle of 2.0-V 1-Hz input power oscillation reduces energy consumption to 1.05 mW. The suspended microheater structure is suitable for preventing heat transfer which contributes to very low energy consumption. The pulsed input power may reduce damages of graphene and structures. The aforementioned low-energy consumption gas sensor platform will be an attractive platform for flexible and flexible mobile electronic devices.

Acknowledgments: The electron beam process and metal evaporation were conducted in the University of Utah NanoFab facility. This work was supported by the KIST Institutional Program (Project No. 2E27050)

Author Contributions: J.K. performed graphene fabrication, experiments and wrote the paper. Q.Z. and J.C. contributed design and analysis for fabrication of the device and wrote the paper.

Conflicts of Interest: The authors declare no conflict of interest.

References

1. Cho, B.; Yoon, J.; Hahm, M.G.; Kim, D.H.; Kim, A.R.; Kahng, Y.H.; Park, S.W.; Lee, Y.J.; Park, S.G.; Kwon, J.D.; et al. Graphene-based gas sensor: metal decoration effect and application to a flexible device. *J. Mater. Chem. C.* **2014**, *2*, 5280. [CrossRef]

2. Lee, K.; Scardaci, V.; Kim, H.Y.; Hallam, T.; Nolan, H.; Bolf, B.E.; Maltbie, G.S.; Abbott, J.E.; Duesberg, G.S. Highly sensitive, transparent, and flexible gas sensors based on gold nanoparticle decorated carbon nanotubes. *Sens.Actuators B Chem.* **2013**, *188*, 571–575. [CrossRef]

3. Schedin, F.; Geim, A.K.; Morozov, S.V.; Hill, E.W.; Blake, P.; Katsnelson, M.I.; Novoselov, K.S. Detection of individual gas molecules adsorbed on graphene. *Nat. Mater.* **2007**, *6*, 652–655. [CrossRef] [PubMed]

4. Sheehan, P.E.; Whitman, L.J. Detection limits for nanoscale biosensors. *Nano Lett.* **2005**, *5*, 803–807. [CrossRef] [PubMed]

5. Leenaerts, O.; Partoens, B.; Peeters, F.M. Adsorption of H_2O, NH_3, CO, NO_2, and NO on graphene: A first-principles study. **2007**, *2*, 1–6.

6. Inoue, T.; Ohtsuka, K.; Yoshida, Y.; Matsuura, Y.; Kajiyama, Y. Metal oxide semiconductor NO_2 sensor. *Sens. Actuators B Chem.* **1995**, *25*, 388–391. [CrossRef]

7. Mitoma, N.; Nouchi, R.; Tanigaki, K. Enhanced sensing response of oxidized graphene formed by UV irradiation in water. *Nanotechnology* **2015**, *26*, 105701. [CrossRef] [PubMed]

8. Chen, g.; Paronyan, T.M.; Pigos, E.M.; Harutyunyan, A.R. Enhanced gas sensing in pristine carbon nanotubes under continuous ultraviolet light illumination. *Sci. Rep.* **2012**, *2*, 343. [CrossRef] [PubMed]

9. Semancik, S.; Cavicchi, R.E.; Wheeler, M.C.; Tiffany, J.E.; Poirier, G.E.; Walton, R.M.; Suehle, J.S.; Panchapakesan, B.; DeVoe, D.L. Microhotplate platforms for chemical sensor research. *Sens. Actuators B Chem.* **2001**, *77*, 579–591. [CrossRef]

10. Cavicchi, R.E.; Suehle, J.S.; Kreider, K.G.; Shomaker, B.L.; Small, J.A.; Gaitan, M.; Chaparala, P. Growth of SnO_2 films on micromachined hotplates. *Appl. Phys. Lett.* **1995**, *66*, 812–814. [CrossRef]

11. Fowler, J.D.; Allen, M.J.; Tung, V.C.; Yang, Y.; Kaner, R.B.; Weiller, B.H. Practical chemical sensors from chemically derived graphene. *ACS Nano.* **2009**, *3*, 301–306. [CrossRef] [PubMed]

12. Liu, L.; Zhao, L.; Sun, H. Simulation of NH_3 temperature-programmed desorption curves using an ab initio force field. *J. Phys. Chem. C.* **2009**, *113*, 16051–16057. [CrossRef]

13. Zhang, Z.; Zhang, X.; Luo, W.; Yang, H.; He, Y.; Liu, Y.; Zhang, X.; Peng, G. Study on adsorption and desorption of ammonia on graphene. *Nanoscale Res. Lett.* **2015**, *10*, 359. [CrossRef] [PubMed]

14. Akinwande, D.; Petrone, N.; Hone, J. Two-dimensional flexible nanoelectronics. *Nat. Commun.* **2014**, *5*, 5678. [CrossRef] [PubMed]

15. Novoselov, K.S.; Geim, A.K.; Morozov, S.V.; Jiang, D.; Zhang, Y.; Dubonos, S.V.; Grigorieva, I.V.; Firsov, A.A. Electric field effect in atomically thin carbon films. *Science* **2004**, *306*, 666–669. [CrossRef] [PubMed]

16. Liu, Y.; Liu, Z.; Lew, W.S.; Wang, Q.J. Temperature dependence of the electrical transport properties in few-layer graphene interconnects. *Nanoscale Res. Lett.* **2013**, *8*, 335. [CrossRef] [PubMed]

17. Silvestrelli, P.L. van der Waals interactions in density functional theory using wannier functions. *J. Phys. Chem. A.* **2009**, *113*, 5224–5234. [CrossRef] [PubMed]

18. Hwang, E.H.; Adam, S.; Das Sarma, S. Transport in chemically doped graphene in the presence of adsorbed molecules. *Phys. Rev. B* **2007**, *76*, 195421. [CrossRef]

19. Lin, X.; Ni, J.; Fang, C. Adsorption capacity of H_2O, NH_3, CO, and NO_2 on the pristine graphene. *J. Appl. Phys.* **2013**, *113*, 34306. [CrossRef]

micromachines

MDPI

Article

One-Dimensional Thermal Analysis of the Flexible Electronic Devices Integrated with Human Skin

Yun Cui [1], Yuhang Li [1,2,3,*], Yufeng Xing [1], Tianzhi Yang [4] and Jizhou Song [2,5,*]

[1] Institute of Solid Mechanics, Beihang University (BUAA), Beijing 100191, China; kyalex@sina.com (Y.C.); xingyf@buaa.edu.cn (Y.X.)

[2] Key Laboratory of Soft Machines and Smart Devices of Zhejiang Province, Zhejiang University, Hangzhou 310027, China

[3] State Key Laboratory of Digital Manufacturing Equipment and Technology, Huazhong University of Science and Technology, Wuhan 430074, China

[4] School of Aerospace Engineering, Shenyang Aerospace University, Shenyang 110136, China; yangtz@sau.edu.cn

[5] Department of Engineering Mechanics and Soft Matter Research Center, Zhejiang University, Hangzhou 310027, China

[*] Correspondence: liyuhang@buaa.edu.cn (Y.L.); jzsong@zju.edu.cn (J.S.);
Tel.: +86-10-8231-7507 (Y.L.); +86-571-8795-2368 (J.S.)

Academic Editors: Seung Hwan Ko, Daeho Lee and Zhigang Wu
Received: 3 September 2016; Accepted: 14 November 2016; Published: 18 November 2016

Abstract: A one-dimensional analytic thermal model for the flexible electronic devices integrated with human skin under a constant and pulsed power is developed. The Fourier heat conduction equation is adopted for the flexible electronics devices while the Pennes bio-heat transfer equation is adopted for the skin tissue. Finite element analysis is performed to validate the analytic model through the comparison of temperature distributions in the system. The influences of geometric and loading parameters on the temperature increase under a pulsed power are investigated. It is shown that a small duty cycle can reduce the temperature increase of the system effectively. A thin substrate can reduce the device temperature but increase the skin surface temperature. The results presented may be helpful to optimize the design of flexible electronic devices to reduce the adverse thermal influences in bio-integrated applications.

Keywords: flexible electronics; thermal analysis; human skin

---

1. Introduction

Recent advances in flexible electronics enable the development of epidermal electronics [1,2], which could be mounted onto the skin and retain conformal contact with the skin under compression and tension. Knowledge on heat transfer within skin tissue due to the physical coupling of epidermal electronics to the skin is critical for the use of epidermal electronics since even only a few degrees in temperature increase may induce uncomfortable feelings.

Many researchers have performed the thermal analysis of flexible electronic devices by adopting the Fourier heat conduction equation. Kim et al. [3], Lu et al. [4], and Cui et al. [5] studied the heat conduction of flexible micro-scale inorganic light-emitting diodes under a constant power experimentally and analytically. Relevant experimental results are presented as surface temperature contours, which can be captured by QFI Infra-Scope Micro-Thermal Imager [3]. Results of experiment agree well with analytic ones. Kim et al. [6] and Li et al. [7] investigated the thermal properties of flexible devices on various substrates under a pulsed power. They developed an analytic axi-symmetric model to explore the dependence of device temperature on the geometric dimensions, material properties and loading parameters. There exist several reviews of thermal and mechanical analysis of flexible electronics [8,9].

The heat transfer in skin tissue is much more complicated than that in flexible electronic devices due to the influences of blood perfusion and metabolism and the complex multilayered structure consisting of four layers: stratum corneum, epidermis, dermis and fat as shown in Figure 1a. The Pennes bioheat equation has been widely used to investigate the thermal response of skin tissue under surface heating. Jiang et al. [10] studied the skin burn process resulting from a high temperature heat source to the skin surface by using finite difference method to solve the Pennes bio-heat equation. Ozen et al. [11] presented a one-dimensional multi-layer model to characterize the temperature rise resulting from skin exposure to microwaves. Im et al. [12] carried out a numerical study on the temperature profiles as a result of local heating of human skin. An extensive review on heat transfer of skin tissue was given by Xu et al. [13].

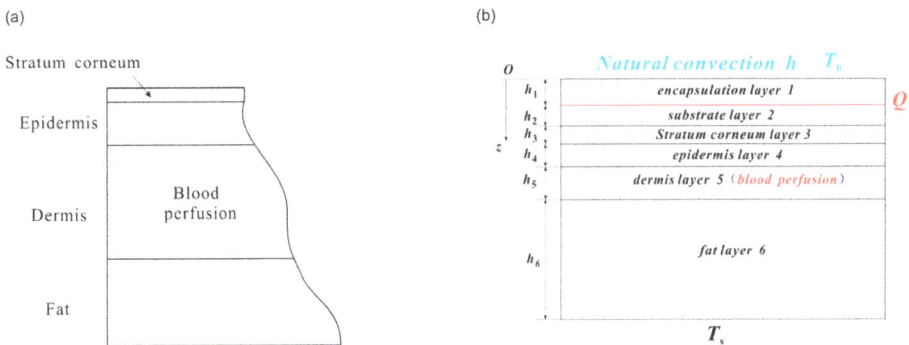

Figure 1. (**a**) Schematic illustration of the skin tissue structure; and (**b**) schematic illustration of one-dimensional geometry of the analytic modeled device-skin system.

The above heat transfer analyses are either for flexible electronic devices or human skin. There are few studies on the system of flexible electronic devices integrated with the human skin. For better understanding of the thermal properties of device and establishing design guidelines for flexible electronic devices to minimize the adverse thermal effect, we aims to develop an analytical model for the flexible electronic devices integrated with human skin under both constant and pulsed power. Basing on the Kim's experimental results and analytical model for the device on top of a metal trunk [3,4], this paper aims at the device integrated with human skin. For flexible electronic devices with the in-plane dimension on the order of a few millimeters, which is much larger than the device thickness, the heat mainly transfers along the thickness direction. It is reasonable to perform a one-dimensional thermal analysis. Such treatment simplifies the theoretical model significantly with enough accuracy. Moreover, the finite element models are also established to validate the analytic solutions.

2. Thermal Analysis under a Constant Power

The human skin is modeled as a multi-layer structure consisting of four layers: stratum corneum, epidermis, dermis and fat. The flexible electronic device consists of the functional component on a flexible substrate (e.g., polydimethylsiloxane (PDMS)) encapsulated by an encapsulation layer (e.g., SU8). The functional component could be modeled as a planar heat source with heat generation power density Q (W/m^2) since its thermal conductivity (~100 W/m/K) is much larger than that of substrate or encapsulation layer. Figure 1b shows the one-dimensional geometry of the analytic model with the flexible device on the human skin. The origin of the coordinate (z) is established on the top surface of encapsulation layer with the positive direction pointing from the flexible device to the skin tissue. The total thickness of the system is denoted by H. h_i denotes the thickness for each layer with the substrate i (i = 1, 2, 3, 4, 5 and 6) for the encapsulation, substrate, stratum corneum, epidermis, dermis and fat layer, respectively. The top surface of encapsulation layer has the natural convection boundary with h as the coefficient of heat convection. The bottom surface of fat layer has constant core body temperature T_s.

The temperature in the device satisfies the Fourier heat conduction equation

$$k_i \frac{d^2 T_i}{dz^2} = 0 \; (i = 1, 2) \tag{1}$$

while the temperature in the skin tissue satisfies the Pennes bio-heat equation [14]

$$\begin{cases} k_i \dfrac{d^2 T_i}{dz^2} + q_{met} = 0 \; (i = 3, 4, 6) \\ k_5 \dfrac{d^2 T_5}{dz^2} - \omega_b \rho_b c_b \, (T_5 - T_s) + q_{met} = 0 \end{cases} \tag{2}$$

where k_i represents the thermal conductivity of corresponding layers; ρ_b and c_b are the density and specific heat of blood, respectively; ω_b is the blood perfusion rate; and q_{met} is the metabolic heat generation. Here, we assumed that the blood temperature is just the same as the core temperature [14]. The main difference between two heat conduction models mentioned above is confined to blood perfusion effect and metabolism. The blood perfusion is effected by ω_b, which varies from 0–1 mL/(mL·s) [15]. While the metabolic heat generation is almost constant for healthy person. It should be noted that the blood perfusion only exists inside the dermis layer [13].

Let ΔT = $T(z)$ − T_s denote the temperature increase from the core temperature. Equations (1) and (2) then become

$$\begin{cases} k_i \dfrac{d^2 \Delta T_i}{dz^2} = 0 \; (i = 1, 2) \\ k_i \dfrac{d^2 \Delta T_i}{dz^2} + q_{met} = 0 \; (i = 3, 4, 6) \\ k_5 \dfrac{d^2 \Delta T_5}{dz^2} - \omega_b \rho_b c_b \Delta T_5 + q_{met} = 0 \end{cases} \tag{3}$$

The temperature increase in each layer can be given

$$\begin{cases} \Delta T_1 = A_1 z + B_1 \\ \Delta T_2 = A_2 z + B_2 \\ \Delta T_3 = -q_{met} z^2 / 2k_3 + B_3 z + C_3 \\ \Delta T_4 = -q_{met} z^2 / 2k_4 + B_4 z + C_4 \\ \Delta T_5 = A_5 \exp\left(z\sqrt{\eta/k_5}\right) + B_5 \exp\left(-z\sqrt{\eta/k_5}\right) + q_{met}/\eta \\ \Delta T_6 = -q_{met} z^2 / 2k_6 + B_6 z + C_6 \end{cases} \tag{4}$$

where $\eta = \omega_b \rho_b c_b$, and the coefficients A_i, B_i and C_i are to be determined by the boundary conditions and the interfacial continuity condition. At the top surface of encapsulation layer (z = 0), the natural convection condition gives

$$-k_1 \left. \frac{d\Delta T}{dz} \right|_{z=0} = -h \cdot (\Delta T - \Delta T_0)|_{z=0} \tag{5}$$

where ΔT_0 = T_0 − T_s, T_0 denotes the ambient temperature, k_1 is the thermal conductivity of encapsulation layer and h is the coefficient of heat convection. At the encapsulation/substrate interface

($z = h_1$), the temperature is continuous while the heat flux satisfies the surface heat source condition, which gives,

$$\Delta T|_{z=h_1^+} = \Delta T|_{z=h_1^-} \text{ and } k_1 \left.\frac{d\Delta T}{dz}\right|_{z=h_1^-} - k_2 \left.\frac{d\Delta T}{dz}\right|_{z=h_1^+} = Q \qquad (6)$$

where k_2 is the thermal conductivity of substrate layer. At the interface between any other two layers, both the temperature and heat flux are continuous, which give

$$\Delta T\Big|_{z=\left(\sum_{i=1}^{n} h_i\right)^+} = \Delta T\Big|_{z=\left(\sum_{i=1}^{n} h_i\right)^-} \text{ and } k_{(n+1)} \left.\frac{d\Delta T}{dz}\right|_{z=\left(\sum_{i=1}^{n} h_i\right)^+} = k_n \left.\frac{d\Delta T}{dz}\right|_{z=\left(\sum_{i=1}^{n} h_i\right)^-} \qquad (7)$$

where n = 2, 3, 4, and k_n is the thermal conductivity of layer n with n = 2 for substrate n = 3 for stratum corneum, n = 4 for epidermis, and n = 5 for dermis.

The ambient temperature at the bottom surface ($z = \sum_{i=1}^{6} h_i$) of fat layer gives

$$\Delta T\Big|_{z=\sum_{i=1}^{6} h_i} = 0 \qquad (8)$$

It should be noted that the stratum corneum usually has the same thermal property as that of epidermis, i.e., $k_3 = k_4$. The coefficients in Equation (4) are then determined by the boundary and continuity conditions in Equations (5)–(8) as

$$
\begin{Bmatrix}
A_1 \\
B_1 \\
A_2 \\
B_2 \\
B_3 \\
C_3 \\
B_4 \\
C_4 \\
A_5 \\
B_5 \\
B_6 \\
C_6
\end{Bmatrix}
=
\begin{Bmatrix}
\dfrac{y\left[-E\sinh(y) + F\cosh(y)\right] - E\cosh(y) + F\sinh(y) - M}{y\left[C\sinh(y) + D\cosh(y)\right] + C\cosh(y) + D\sinh(y)} \\[2mm]
\dfrac{k_1}{h}A_1 + \Delta T_0 \\[2mm]
(k_1 A_1 - Q)/k_2 \\[1mm]
h_1 A_1 + B_1 - h_1 A_2 \\[1mm]
\left[k_1 A_1 - Q + q_{met}(h_1 + h_2)\right]/k_3 \\[1mm]
A_2(h_1 + h_2) + h_1 A_1 + B_1 - h_1 A_2 + q_{met}(h_1 + h_2)^2/(2k_3) - B_3(h_1 + h_2) \\[1mm]
\left[k_1 A_1 - Q + q_{met}(h_1 + h_2)\right]/k_3 \\[1mm]
A_2(h_1 + h_2) + h_1 A_1 + B_1 - h_1 A_2 + q_{met}(h_1 + h_2)^2/(2k_3) - B_3(h_1 + h_2) \\[1mm]
\left[(C+D)A_1 + E - F\right]e^{-(h_1+h_2+h_3+h_4)\sqrt{\eta/k_5}}/2 \\[1mm]
\left[(C-D)A_1 + E + F\right]e^{(h_1+h_2+h_3+h_4)\sqrt{\eta/k_5}}/2 \\[1mm]
\begin{bmatrix} y(CA_1 + E)\sinh(y)/h_6 + y(DA_1 - F)\cosh(y)/h_6 \\ +q_{met}(h_1 + h_2 + h_3 + h_4 + h_5)/k_6 \end{bmatrix} \\[3mm]
\begin{bmatrix} (CA_1 + E)\cosh(y) + (DA_1 - F)\sinh(y) + q_{met}/\eta \\ +q_{met}(h_1 + h_2 + h_3 + h_4 + h_5)^2/2k_6 - B_6(h_1 + h_2 + h_3 + h_4 + h_5) \end{bmatrix}
\end{Bmatrix}
\qquad (9)
$$

where,

$$C = h_1 + k_1/h + k_1 h_2/k_2 + k_1(h_3 + h_4)/k_3$$
$$D = k_1/\sqrt{k_5\eta}$$
$$E = -(h_2/k_2 + (h_3 + h_4)/k_3)Q - q_{met}(h_3 + h_4)^2/(2k_3) - q_{met}/\eta + \Delta T_0$$
$$F = (Q + q_{met}(h_3 + h_4))/\sqrt{k_5\eta}$$
$$M = q_{met}/\eta - q_{met}h_6^2/(2k_6)$$
$$y = h_5\sqrt{\eta/k_5}$$

In order to validate the analytical solutions in Equation (9), finite element analysis (FEA) is performed using ABAQUS software (6.13-1, Dassault Simulia, Waltham, MA, USA) to study the thermal properties of this system with continuum element DC3D8. The thicknesses of encapsulation, substrate, stratum corneum, epidermis, dermis, and fat are taken as 7 μm, 2 mm, 0.02 mm, 0.08 mm, 1.5 mm and 4.4 mm, respectively [15]. The thermal conductivities of encapsulation, substrate, stratum corneum, epidermis, dermis, and fat are 0.2 W/(m·K), 0.15 W/(m·K), 0.21 W/(m·K), 0.21 W/(m·K), 0.37 W/(m·K) and 0.16 W/(m·K) [4,15]. The top surface of encapsulation layer has a natural convection boundary with the coefficient of heat convection h = 25 W/(m²·K) [2,16]. The ambient temperature is set as 25 °C. The core body temperature at the bottom surface of substrate is 37 °C. The metabolic heat generation in the skin tissue is 368 W/m³ [15]. The product of mass density and specific heat capacity of the blood is 4.218 × 10⁶ J/(m³·K) [17]. The blood perfusion rate inside the dermis layer is 0.03 mL/(mL·s) [15]. Figure 2 shows the comparison of temperature increase along the thickness direction between the analytic predictions and FEA under the input power density 2500 W/m². The good agreement validates the analytic model. With the distance to the top surface of encapsulation layer increasing, the temperature increase first increases inside the encapsulation layer then decreases in other layers. The temperature increase of the heat source can reach about 28 °C while the temperature increase at the device/skin interface is about 9 °C. Compared to the temperature increase of skin away from electronic device, which is about −1.5 °C, the constant input power causes a remarkable temperature increase.

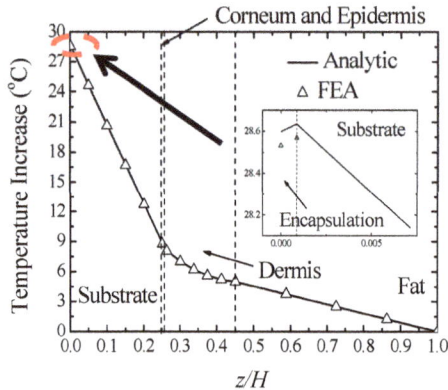

Figure 2. The comparison of temperature increase along the thickness direction between the analytic prediction and finite element analysis with the total thickness H = 8.007 mm.

3. Thermal Analysis under a Pulsed Power

As shown in Figure 2, the temperature increase of the device/skin interface is much more than that the human can stand. It may induce uncomfortable feelings or even tissue lesion in bio-integrated applications. Clearly, the temperature increase can be reduced by decreasing the input power density. However, the devices can hardly operate properly under small input power. In order to solve this dilemma, Kim et al. [6] and Li et al. [7] changed the input loading method from constant power to pulsed protocol, which has proven pretty effective on modern devices. The operation of flexible electronic devices (e.g., flexible light-emitting diodes) in a pulsed mode could significantly reduce the temperature increase to reach the goal in thermal management. In this section, an analytic model is developed to investigate the thermal response of flexible electronic devices on human skin.

Under a pulsed power, the temperature in the system increases to saturation in a fluctuation way [12]. We are interested in the saturated temperature since it gives the maximum temperature that could reach in the system. The pulsed power $Q(t)$ is defined in Figure 3 with Q_0 as the peak value and τ as the heating duration in one period t_0. The duty cycle D is given by $D = \tau/t_0$. Under a pulsed power density $Q(t)$, the temperature increase from the core body temperature in the system satisfies,

$$
\begin{cases}
k_i \dfrac{\partial^2 \Delta T_i}{\partial z^2} = \rho_i c_i \dfrac{\partial \Delta T_i}{\partial t} \ (i = 1, 2) \\[2mm]
k_i \dfrac{\partial^2 \Delta T_i}{\partial z^2} + q_{met} = \rho_i c_i \dfrac{\partial \Delta T_i}{\partial t} \ (i = 3, 4, 6) \\[2mm]
k_5 \dfrac{\partial^2 \Delta T_5}{\partial z^2} - \omega_b \rho_b c_b \Delta T_5 + q_{met} = \rho_5 c_5 \dfrac{\partial \Delta T_5}{\partial t}
\end{cases}
\tag{10}
$$

where ρ_i and c_i are the mass density and the specific heat capacity for each layer. The boundary and continuity conditions are the same as those for the case under a constant power in Equations (5)–(8) except the constant power density Q in Equation (6) should be changed to $Q(t)$, i.e.,

$$
\Delta T|_{z=h_1^+} = \Delta T|_{z=h_1^-} \text{ and } k_1 \left.\frac{\partial \Delta T}{\partial z}\right|_{z=h_1^-} - k_2 \left.\frac{\partial \Delta T}{\partial z}\right|_{z=h_1^+} = Q(t)
\tag{11}
$$

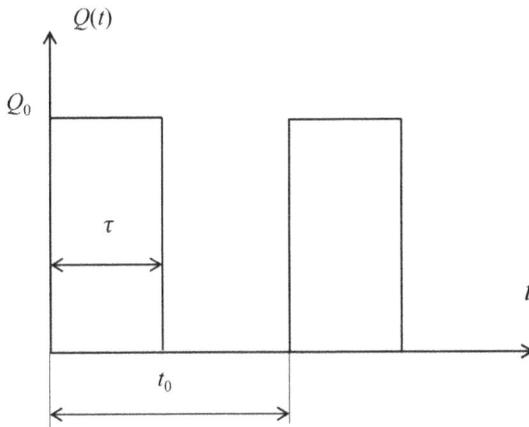

Figure 3. The pulsed power density $Q(t)$ with Q_0 as the peak power density, t as the during time and t_0 as the period.

The solution for Equation (10) is the summation of the specific solution, which is given in Section 2 without the input power, and the homogeneous solution, which is to be determined below. The homogenous solution for Equation (10) satisfies

$$
\begin{cases}
\dfrac{\partial \Delta T_i}{\partial t} - \lambda_i \dfrac{\partial^2 \Delta T_i}{\partial z^2} = 0 \ (i = 1, 2, 3, 4, 6) \\[2mm]
\dfrac{\partial \Delta T_5}{\partial t} - \lambda_5 \dfrac{\partial^2 \Delta T_5}{\partial z^2} + \omega_b \Delta T_5 = 0
\end{cases}
\tag{12}
$$

where $\lambda_i = k_i/(c_i \rho_i)$ is the thermal diffusivity with c, ρ and k as the specific heat capacity, mass density and thermal conductivity, respectively.

The method of superposition is adopted to obtain the temperature increase after saturation (i.e., homogeneous solution) under a pulsed power density, which could be written via Fourier series by

$$
Q(t) = Q_0 \begin{cases} 1 & 0 < t \le \tau \\ 0 & \tau < t \le t_0 \end{cases} = Q_0 \left[a_0 + \sum_{n=1}^{\infty} (a_n \cos n\omega t + b_n \sin n\omega t) \right]
\tag{13}
$$

where $w = 2\pi/t_0$, $a_0 = D = \tau/t_0$, $a_n = \sin(2n\pi D)/(n\pi)$, and $b_n = \frac{[1-\cos(2n\pi D)]}{n\pi}$. The temperature increase after saturation due to each sinusoidal power density $Q_0\cos n wt$ (or $Q_0 \sin n wt$) in Equation (13) corresponds to the real (or imaginary part) of the solution due to a power of $Q_0 e^{nwt\cdot i}$. The power of $Q_0 e^{nwt\cdot i}$ yields the temperature increase as $\theta(z; nw) e^{nwt\cdot i}$, which gives the real and imaginary part as $|\theta(z; nw)| \cos(nwt + \beta_n)$ and $|\theta(z; nw)| \sin(nwt + \beta_n)$, respectively. Here $\beta_n(nw)$ is the phase angle of $\theta(z; nw)$. Therefore, the temperature increase due to the pulsed power can be obtained by

$$\Delta T(z, t) = D\theta(z; 0) + \sum_{n=1}^{\infty} |\theta(z; nw)| \cdot \left[\begin{array}{c} \dfrac{\sin(2n\pi D)}{n\pi} \cos(nwt + \beta_n) \\ + \dfrac{1 - \cos(2n\pi D)}{n\pi} \sin(nwt + \beta_n) \end{array} \right] \tag{14}$$

where $\theta(z; nw)$ is to be determined from Equation (11) and boundary conditions. The substitution of $\theta(z; nw) e^{nwt\cdot i}$ into Equation (11) gives the governing equation of $\theta(z; nw)$ as

$$\begin{cases} \dfrac{d^2\theta_i}{dz^2} - q_i^2 \theta_i = 0 \ (i = 1, 2, 3, 4, 6) \\ \dfrac{d^2\theta_5}{dz^2} - (q_5^2 + \eta/k_5)\,\theta_5 = 0 \end{cases} \tag{15}$$

where the $q^2 = n w i/\lambda$. $\theta(z; nw)$ satisfies the following boundary and continuity conditions: $-k_1 \partial\theta/\partial z|_{z=0} = -h \cdot \theta|_{z=0}$, $\theta|_{z=h_1^+} = \theta|_{z=h_1^-}$, $k_1 \partial\theta/\partial z|_{z=h_1^-} - k_2 \partial\theta/\partial z|_{z=h_1^+} = Q_0$, $\theta|_{z=(\sum\limits_{i=1}^{n} h_i)^+} =$

$\theta|_{z=(\sum\limits_{i=1}^{n} h_i)^-}$, $k_{(n+1)} \partial\theta/\partial z|_{z=(\sum\limits_{i=1}^{n} h_i)^+} = k_n \partial\theta/\partial z|_{z=(\sum\limits_{i=1}^{n} h_i)^-}$, $\theta|_{z=\sum\limits_{i=1}^{6} h_i} = 0.$

The solution of Equation (15) takes the form of

$$\begin{cases} \theta_i = A_i \exp\left(z\sqrt{q_i^2}\right) + B_i \exp\left(-z\sqrt{q_i^2}\right) \ (i = 1, 2, 3, 4, 6) \\ \theta_5 = A_5 \exp\left(z\sqrt{q_5^2 + \eta/k_5}\right) + B_5 \exp\left(-z\sqrt{q_5^2 + \eta/k_5}\right) \end{cases} \tag{16}$$

where the unknown coefficients in Equation (16) are to be determined by the boundary and continuity conditions as

$$\left\{ \begin{array}{c} A_1' \\ B_1' \\ A_2' \\ B_2' \\ A_3' \\ B_3' \\ A_4' \\ B_4' \\ A_5' \\ B_5' \\ A_6' \\ B_6' \end{array} \right\} = \left\{ \begin{array}{c} \dfrac{T_1 R \cosh\left(h_6\sqrt{q_6^2}\right) + m_5 R T_2 \sinh\left(h_6\sqrt{q_6^2}\right)}{(P_1 S_1 + m_1 P_2 T_1)\cosh\left(h_6\sqrt{q_6^2}\right) + m_5 (P_1 S_2 + m_1 P_2 T_2)\sinh\left(h_6\sqrt{q_6^2}\right)} \\ \left(k_1\sqrt{q_1^2} - h\right) A_1 / \left(k_1\sqrt{q_1^2} + h\right) \\ [(P_1 + m_1 P_2) A_1 - R]\, e^{-h_1\sqrt{q_2^2}} \\ [(P_1 - m_1 P_2) A_1 + R]\, e^{h_1\sqrt{q_2^2}} \\ (G + H)\, e^{-(h_1+h_2)\sqrt{q_3^2}} \\ (G - H)\, e^{(h_1+h_2)\sqrt{q_3^2}} \\ (G + H)\, e^{-(h_1+h_2)\sqrt{q_3^2}} \\ (G - H)\, e^{(h_1+h_2)\sqrt{q_3^2}} \\ (I + J)\, e^{-(h_1+h_2+h_3+h_4)\sqrt{q_5^2+\eta/k_5}} \\ (I - J)\, e^{(h_1+h_2+h_3+h_4)\sqrt{q_5^2+\eta/k_5}} \\ (K + L)\, e^{-(h_1+h_2+h_3+h_4+h_5)\sqrt{q_6^2}} \\ (K - L)\, e^{(h_1+h_2+h_3+h_4+h_5)\sqrt{q_6^2}} \end{array} \right\} \tag{17}$$

where

$$P_1 = \frac{k_1\sqrt{q_1^2}\cosh\left(h_1\sqrt{q_1^2}\right) + h\sinh\left(h_1\sqrt{q_1^2}\right)}{k_1\sqrt{q_1^2} + h}$$

$$P_2 = \frac{k_1\sqrt{q_1^2}\sinh\left(h_1\sqrt{q_1^2}\right) + h\cosh\left(h_1\sqrt{q_1^2}\right)}{k_1\sqrt{q_1^2} + h}$$

$$R = \frac{Q_0}{2k_2\sqrt{q_2^2}}$$

$$S_1 = \begin{bmatrix} \cosh\left(h_2\sqrt{q_2^2}\right)\cosh\left((h_3+h_4)\sqrt{q_3^2}\right)\cosh\left(h_5\sqrt{q_5^2+\eta/k_5}\right) \\ +m_2\sinh\left(h_2\sqrt{q_2^2}\right)\sinh\left((h_3+h_4)\sqrt{q_3^2}\right)\cosh\left(h_5\sqrt{q_5^2+\eta/k_5}\right) \\ +m_4\cosh\left(h_2\sqrt{q_2^2}\right)\sinh\left((h_3+h_4)\sqrt{q_3^2}\right)\sinh\left(h_5\sqrt{q_5^2+\eta/k_5}\right) \\ +m_2m_4\sinh\left(h_2\sqrt{q_2^2}\right)\cosh\left((h_3+h_4)\sqrt{q_3^2}\right)\sinh\left(h_5\sqrt{q_5^2+\eta/k_5}\right) \end{bmatrix}$$

$$T_1 = \begin{bmatrix} \sinh\left(h_2\sqrt{q_2^2}\right)\cosh\left((h_3+h_4)\sqrt{q_3^2}\right)\cosh\left(h_5\sqrt{q_5^2+\eta/k_5}\right) \\ +m_2\cosh\left(h_2\sqrt{q_2^2}\right)\sinh\left((h_3+h_4)\sqrt{q_3^2}\right)\cosh\left(h_5\sqrt{q_5^2+\eta/k_5}\right) \\ +m_4\sinh\left(h_2\sqrt{q_2^2}\right)\sinh\left((h_3+h_4)\sqrt{q_3^2}\right)\sinh\left(h_5\sqrt{q_5^2+\eta/k_5}\right) \\ +m_2m_4\cosh\left(h_2\sqrt{q_2^2}\right)\cosh\left((h_3+h_4)\sqrt{q_3^2}\right)\sinh\left(h_5\sqrt{q_5^2+\eta/k_5}\right) \end{bmatrix}$$

$$S_2 = \begin{bmatrix} \cosh\left(h_2\sqrt{q_2^2}\right)\cosh\left((h_3+h_4)\sqrt{q_3^2}\right)\sinh\left(h_5\sqrt{q_5^2+\eta/k_5}\right) \\ +m_2\sinh\left(h_2\sqrt{q_2^2}\right)\sinh\left((h_3+h_4)\sqrt{q_3^2}\right)\sinh\left(h_5\sqrt{q_5^2+\eta/k_5}\right) \\ +m_4\cosh\left(h_2\sqrt{q_2^2}\right)\sinh\left((h_3+h_4)\sqrt{q_3^2}\right)\cosh\left(h_5\sqrt{q_5^2+\eta/k_5}\right) \\ +m_2m_4\sinh\left(h_2\sqrt{q_2^2}\right)\cosh\left((h_3+h_4)\sqrt{q_3^2}\right)\cosh\left(h_5\sqrt{q_5^2+\eta/k_5}\right) \end{bmatrix}$$

$$T_2 = \begin{bmatrix} \sinh\left(h_2\sqrt{q_2^2}\right)\cosh\left((h_3+h_4)\sqrt{q_3^2}\right)\sinh\left(h_5\sqrt{q_5^2+\eta/k_5}\right) \\ +m_2\cosh\left(h_2\sqrt{q_2^2}\right)\sinh\left((h_3+h_4)\sqrt{q_3^2}\right)\sinh\left(h_5\sqrt{q_5^2+\eta/k_5}\right) \\ +m_4\sinh\left(h_2\sqrt{q_2^2}\right)\sinh\left((h_3+h_4)\sqrt{q_3^2}\right)\cosh\left(h_5\sqrt{q_5^2+\eta/k_5}\right) \\ +m_2m_4\cosh\left(h_2\sqrt{q_2^2}\right)\cosh\left((h_3+h_4)\sqrt{q_3^2}\right)\cosh\left(h_5\sqrt{q_5^2+\eta/k_5}\right) \end{bmatrix}$$

$$m_i = \frac{k_i\sqrt{q_i^2}}{k_{i+1}\sqrt{q_{i+1}^2}} \quad (i = 1,2,3)$$

$$m_4 = \frac{k_4\sqrt{q_4^2}}{k_5\sqrt{q_5^2+\eta/k_5}}$$

$$m_5 = \frac{k_5\sqrt{q_5^2+\eta/k_5}}{k_6\sqrt{q_6^2}}$$

$$G = \left[P_1\cosh\left(h_2\sqrt{q_2^2}\right) + m_1P_2\sinh\left(h_2\sqrt{q_2^2}\right)\right]A_1' - R\sinh\left(h_2\sqrt{q_2^2}\right)$$

$$H = m_2\left\{\left[P_1\sinh\left(h_2\sqrt{q_2^2}\right) + m_1P_2\cosh\left(h_2\sqrt{q_2^2}\right)\right]A_1' - R\cosh\left(h_2\sqrt{q_2^2}\right)\right\}$$

$$I = G\cosh\left[(h_3+h_4)\sqrt{q_3^2}\right] + H\sinh\left[(h_3+h_4)\sqrt{q_3^2}\right]$$

$$J = m_4 \left\{ G\sinh\left[(h_3+h_4)\sqrt{q_3^2}\right] + H\cosh\left[(h_3+h_4)\sqrt{q_3^2}\right] \right\}$$

$$K = I\cosh\left(h_5\sqrt{q_5^2+\eta/k_5}\right) + J\sinh\left(h_5\sqrt{q_5^2+\eta/k_5}\right)$$

$$L = m_5\left[I\sinh\left(h_5\sqrt{q_5^2+\eta/k_5}\right) + J\cosh\left(h_5\sqrt{q_5^2+\eta/k_5}\right)\right]$$

Thus, the temperature increase of heat source and device/skin interface can be given as follows,

$$\Delta T_{\text{device}}(t;\omega) = \Delta T_{\text{initial}}(h_1) + D\theta_1(h_1;0)$$
$$+ \sum_{n=1}^{\infty} |\theta_1(h_1;n\omega)| \cdot \left[\begin{array}{l} \frac{\sin(2n\pi D)}{n\pi}\cos\left(n\omega t + \beta_n^{\text{surface}}\right) \\ +\frac{1-\cos(2n\pi D)}{n\pi}\sin\left(n\omega t + \beta_n^{\text{surface}}\right) \end{array} \right] \tag{18}$$

$$\Delta T_{\text{interface}}(t;\omega) = \Delta T_{\text{initial}}(h_1+h_2) + D\theta_2(h_1+h_2;0)$$
$$+ \sum_{n=1}^{\infty} |\theta_2(h_1+h_2;n\omega)| \cdot \left[\begin{array}{l} \frac{\sin(2n\pi D)}{n\pi}\cos\left(n\omega t + \beta_n^{\text{surface}}\right) \\ +\frac{1-\cos(2n\pi D)}{n\pi}\sin\left(n\omega t + \beta_n^{\text{surface}}\right) \end{array} \right] \tag{19}$$

where $\Delta T_{\text{initial}}$ denotes the specific solution [initial temperature increase field in Equation (4) when there is no input power, given by

$$\Delta T_{\text{initial}}(h_1) = \left(\frac{k_1}{h}+h_1\right)A_1 + \Delta T_0 \tag{20}$$

and

$$\Delta T_{\text{initial}}(h_1+h_2) = \left(\frac{k_1}{h}+h_1+\frac{k_1 h_2}{k_2}\right)A_1 + \Delta T_0 \tag{21}$$

Here $\theta_1(h_1;n\omega)$ and $\theta_2(h_1+h_2;n\omega)$ can be obtained from Equations (16) and (17) as

$$\theta_1(h_1;n\omega) = 2\left[k_1\sqrt{q_1^2}\cosh\left(h_1\sqrt{q_1^2}\right) + h\sinh\left(h_1\sqrt{q_1^2}\right)\right]A_1' / \left(k_1\sqrt{q_1^2}+h\right) \tag{22}$$

and

$$\theta_2(h_1+h_2;n\omega) = 2\left[P_1 A_1'\cosh\left(-h_2\sqrt{q_2^2}\right) + (m_1 P_2 - R)\sinh\left(-h_2\sqrt{q_2^2}\right)\right] \tag{23}$$

To validate the analytic model under a pulsed power, we also performed FEA to obtain the temperature increase after saturation due to the pulsed power. The continuum element DC3D8 in ABAQUS is used to discretize the geometry. The material and geometry parameters such as thicknesses and thermal conductivity of each layer are the same as those in Section 2. The thermal diffusivity of encapsulation, substrate, stratum corneum, epidermis, dermis and fat are 1.4×10^{-7} m^2/s, 1.1×10^{-7} m^2/s, 6.6×10^{-8} m^2/s, 6.6×10^{-8} m^2/s, 1.3×10^{-7} m^2/s, 8.1×10^{-8} m^2/s, respectively [6,14]. The metabolic heat generation is set as 368 W/m^3 [14]. The natural convection boundary with the coefficient of heat convection 25 W/(m^2·K) is applied on the top surface of encapsulation layer. The ambient temperature is set as 25 °C. The core body temperature at the bottom surface of substrate is 37 °C. The thermal properties of blood and skin tissue can be found in Section 2 or in reference [14].

Figure 4 shows the influence of duty cycle on the maximum and minimum heat source temperature increase under the pulsed peak power density of 2500 W/m^2 (here we fix this peak power density to ensure that devices operate properly) with period as 500 ms. The analytic prediction agrees very well with FEA. Both the maximum temperature increase and minimum temperature increase of the heat source decrease as the duty cycle decreases. This trend can qualitatively be understood since the average emitted heat from the electronic device is reduced. For the case of constant power density corresponding to a duty cycle of 100%, the maximum temperature increase of heat source is 28.6 °C. As the duty cycle decreases to a smaller duty cycle of 10%, the maximum temperature increase of the heat source drops rapidly to about 0 °C. When the duty cycle is smaller than 10%, the temperature increase may become negative. This is because the temperature increase is defined from the core body temperature instead of the ambient temperature. A negative temperature increase means that the temperature is lower than the core body temperature.

It is necessary to mention that time period also has an influence on the results in a pulsed protocol. According to a transient heat transfer FEA, we obtained that the characteristic time of the temperature profile to reach a steady state is about 280 s. When the time periods used in the pulsed protocol are much smaller than the characteristic time, just as 500 ms we set above, the changes in time periods have only minor effects on the maximal heat source temperature. However, if the duration (τ) of the pulse is much longer than the characteristic time, the maximal heat source temperature is expected to be the same as if the electronic device is always turned on (constant protocol).

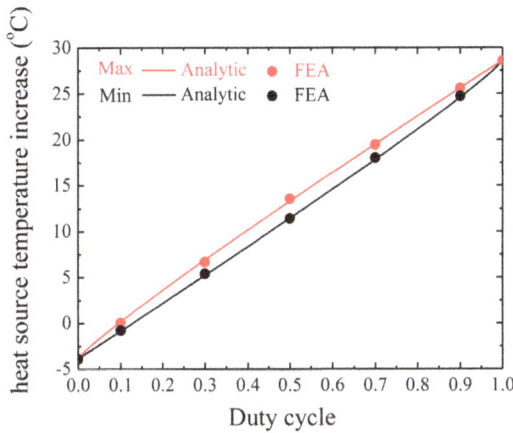

Figure 4. The maximum and minimum heat source temperature increase after saturation versus duty cycle.

Figure 5a compares the temperature increase of heat source after saturation from the analytic model in Equation (18) and FEA under the pulsed peak power density of 2500 W/m^2 with duty cycle D as 50% and period as 500 ms. The good agreement between the analytical prediction and FEA validates the analytical model. The temperature increase of heat source decreases with the increase of substrate thickness. Figure 5b compares the temperature increase of skin surface after saturation versus time from the analytic model in Equation (19) and FEA under the same condition in Figure 5a. With the change of substrate thickness, the temperature increase of skin surface shows the opposite trend to Figure 5a, i.e., the temperature increase of skin surface increases with the increase of the substrate thickness. These results clearly show that the pulsed operation is very effective at reducing the temperature increase in the system. A thin substrate can reduce the device temperature increase due to the heat sink effect of skin while increase the temperature increase of skin surface due to the influence of heat source.

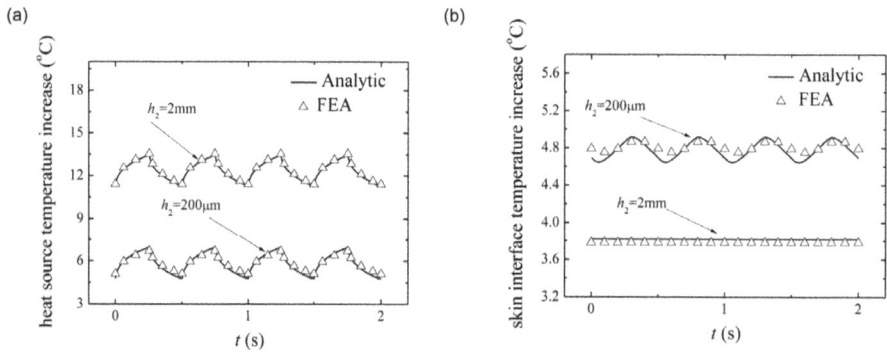

Figure 5. Temperature increase comparison between the analytic prediction and finite element analysis for the pulsed peak power density 2500 W/m^2 with 50% duty cycle and period 500 ms: (**a**) heat source; and (**b**) skin surface.

4. Conclusions

A one-dimensional analytic thermal model, as validated by the finite element analysis, for the flexible electronic devices integrated with human skin under a constant power and pulsed power is presented in this paper. This model combines the Fourier heat conduction equation for the flexible electronic devices and the Pennes bio-heat transfer equation for the skin tissue. The influences of geometric and loading parameters on the temperature increase under a pulsed power are investigated. The results could provide design guidelines for flexible electronic devices to minimize the adverse thermal effect in bio-integrated applications.

Acknowledgments: This work is supported by the National Basic Research Program of China (Grant No. 2015CB351901), the National Natural Science Foundation of China (Grant Nos. 11502009, 11172028 and 11372021), the Fundamental Research Funds for the Central Universities (Grant No. 2016XZZX001-05), the Research Fund for the Doctoral Program of Higher Education of China (20131102110039) and the open fund of Key Laboratory of Soft Machines and Smart Devices of Zhejiang Province.

Author Contributions: Yuhang Li and Jizhou Song conceived and designed the paper construction; Yun Cui performed the FEA simulations and derived the analytical solutions; Yufeng Xing and Tianzhi Yang provided the discussion about the analysis results; Yun Cui and Yuhang Li and Jizhou Song wrote the paper.

Conflicts of Interest: The authors declare no conflict of interest.

References

1. Kim, D.H.; Lu, N.; Ma, R.; Kim, Y.S.; Kim, R.H.; Wang, S.; Wu, J.; Won, S.M.; Tao, H.; Islam, A.; et al. Epidermal electronics. *Science* **2011**, *333*, 838–843. [CrossRef] [PubMed]
2. Webb, R.C.; Bonifas, A.P.; Behnaz, A.; Zhang, Y.; Yu, K.J.; Cheng, H.; Shi, M.; Bian, Z.; Liu, Z.; Kim, Y.S.; et al. Ultrathin conformal devices for precise and continuous thermal characterization of human skin. *Nat. Mater.* **2013**, *12*, 938–944. [CrossRef] [PubMed]
3. Kim, H.S.; Brueckner, E.; Song, J.Z.; Li, Y.H.; Kim, S.; Lu, C.F.; Sulkin, J.; Choquette, K.; Huang, Y.G.; Nuzzo, R.G.; et al. Unusual strategies for using indium gallium nitride grown on silicon (111) for solid-state lighting. *Proc. Natl. Acad. Sci. USA* **2011**, *108*, 10072–10077. [CrossRef] [PubMed]
4. Lu, C.F.; Li, Y.H.; Song, J.Z.; Kim, H.S.; Brueckner, E.; Fang, B.; Hwang, K.C.; Huang, Y.G.; Nuzzo, R.G.; Rogers, J.A. A thermal analysis of the operation of microscale, inorganic light-emitting diodes. *Proc. R. Soc. A Math. Phys. Eng. Sci.* **2012**, *468*, 3215–3223. [CrossRef]
5. Cui, Y.; Xing, Y.F.; Li, Y.H.; Song, J.Z. Thermal management for microscale inorganic light-emitting diodes. *Sci. Sin. Phys. Mech. Astron.* **2016**, *46*, 044612. (In Chinese) [CrossRef]

6. Kim, T.I.; Jung, Y.H.; Song, J.Z.; Kim, D.; Li, Y.H.; Kim, H.S.; Song, I.S.; Wierer, J.J.; Pao, H.A.; Huang, Y.G.; et al. High-efficiency, microscale GaN Light-emitting diodes and their thermal properties on unusual substrates. *Small* **2012**, *8*, 1643–1649. [CrossRef] [PubMed]

7. Li, Y.H.; Shi, Y.; Song, J.Z.; Lu, C.F.; Kim, T.I.; Rogers, J.A.; Huang, Y.G. Thermal properties of microscale inorganic light-emitting diodes in a pulsed operation. *J. Appl. Phys.* **2013**, *113*, 144505. [CrossRef]

8. Song, J.; Feng, X.; Huang, Y. Mechanics and thermal management of stretchable inorganic electronics. *Nat. Sci. Rev.* **2016**, *3*, 128–143. [CrossRef] [PubMed]

9. Li, Y.; Gao, Y.; Song, J. Recent advances on thermal analysis of stretchable electronics. *Theor. Appl. Mech. Lett.* **2016**, *6*, 32–37. [CrossRef]

10. Jiang, S.C.; Ma, N.; Li, H.J.; Zhang, X.X. Effects of thermal properties and geometrical dimensions on skin burn injuries. *Burns* **2002**, *28*, 713–717. [CrossRef]

11. Ozen, S.; Helhel, S.; Bilgin, S. Temperature and burn injury prediction of human skin exposed to microwaves: A model analysis. *Radiat. Environ. Biophys.* **2011**, *50*, 484–489. [CrossRef] [PubMed]

12. Im, I.; Youn, S.B.; Kim, K. Numerical study on the temperature profiles and degree of burns in human skin tissue during combined thermal therapy. *Numer. Heat Transf. A Appl.* **2015**, *67*, 921–933. [CrossRef]

13. Xu, F.; Lu, T.J.; Seffen, K.A.; Ng, E.Y.K. Mathematical modeling of skin bioheat transfer. *Appl. Mech. Rev.* **2009**, *62*, 050801. [CrossRef]

14. Pennes, H.H. Analysis of tissue and arterial blood temperatures in the resting human forearm. *J. Appl. Physiol.* **1948**, *1*, 93–122. [PubMed]

15. Xu, F.; Lu, T.J.; Seffen, K.A. Biothermomechanics of skin tissues. *J. Mech. Phys. Solids* **2008**, *56*, 1852–1884. [CrossRef]

16. Incropera, F.P.; Dewitt, D.P. *Fundamental of Heat and Mass Transfer*; Wiley: Hoboken, NJ, USA, 1996.

17. Zhao, G.; Liu, Z.F.; Zhang, A.L.; Zhang, H.F.; Cheng, S.X. Theoretical analyses of thermal stress of blood vessel during cryopreservation. *Cryo Lett.* **2005**, *26*, 239–250.

micromachines

MDPI

Review

Mechanisms and Materials of Flexible and Stretchable Skin Sensors

Yicong Zhao and Xian Huang *

Biomedical Engineering, School of Precision Instrument and Opto-Electronics Engineering, Tianjin University, Tianjin 300072, China; zhaoyicong@tju.edu.cn
* Correspondence: huangxian@tju.edu.cn; Tel.: +86-22-2740-9253

Academic Editors: Seung Hwan Ko, Daeho Lee and Zhigang Wu
Received: 9 January 2017; Accepted: 21 February 2017; Published: 28 February 2017

Abstract: Wearable technology has attracted significant public attention and has generated huge societal and economic impact, leading to changes of both personal lifestyles and formats of healthcare. An important type of devices in wearable technology is flexible and stretchable skin sensors used primarily for biophysiological signal sensing and biomolecule analysis on skin. These sensors offer mechanical compatibility to human skin and maximum compliance to skin morphology and motion, demonstrating great potential as promising alternatives to current wearable electronic devices based on rigid substrates and packages. The mechanisms behind the design and applications of these sensors are numerous, involving profound knowledge about the physical and chemical properties of the sensors and the skin. The corresponding materials are diverse, featuring thin elastic films and unique stretchable structures based on traditional hard or ductile materials. In addition, the fabrication techniques that range from complementary metal-oxide semiconductor (CMOS) fabrication to innovative additive manufacturing have led to various sensor formats. This paper reviews mechanisms, materials, fabrication techniques, and representative applications of flexible and stretchable skin sensors, and provides perspective of future trends of the sensors in improving biomedical sensing, human machine interfacing, and quality of life.

Keywords: flexible electronics; stretchable electronics; skin sensors; precision medicine; health monitoring; wearable technology

---

1. Introduction

Rapid growth in electronic technology yields miniaturized electronic devices and recent evolution of wearable electronic technology that can be integrated on human bodies and conduct diverse functions, such as mobile computation [1], health monitoring [2,3], activity tracking [4–6], and rehabilitation [7]. Wearable electronic devices can combine with portable electronic gadgets such as cell phones, laptops, and tablets to offer access to remote resources and enable data exchange, analysis and diagnosis. The wearable devices demonstrated both by various commercial available devices [8] as well as devices under exploration [9] have shown great promise to enrich personal health records and facilitate biomedical informatics, both of which are considered essential elements in the newly proposed precision medicine [10,11]. However, current wearable devices are predominately realized by encapsulating integrated circuits on solid substrates in rigid packages, which are mechanically incompatible with soft and curvilinear human body, resulting in unreliable and unrepeatable measurement results due to unreliable skin contact and changing measurement locations.

Some wearable devices are based on flexible and stretchable skin sensors, which are used primarily for biophysiological signal sensing and biomolecule analyzing on skin. These sensors can serve as activity tracking devices to record basic biophysiological parameters, or used for facilitating diagnosis and treatment of certain diseases such as diabetes [12], cystic fibrosis [13], dermatitis [14,15], and peripheral vascular disease [16]. In addition, they can be used as human–computer interface to assist human with speech and action disorders [17,18]. Additional use of skin sensors may involve monitoring exogenous parameters such as air qualities, environmental temperature, ultraviolet (UV) exposure, and humidity, allowing comprehensive assessment of health-related issues by considering both environmental and internal effects. The skin sensors offer mechanical compatibility to human skin and maximum compliance to skin morphology and motion. Stretchability is essential for these skin sensors, as the sensing precision, repeatability, stability and adhesion to skin are all determined by the capability of the sensors in following the skin motion, which causes skin deformation up to 30% [19]. The skin sensors contain unique structures constructed by either intrinsically soft materials or thin film materials on elastomer substrates. They can be simply mounted on bodies using fixtures such as bandages and body straps or use improved approaches that allow spontaneous skin attachment by van der Waals force using ultrathin and soft materials [14,18,20]. In addition, pressure sensitive silicon adhesive [21,22] can also be used to enhance the interface between the sensors and the skin, and offer reversible adhesion for long-term skin integration. Although the underlying mechanisms and relevant techniques of the skin sensors have been studied in many research papers that focus on various aspects [20,23,24], it will be beneficial if a systematic summary can be offered with comprehensive review of the state-of-the-art technology in flexible and stretchable skin sensor development.

This paper reviews some essential elements of flexible and stretchable skin sensors, including their mechanisms, materials, fabrication techniques, and applications, all of which represent recent progress in both theoretical and applied research of skin sensors. The fundamental mechanisms that determine the stretchability of the sensors are first presented, followed by materials used in skin sensors and their processing techniques. Finally, representative applications of the skin sensors are presented to demonstrate their capability in the areas of biomedical sensing and daily activity tracking. Flexible and stretchable skin sensors hold the promise to replace current wearable sensors based on rigid substrates and packages, and may eventually lead to the revolutionary changes in the formats of continuous, long-term health monitoring devices to improve social health levels.

2. Mechanisms of Flexible and Stretchable Skin Sensors

Flexible and stretchable skin sensors offer maximum compliance to the skin, and, thus, minimum reaction force from the sensor in response to the deformation, allowing less influence to the normal functions of the skin. The stretchability of the skin sensors can be achieved at the structural and material levels. The former refers to unique designed structures that offer tolerance to certain levels of deformation within the limits of the fracture strain of the constituent materials, while the latter can be attributed to intrinsically soft materials that are mechanically elastic to allow reversible extension and compression in response to external forces. The following section summarizes details of these two approaches.

2.1. Stretchable Structures

Materials used in skin sensors follow a simply rule that the bending strain of the materials decreases linearly with thickness of materials [23]. As a result, composition materials of skin sensors such as semiconductors, polymers, and metals are used in formats of ribbons, wires, and membranes with thickness in the scale of tens of nanometers to a few micrometers. They can be readily bended to reach a radius of curvature of ~150 mm with ~0.1% strains [25,26], which is less than the facture strain of these materials [26]. Two design approaches have been developed to make intrinsically rigid materials stretchable on elastomeric substrates. The former design uses out-of-plane buckling of ultra-thin nanoscale wires, ribbons, or membranes to release stress caused by in-plane prestrain applied to the substrates. The latter uses stretchable interconnects as bridges to connect with rigid islands, which typically contain functional components such as sensors [27,28], electronics [29–31], and commercial off-the-shelf components [32]. Both of these strategies have been widely used in layout of stretchable skin sensors.

2.1.1. Out-of-Plane Design

The formation of out-of-plane design with buckling structure is illustrated in Figure 1a. The ultrathin ribbons can be fabricated using conventional lithography process, followed by bonding the nanoribbons on to a prestrained elastomeric substrate. Releasing of the prestrain leads to periodical wavy structures on both ribbons and the substrate. The wavelength (λ) and magnitude (A) of the wavy structures can be determined by following equation.

$$\lambda = 2\pi h_f \left(\frac{\overline{E}_f}{3\overline{E}_s}\right)^{1/3}, A = h_f \sqrt{\frac{\varepsilon_{pre} - \varepsilon_{applied}}{\varepsilon_c} - 1} \tag{1}$$

in which h_f is the thickness of the stiff ribbons, $\overline{E}_f$ the Young's modulus of the elastic substrate, $\varepsilon_{applied}$ the applied strain, ε_{pre} the prestrain level, and ε_c the critical strain for the buckled ribbons. The peak strain in the ribbon is approximately equal to

$$\varepsilon_{peak} \approx 2\sqrt{\left(\varepsilon_{pre} - \varepsilon_{applied}\right)\varepsilon_c} \tag{2}$$

As a result, the maximum stretchability of the wavy structures can be determined by equating ε_{peak} to facture strain of the ribbon materials.

Recent development of wavy design has led to more complicated three-dimensional (3D) structures that buckle at higher orders, indicating potential applications of these technologies to form miniaturized flexible and stretchable electronics with highly spatial complexity and capability to achieve predefined shape shifting and function alternation [33]. The wavy design has been wildly available to integrate silicon [24], carbon nanotubes [34–36], graphene [37] and ferroelectrics [38] in formats of nanoribbons (Figure 1b) [24], nanowires (Figure 1c,e) [39], and nanomembranes (Figure 1d) [40]. However, the stretchability of the wavy design is determined by the prestrain levels of the substrates and the bending curvature of the materials as shown in Equations (1) and (2), limiting the applications of this technology in situations that require larger stretchability and less complex fabrication processes. As a result, island–bridge configurations have been developed to offer improved stretchability with both out-of-plane and in-plane structures in which islands based on functional sensing and circuit elements are mechanically and electrically connected with bridges made of narrow polymeric and metallic strips. The bridges, which contain either straight or serpentine interconnects, are freely suspended or bond on substrates between two islands. The stretchability of such structures are achieved either by deformation of the spatially buckled bridges (Figure 2a) or planar deformation of the interconnects (Figure 3c).

Novel substrate-free spatial helical structures have also been developed using diverse twisting modes, and have been used to act as sensors and power harvesters. Shang et al. have explored a carbon nanotube (CNT) yarn supercapacitor utilizing the helical loop structure. The entire structure can withstand strain of ~150% and repeated high frequency stretching (up to 10 Hz over 10,000 cycles). A helical spring based on copper nanowire [41], as shown in Figure 2b, offers higher stretchability (~700%). Similar structures can be formed by other metallic nanowires, showing variety of potential applications in wearable sensors and interconnects that can deform with the gradual growth of body parts.

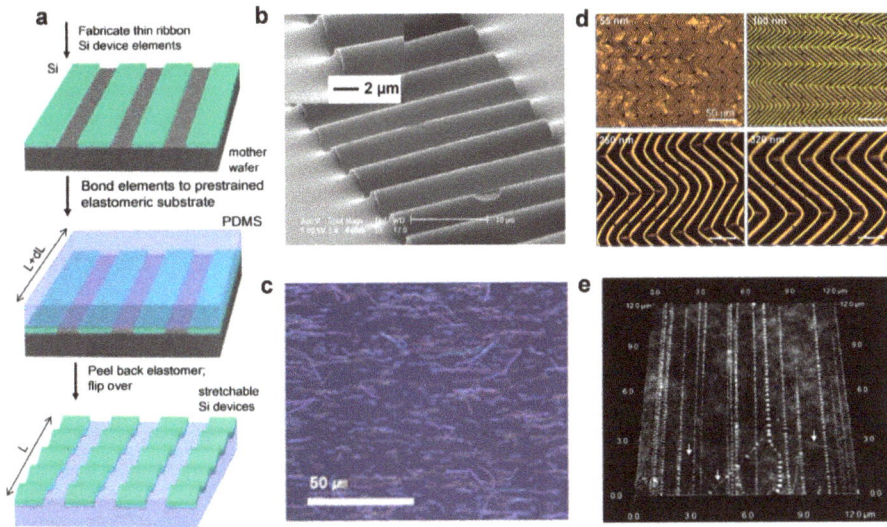

Figure 1. Mechanism of out-of-plane stretchable structures: (**a**) formation of out-of-plane nanoribbons (Reprinted with permission from Ref. [30] Copyright 2006 American Association for the Advancement of Science); (**b**) scanning electron microscope (SEM) images of a wavy nanoribbon (Reprinted with permission from Ref. [24] Copyright 2010 American Association for the Advancement of Science); (**c**) a large area optical micrograph of silicon nanowires (Reprinted with permission from Ref. [39] Copyright 2009 American Chemical Society); (**d**) optical micrographs of 2D wavy Si nanomembranes with various thickness (55, 100, 260, 320 nm) on polydimethylsiloxane (PDMS), formed with a thermal prestrain of 3.8% (Reprinted with permission from Ref. [40] Copyright 2007, American Chemical Society); and (**e**) an atomic force microscopic image of wavy SWNTs on a PDMS substrate (Reprinted with permission from Ref. [34] Copyright 2008 American Chemical Society).

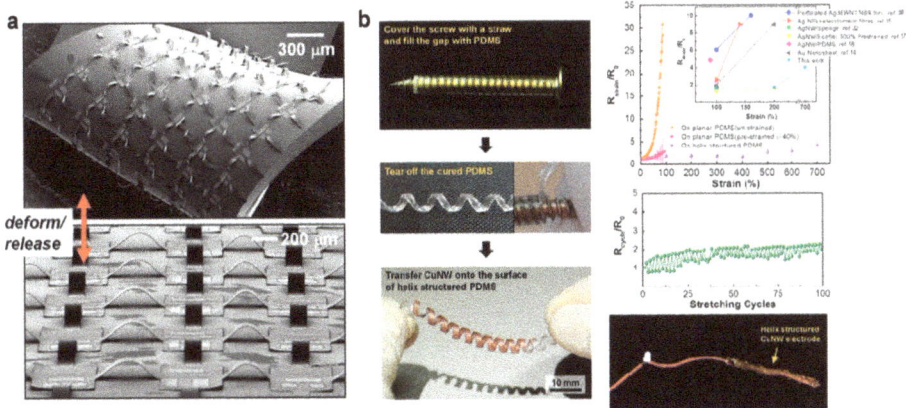

Figure 2. Examples of out-of-plane strcutres with high stretchability: (**a**) SEM images of arrays of complementary metal-oxide semiconductor (CMOS) inverters with spatially buckled bridges (Reprinted with permission from Ref. [29] Copyright 2008 National Academy of Sciences); and (**b**) helical-structured copper nanowire (CuNW)-based electrodes (Reprinted with permission from Ref. [41] Copyright 2014 Nature Publishing Group).

2.1.2. In-Plane Design

Improved deign without using prestrained substrates can be achieved by in-plane island–bridge design. The polymeric and metallic interconnects are typically in forms of serpentine or fractal [42–46] meshes that are completely bonded onto elastomeric substrates. In some cases, the island structures can be further omitted, resulting in continuous self-similar serpentine or fractal structures as both sensors and interconnects. Compared to wavy structures, these planar serpentine or fractal design effectively accommodate much larger applied strain through in-plane structural deformation without requirement of prestrain on substrates, and eliminate the concern of delicate spatially-buckled structures that can be easily broken under external scratch.

However, design of serpentine interconnect is still largely empirical, only a few theoretical models have been developed to analyze the deformation and stretchability of serpentine geometry. Fan et al. formulated an analytic model of in-planar serpentine interconnects based on finite deformation theory [47]. As illustrated in Figure 3a, a serpentine interconnect is simplified as three straight wires with length L or $L/2$ connected with two arcs with an identical radius R and an arc angle α. Three dimensionless parameters, width/radius ratio $\overline{w} = w/R$, arm length/radius ratio $\overline{L} = L/R$ and arc angle α, can then be used to represent the shape of the serpentine interconnect. As $\overline{w}$ of the non-buckled interconnect is usually much smaller than 0.5, such interconnect can be modeled as a curved, Euler–Bernoulli beam. When the serpentine interconnect is subjected to a tensile displacement of $U_{app}/2$ at the end, the effective applied strain ε_{app} of the serpentine interconnect can be represented by

$$\varepsilon_{app} = \frac{U_{app}}{4R\sin(\alpha/2) + 2L\cos(\alpha/2)} \tag{3}$$

The peak value of maximum principal strain in the serpentine interconnect can be related to the applied strain and the geometric parameters by

$$\varepsilon_{max-nonlinear} = \overline{w}F_2\left(\overline{L}, \alpha, \varepsilon_{app}\right) \tag{4}$$

where $F_2\left(\overline{L}, \alpha, \varepsilon_{app}\right)$ is a function that can be determined numerically using an approximate model based on finite deformation theory.

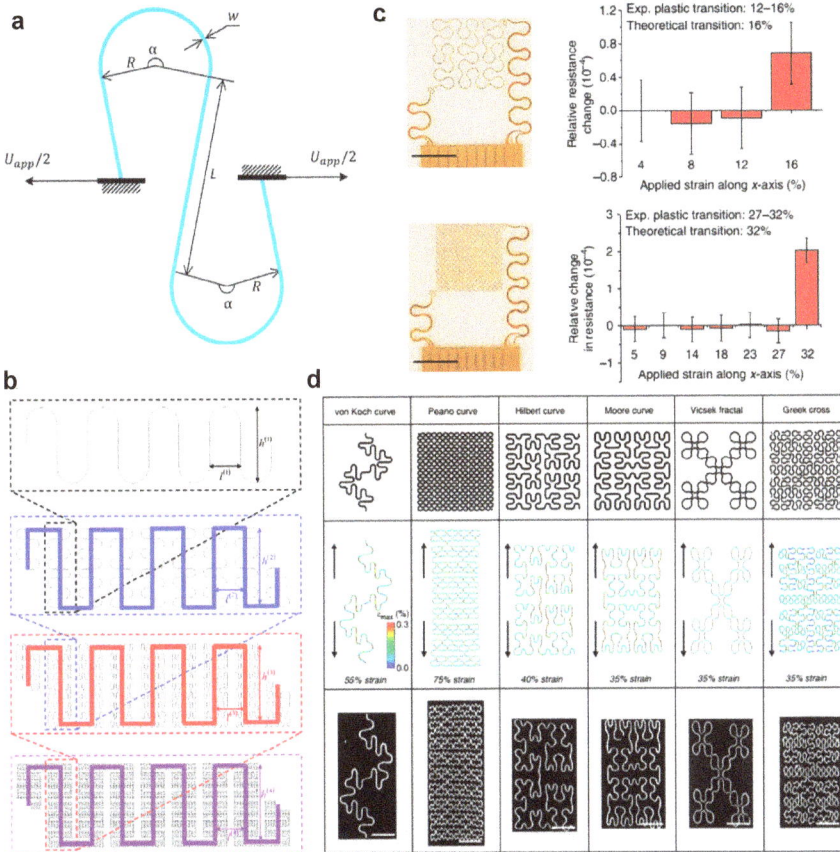

Figure 3. Mechanisms of serpentine and fractal structures: (**a**) a serpentine interconnect subjected to an axial stretching (U_{app}) at the two ends (Reprinted with permission from Ref. [47] Copyright 2016 Elsevier); (**b**) schematic illustration on the geometric construction of self-similar serpentine interconnects (Reprinted with permission from Ref. [48] Copyright 2013 Elsevier); and (**c**,**d**) representatives of fractal structures (Reprinted with permission from Ref. [45] Copyright 2014 Nature Publishing Group).

A fractal design concept that allows formation of stretchable layouts through stepwise iterations of basic shape units has been introduced to realize highly stretchable lithium-ion batteries [43] as well as several epidermal sensors [31,44,49]. As illustrated in Figure 3b, a fractal-based layout is created from the first order of serpentine geometry, and then constructed by connecting multiple copies of the unit cell forming self-similar design that offers increased area coverage and improved stretchability [48]. Theoretical models have been developed to analyze the deformation and stretchability of the fractal geometry. Fan et al. [45] have studied the deformations of various fractal layouts (Figure 3d), using a finite element method (FEM) followed by experimental evaluation. They also introduced a high precision approach to measure the elastic-plastic transition (or the elastic stretchability) through measuring differential resistances of the fractal interconnects, showing reasonable consistency with numerical analysis (Figure 3c). Several analytical models have been developed to determine elasticity for fractal interconnects. For example, Zhang et al. [48] have developed analytical models of flexibility and elastic stretchability, through establishing recursive formulae at different fractal orders. The analytical models show that the stretchability of system increases with the order of self-similar interconnect, and a surface filling ratios of 50% would yield 70% stretchability. In addition, the tensile stiffness for fractal interconnects has been determined by analytic approach, and has been verified by finite element analysis and experiments [50].

2.2. Intrinsic Elastic Materials

The stretchability of skin sensors can be achieved at the material levels using intrinsic elastic materials. Major materials in this category include elastomers and liquid metals, all of which adopt the stretchability due either to the flexible and long polymer chains or to weak intermolecular forces. This section offers a general review of the mechanism of these materials, while a detailed list of these materials will be given in next section.

One of the most widely used materials in stretchable skin sensors is silicone-based elastomers represented by polydimethylsiloxane (PDMS), fluorosilicone and various commercially available products under different tradenames such as Ecoflex, Dragon Skin, and Solaris. Silicone-based elastomers are notable for their high electrical resistivity (e.g., 2.9×10^{14} Ω·cm for PDMS), low glass transition temperatures (e.g., -125 °C for PDMS), large thermal coefficient of expansion (typically 4.8×10^{-4} K^{-1}) and high flexibility (with Young's module of 1 MPa). The polymer chains of the silicone contain siloxane backbones that consist of alterative sequences of silicon and oxygen and two organic substituents connected to each silicon atom, resulting in various properties (e.g., chemical resistance, elasticity, and phase) and processibility (e.g., curing time, and curing temperature). The elastomers obtain their stretchability through highly flexible siloxane backbones, which can be stretched under external forces. Another material that widely used in stretchable skin sensors is polyurethane (PU), which contains urethane groups connected with other groups such as ester, ether, amine and urea. The PU elastomers adopt their elasticity from the elastic polyol parts in the polymer chains, and offer large tear strength and abrasion resistant than silicone rubbers, making them ideal materials to construct substrates for skin sensors when frequent surface scratch and impact are expected.

3. Materials in Skin Sensors

The major materials of stretchable skin sensors can be classified into two categories. One involves various intrinsically stretchable materials, such as elastomers, liquid metals, and composite materials. The other includes materials such as solid metals, semiconductors, polymers, and inorganic compounds, which are rigid as bulk materials, but can be used as ultrathin films or membranes designed into special stretchable structures with thickness ranging from tens of nanometers to tens of micrometers. Therefore, the Young's modulus of the materials for stretchable skin sensors are wildly distributed from 0 to 10^{12} Pa [51].

3.1. Physically Soft and Stretchable Materials

3.1.1. Elastomers

Elastomers are available in different compositions with varied stretchability. As the fundamental materials in stretchable skin sensors, elastomers are mainly used as substrates, binders and adhesion layers. Among the available elastomers, PDMS is most commonly used. The mechanical properties of PDMS can be tuned by varying the curing conditions such as chemical ratios, temperature, and time, resulting in a Young's modulus in a controllable range from 1 to 150 MPa, and a stretchability up to 100%. By alternating the side groups as well as the lengths of the polymer chains, it is possible to obtain different types of elastomers with different physical and chemical properties. For example, PU and acrylic elastomer are two alternatives for skin sensor substrates, and are softer than PDMS due to its low Young's modules. The maximum stretchability that can be achieved is 300% [52]. The low-temperature curing silicone is a type of adhesive that can provide pressure sensitive reversible bonding between the skin and the devices. Besides of mold casting, silicone such as PDMS and polyurethane acrylate (PUA) can be photocurable to allow pattern definition through traditional photolithography processes [53,54] or even 3D printing techniques [55].

3.1.2. Liquid Metals

Liquid metals such as eutectic gallium-indium (eGaIn) and gallium-indium-tin (Galinstan) are intrinsically elastic with low resistivity (~2.9 × 10^{-7} Ω·m), low viscosity (~2 × 10^{-3} Pa·s), and low toxicity [56]. Their melting points are 15.5 °C [57] (75 wt % Ga and 25 wt % In) and −19 °C [58] (68.5 wt % Ga, 21.5% In, and 10.0% Sn), respectively, resulting in their liquid states at room temperature. Various functional components such as pressure sensors [59,60], strain sensors [61], antennas [62,63], and soft wires [64] have been fabricated by injecting liquid metals into microfluidic channels. Devices made of liquid metals can withstand deformation of microchannels at very high strain (up to 800%) [65].

3.1.3. Conductive Polymers

Conductive polymers (CPs) applied in stretchable electronics can be achieved by intrinsically conductive polymers (ICPs) [66] or conductive polymer composites [67]. Intrinsically conductive polymer materials such as synthetic poly(acetylene) (PA), poly(pyrrole) (PPy), poly(thiophene) (PT), poly(aniline) (PANI), and poly-(3,4-ethylenedioxythiophene) (PEDOT) can be realized by conjugation of polymer backbone, forming high energy orbitals with loosely bonded electrons to corresponding atoms, allowing maximum facture strain at the level of 1000% [68]. The conductive polymer composites are composed of polymers and conductive fillers (e.g., metal nanoparticles, metal nanowires, graphite, carbon nanotubes, and graphene). Conductive polymers are subject to influence of strain, which may lead to increased resistivity with strain. Park et al. have demonstrated a conductive composite mat using electrospun poly(styrene-block-butadiene-blocks-tyrene) (SBS) rubber fibers embedded with silver nanoparticles, leading to high conductivity even under large deformations (σ ≈ 2200 S·cm^{-1} at 100% strain) [69]. Shang et al. have achieved an elastic conductive nanocomposite composed of multiwall carbon nanotubes (MWNTs) and polyurethane (PU), which has initial conductivity of more than 5.3 S·cm^{-1} and stretchability of more than 100% [70]. Niu et al. have realized buckled single-wall carbon nanotube (SWCNT) electrodes by fabricating directly grown SWCNT films with continuous reticulate architecture on pre-strained PDMS [71]. The electrodes can stretch under a strain of 140% without significant change of resistance.

3.1.4. 1D and 2D Materials

The applications of one-dimensional (1D) and two-dimensional (2D) materials to construct stretchable electronics represent important trends in constructing stretchable electronics. Representative 1D and 2D materials include multi-walled or single-walled carbon nanotubes [72,73], silicon nanowires [74], metal nanowires [75,76], graphene [77], and transition metal dichalcogenides (TMDCs) [78]. Among them, both carbon nanotubes and graphene possess high electron mobility (~10^5 cm^2·v^{-1}·s^{-1} for carbon nanotube [79] and ~2 × 10^5 cm^2·v^{-1}·s^{-1} for graphene [80] at room temperature) and excellent mechanical flexibility (~1 Tpa Young's modulus) [81], making them promising materials for high performance electronic devices, such as top-gated transistors [82–84]. When used as sensors, the large surface-to-volume ratios of the 1D and 2D materials can lead to improved capabilities, such as highly sensitive biochemical sensing and large interfacial adhesion [85]. In addition, the optical transparency of graphene and carbon nanotubes allows construction of fully transparent sensors that possess high flexibility and softness [86,87]. Some excellent reviews about the 1D and 2D materials used in flexible and stretchable electronics have been provided by the following articles [88–90].

3.2. Unique Stretchable Structures

As mentioned in the previous section, raw materials that are rigid in their bulky formats can also offer stretchability in ultrathin configurations and unique stretchable design. Some major materials used for constructing skin sensors include metals, semiconductors, polymers, and inorganic compounds.

3.2.1. Solid Metals

Solid metals are intrinsically hard conductive materials that would become flexible when appear as thin films. Dominant metals used in skin sensors include Au, Cu, Al, Cr, Ti and Pt, which are used for conductive interconnects, electrodes, sensors, contact pads and other circuit components (e.g., resistor, inductors, and capacitors). These metals are typically tens of nanometers to a few micrometers in thickness, and are deposited on target substrates through physical deposition, electrochemical plating, and direct printing approaches. Many of these metals are ductile with a fracture strain of less than 1%. However, the stretchability of the metallic thin films can reach more than 100% when designed into special formats such as self-similar serpentine [91], fractal [28], helical [92] and prestrained bulking [29].

3.2.2. Semiconductors

Various active components such as diodes, transistors, and light emitting diodes (LED) can be made of inorganic semiconductor materials (e.g., silicon [30], GaAs [93], ZnO [94], InP [95], GaN [96]) as well as organic semiconductor materials (e.g., poly(3-hexylthiophene) (P3HT) [97], Poly(p-phenylene)vinylene [98], and Poly(2,5-bis(3-hexadecylthiophen-2-yl)thieno[3,2-b]thiophene) (pBTTT)) [99]. The bending stiffness and bending-induced strain of these rigid semiconductor materials can be exceptionally small due to cubic and linear scaling of these quantities with thickness of the materials. These semiconductor materials can be patterned into nanomembranes [100], nanoribbons [101], and nanowires [95] through complementary metal-oxide semiconductor (CMOS) fabrication processes.

3.2.3. Polymers

Polymers offer mechanical and electrical supports to skin sensors. They can be used as structural layers, electrical insulation layers, and dielectric layers in the skin sensors. Many polymers have been used to construct skin sensors, including some most prominent ones such as polyimide, poly(methyl methacrylate) (PMMA) and parylene. These polymers offer high mechanical strength that are ideal as structural layers to support the skin sensors. In addition, these polymers are typically thermal setting materials that can be easily obtained through spin-coating and dipping followed by curing at escalated temperature. Parylene is an excellent dielectric material, its fabrication process involves chemical vapor deposition (CVD), allowing pinhole-free uniform layers on curved or irregular surface.

4. Fabrication Techniques

Fabrication of skin sensors involve a series of techniques that combine conventional fabrication methods such as microelectromechanical systems (MEMS) technology, CMOS process, and mechanical milling with emerging techniques such as printable electronic, additive manufacturing, and laser process. Under the support from diverse fabrication techniques, many materials can be processed to yield structures in stretchable skin sensors.

4.1. Conventional Microfabrication Processes

Fabrications of active and passive components in flexible and stretchable electronic skin sensors can be achieved by MEMS and CMOS technology. The fundamental challenge of using MEMS and CMOS technology to make skin sensors involve application of ultrathin membranes as compared with rigid or brittle materials used in traditional MEMS and CMOS fabrication processes. Processes for fabricating flexible and stretchable skin sensors can be categorized into device-last and device-first approaches. The former involves fabricating active components on silicon membranes on silicon-on-insulator substrates, and then transfer-printing the membranes to destination substrates for further processing. While the latter refers to thinning down a thick semiconductor substrates integrated with functional components using physical sanding or chemical etching methods. The device-last approaches have been described by several research works. For example, Kim et al. [26] have developed stretchable integrated circuits, which combine multilayer neutral mechanical plane layouts and "wavy" structural configurations, including logic gates, ring oscillators, and differential amplifiers on silicon nanomembrane with thickness of 250 nm (Figure 4a). The device-first approaches have been demonstrated by MOS capacitors [102,103], memory cells [104,105], batteries [106], and MEMS switches [107].

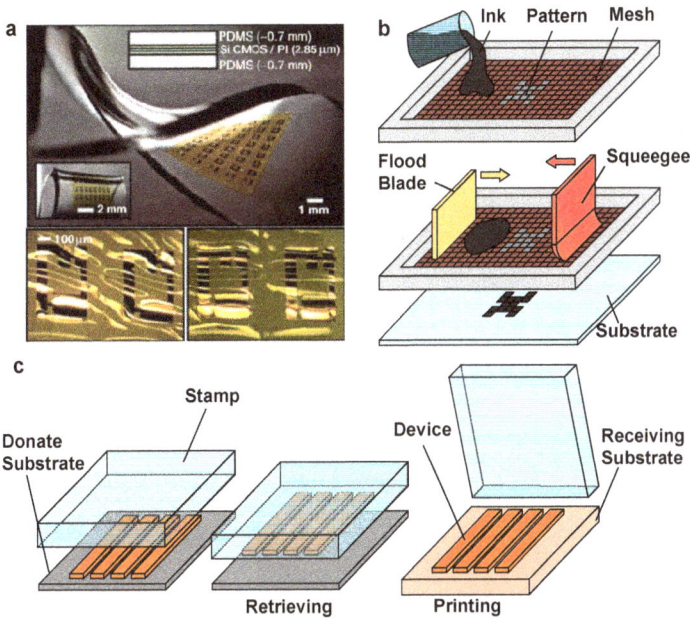

Figure 4. (**a**) Stretchable integrated circuits fabricated based on CMOS technology (Reprinted with permission from Ref. [26] Copyright 2008 American Association for the Advancement of Science); (**b**) a schematics of the screen printing process; and (**c**) a schematics of the transfer printing process.

Simplified approaches for making active and passive components involve using both organic materials as conductors, semiconductors, and dielectric to construct electronic devices. The major fabrication processes include chemical or physical vapor deposition, spin-coating and dip coating, which are much convenient than conventional microfabrication processes of inorganic materials, whose fabrications typically require thermal diffusion, ion implantation, and highly corrosive acid etchant used in inorganic materials. However, drawbacks of using organic materials are their low conductivity (1000 $S \cdot cm^{-1}$) and low charge mobility (approximately 10^{-2} to 10^2 $cm^2 \cdot v^{-1} \cdot s^{-1}$) [108,109], which hinder the applications of organic materials in high performance electronics.

4.2. Printable Electronics

Continuous development of printing electronic technology enables the application of such technology in developing flexible and stretchable electronics. Printable electronic technology allows direct generation of patterns on soft substrates through additive manufacturing technology such as screen-printing, slot-die coating and inkjet printing. In addition, a special printing approach used primarily in flexible and stretchable electronics is based on transfer printing, which transfer components fabricated on donate substrates onto target substrates that are typically thin plastic films, metal foils, and elastomer membranes.

4.2.1. Screen Printing

Screen printing is a low-cost, high-throughput printing technique used to construct skin sensors. The working principle of screen printing system is illustrated in Figure 4b. A screen printing system typically contains a flood blade that moves across a screen with open meshes with pore sizes ranging from 10 to 200 μm [110,111] and fills the meshes with ink. A squeegee is then moved in an opposite direction to push the inks in the meshes towards the substrates. Eventually, the adhesion force from the substrates pulls the inks down in a close distance, resulting in pattern formation that is determined both by the properties of the inks and the size of the meshes. Screen printing is notable for its low fabrication cost and capability to print single or stacked layers onto variety of soft materials such as fabrics and plastic films. The ability of screen printing technique has been demonstrated majorly in the area of printing organic devices such as organic light-emitting diode (OLEDs) [112], organic field-effect transistor (OFETs) [113,114], thin film batteries [115–117], and organic solar cells [118,119]. In addition, it has been widely used to fabrication various flexible and stretchable electronic components, including antennas [120], metal conductors [121,122], and thin-film transistors [123–125]. Limitations of screen printing include limited selection of ink materials, short processing time influenced by solvent evaporation, and low printing resolution (>10 μm).

4.2.2. Inkjet Printing

Inkjet printing is considered as an additive manufacturing technique that is attractive for making electronic components on flexible substrates without using any photomask. Inks that contain either fully dissolved chemicals [126,127] or nanoparticles [128–131] can be deposited through inkjet nozzles, which are actuated through a number of mechanisms such as piezoelectricity and aerosol. Post sintering processing after inkjet printing using direct heating, microwave, laser, and pulsed light can enhance the performance of the printed patterns by converting individual nanoparticles into connected matrixes. Inkjet printing techniques have been used to fabricate various electronic devices. Passive components such as resistors [132–134], capacitors [135–137] and inductors [138] have been printed on polymer substrates with various functional inks. Active components such as thin film transistors [139–141] and LED [142–144] have also been fabricated using inkjet printing methods. Development of colloidal solution for proper ejection of droplets on a targeted area by keeping an acceptable quality of the printed circuits is challenging due to the influence of evaporation rate of the solvents and orientation of the active particles. Despite the advancement of both control electronics and nozzle technology in inkjet printing, its printing speed (at the scale of 10 mm·s^{-1}) are still low as compared with screen printing methods, and its capability in printing complex structures such as serpentine and meander is still demanding further improvement. In addition, possibility of nozzle clogging and limited numbers of nozzles that can work simultaneously making inkjet printing methods more a rapid prototyping tool in labs rather than an acceptable mass fabrication method for industry. The spreading of the printed ink on target substrates and chaotic behavior of droplets during the time of flight further add to the issues of inkjet systems.

4.2.3. Transfer Printing

Transfer printing is an essential procedure to obtain flexible CMOS/MEMS devices. The target devices can be firstly fabricated on a donate substrate and then transferred to a receiving substrate using a viscoelastic stamp (usually a PDMS stamp) (Figure 4c). In this process, the adhesive strength is directly proportional to separation speed of the stamp from a surface. The effective separation speed between the stamp and the substrates is approximately 10 cm·s^{-1} or greater during a retrieval process, and is a few mm·s^{-1} or less for the printing process. Transfer printing technique has been extensively exploited to assemble diverse classes of materials (e.g., semiconductors, metals, carbon, and organic), thereby providing an effective method to fabricate various devices ranging from simple light emitting diodes [145], transistors [146,147] and sensor elements to fully integrated circuits [148].

5. Applications of Flexible and Stretchable Skin Sensors

Mechanisms, materials and fabrication approaches mentioned above can be used to construct diverse skin sensors that offer broad applications in health monitoring, daily activity tracking, and rehabilitation. These skin sensors have been used to record biophysical signals such as biopotential, skin strain, temperature, and hydration. In addition, initial efforts have been made to conduct specific biomolecule analysis using body fluids (e.g., sweat and blood) through direct contact or transdermal sensing approaches.

5.1. Biopotential Measurement

Biopotential measurement using skin sensors represents one of the most important applications of skin sensors. Due to the capability to be mounted on different locations of human skin, electrodes based on stretchable conductive meshes typically made of copper and gold have been used to conduct electroencephalogram (EEG), electrooculogram (EOG), and electrocardiogram (ECG) measurements on body. Yeo et al. [149] have introduced a multifunctional epidermal electronic systems measuring electrophysiological based on skin-contacted metallic electrodes or meshes made of gold and copper, which can measure ECG, EMG, temperature and strain (Figure 5a). Due to close skin contact, the flexible and stretchable skin electrodes can directly contact with skin with a contact resistance at a scale of 35 kΩ, which is compared smaller than conventional dry electrodes (~40 kΩ). Non-contact biopotential sensing is also feasible. Jeong et al. have demonstrated capacitive electrodes that can measure biopotential signals without direct contact with skin [18] (Figure 5b). They have also presented another skin sensor that measures EMG signal induced by arm and wrist movement to control unman drone.

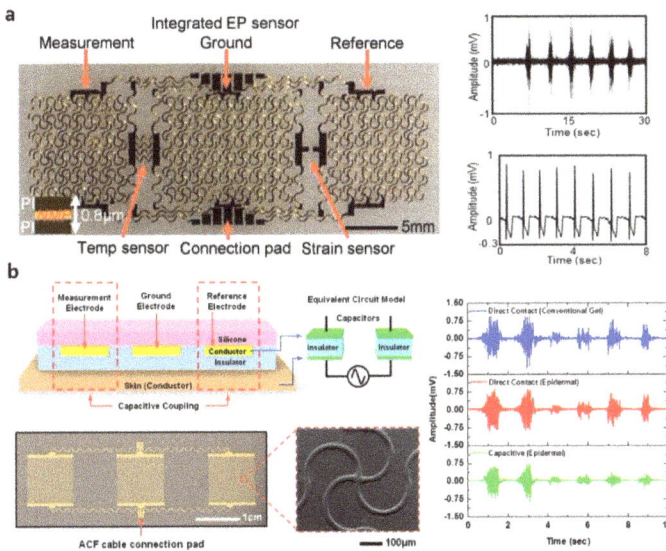

Figure 5. Examples of skin sensors for biopotential measurement: (**a**) biopotential measurement using skin-contacted metallic electrodes (Reprinted with permission from Ref. [91] Copyright 2013 John Wiley and Sons); and (**b**) an epidermal electronic system (EES) with a capacitive sensor for electrophysiological (EP) measurement (Reprinted with permission from Ref. [18] Copyright 2014 John Wiley and Sons).

5.2. Strain Sensing

Strain sensors can be directly attached on skin to measure strain induced by skin deformations caused by respiration, heartbeat, bending of body joints and muscle activities. Strain sensors can be fabricated by diverse materials and methods [17,150,151] with applications ranging from personalized health-monitoring [152,153] to human-machine interfaces [154,155] and soft robotics [156–158]. Park et al. [17] have achieved a stretchable graphene strain sensor using a layer-by-layer assembly method, in which stretchable yarns are repeatedly dip-coated with poly(vinyl alcohol) and graphene nanoplatelets bilayer. This strain sensor can be attached to throat and monitors the motions caused by speaking with maximum stretchability of ~100% (Figure 6a). Surface matrixes made of carbon nanotube and silicone rubber have been used to make strain sensors (Figure 6b). Carbon nanotubes were first coated onto a patterned polyimide film through air spraying, and were then transferred onto an Ecoflex film by casting uncured Ecoflex onto a polyimide film. The carbon nanotubes and Ecoflex form surface matrix that can be separated from the polyimide after the curing of the Ecoflex, resulting in conductive composite with high stretchability (~500%), linear temperature response ($R^2 = 1$) and fast time response (~332 ms) [150]. Majidi et al. [151] have developed thin-film curvature sensors composed of microfluidic channels filled with liquid metal (eGaIn) embedded in PDMS or Ecoflex substrates. The sensors can offer up to 1000% stretchability and measure both bending curvature and strain within the substrates with a gauge factor of 2 and a Young's modulus of 0.1~1 Mpa. Roh et al. [87] have realized a transparent and patchable strain sensor that is made of a sandwich-like stacked piezoresistive nanohybrid film of single-wall carbon nanotubes (SWCNTs) and a conductive elastomeric composite of polyurethane (PU)-poly(3,4-ethylenedioxythiophene) polystyrenesulfonate (PEDOT:PSS). This sensor can offer stretchability of up to 100% and optical transparency of 62%, which can detect small strains on human skin (Figure 6c).

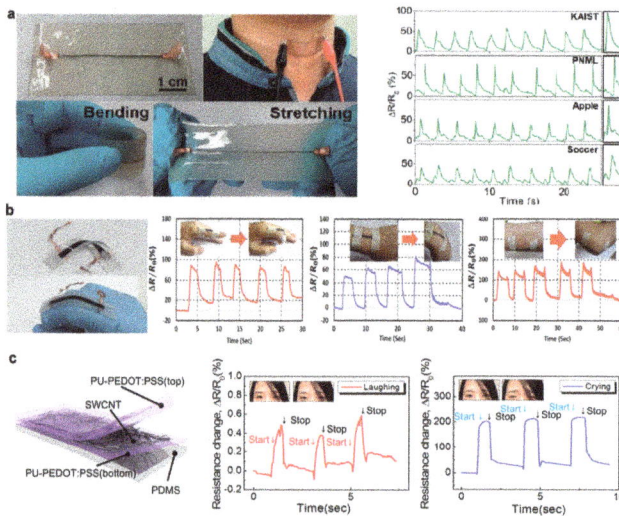

Figure 6. Examples of skin sensors for strain sensing: (**a**) graphene strain sensor embedded in an elastomeric patch that is bendable and stretchable, for detection of the motions of throat (Reprinted with permission from Ref. [17] Copyright 2015 American Chemical Society); (**b**) application of the CNT–Ecoflex nanocomposite based strain sensors to human motion detection (Reprinted with permission from Ref. [150] Copyright 2015 IOP Publishing); and (**c**) a transparent strain sensor consisting of three-layer stacked nanohybrid structure of PU-PEDOT:PSS/SWCNT/PU-PEDOT:PSS on a PDMS substrate (Reprinted with permission from Ref. [87] Copyright 2015 American Chemical Society).

5.3. Skin Temperature Monitoring

Temperature sensors are essential components for many health monitoring systems to determine both physiological and psychological conditions associated with cardiovascular health, cognitive state and malignancy. Skin temperature sensors can be conformably attached to skin surface, and, thus, can accurately measure body temperature with minimized influence from the environmental temperature. Examples of skin temperature sensors include arrays of meander metal wires that determine body temperature through measurement of spatial mapping, and temperature mapping devices based on PIN diodes made of silicon nanomembranes [159] (Figure 7a). To improve capability of long-term integration without disturbing the functions of skin, skin temperature sensors can be integrated with breathable substrates made of porous, semipermeable PU films (Figure 7b). The entire sensor is permeable to air and waterproof. It can realize continuous body temperature measuring for up to 24 h [28]. Organic materials can also be used to construct temperature sensors. Trung et al. [160] have realized a resistive and gated temperature sensor array purely by elastic organic materials with stretchability of ~70% and sensitivity of ~1.34% resistance change per degree Celsius. The sensing layer of this device can be formed by imbedding conductive and graphene oxide nanosheets into an elastomeric PU matrix (Figure 7c).

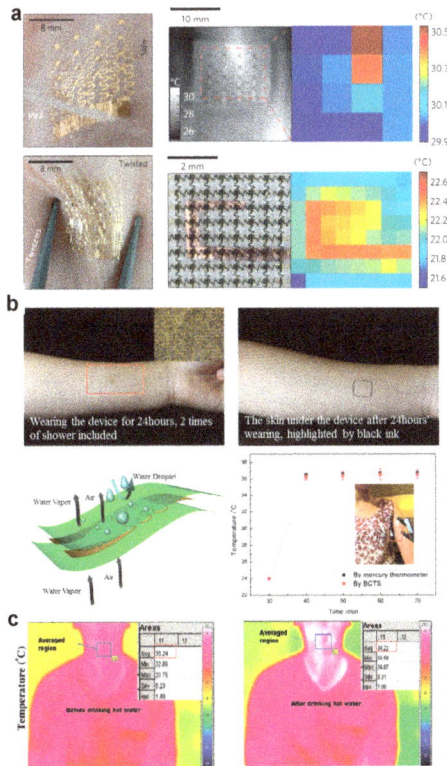

Figure 7. Examples of skin sensors for temepature measurement: (**a**) epidermal sensors that can monitor skin temperature using metallic and semiconductor sensors (Reprinted with permission from Ref. [159] Copyright 2013 Nature Publication Group); (**b**) breathable and stretchable temperature sensors (Reprinted with permission from Ref. [28] Copyright 2015 Nature Publication Group); and (**c**) transparent and stretchable temperature sensors (Reprinted with permission from Ref. [160] Copyright 2015 John Wiley and Sons).

5.4. Hydration Sensing

Accurately measurement of skin hydration levels is important for analyzing various diseases (e.g., dermatitis [161], psoriasis [162], eczema [163] and pruritus [164]) in the fields of dermatology and cosmetology, and evaluating factors (e.g., environmental [165], age [166], and hormone [167]) related to abnormal skin responses. In addition, hydration can also be used for assessing effectiveness of anti-aging treatment, moisturizing treatments and other medical therapies.

Skin hydration can be determined by measurements of electrical impedance, thermal conductivity, spectroscopic property, and mechanical characteristic in conventional approaches. The application of epidermal electronic techniques gives hydration sensing many advantages over traditional methods. Huang et al. have realized several types of epidermal hydration sensors based on detection of skin electrical impedance. The sensors consist of two electrodes connecting with a data acquisition system, which provides alternating electrical current at frequencies between 1 and 100 kHz. The skin electronic impedance can be reflected by resulting attenuation and phase shift of the electrical current. Devices capable of conducting differential monitoring [14] (Figure 8a), regional mapping [27] (Figure 8b), and wireless sensing [15] (Figure 8c) have been developed based on the impedance detection. In addition, hydration can also be assessed through measurements of skin thermal conductivity, which can be determined by time response of skin to constant thermal energy input [159].

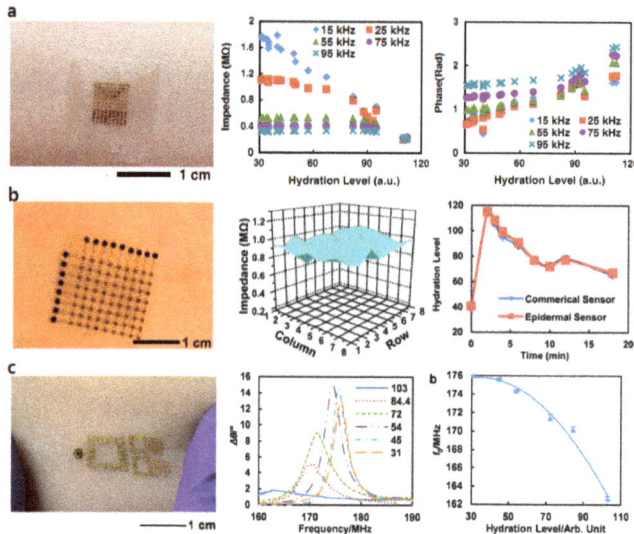

Figure 8. Examples of skin sensors used for hydration sensing: (**a**) epidermal sensor that can monitor biopotential on skin using metallic meshes (Reprinted with permission from Ref. [14] Copyright 2013 John Wiley and Sons); (**b**) hydration sensor that can conduct regional mapping based on the impedance detection (Reprinted with permission from Ref. [27] Copyright 2014 IEEE); and (**c**) hydration sensor capable of passive wireless detection (Reprinted with permission from Ref. [15] Copyright 2014 John Wiley and Sons).

5.5. Biomolecule Analysis

Flexible and stretchable skin sensors can be utilized for biomolecule analysis. Various biomolecules in sweat (e.g., sodium [12,168–170], potassium [12,169], ammonium [171,172], glucose [173], and lactate [12,174]) have been regarded as indicators for human physiological health. Huang et al. have explored materials and design strategies for integrating stretchable wireless sensors on porous sponge-like elastomeric substrates for epidermal analysis of biomolecules in sweat

(Figure 9a). The porous substrates allow sweat collection through capillary forces, without need for complex microfluidic handling systems. Colorimetric measurement is achieved in the same system by introducing indicator compounds into the substrates for sensing specific components (OH^-, H^+, Cu^+, and Fe^{2+}) in sweat [13]. Bandodkar et al. [175] have developed an epidermal tattoo-like sensor using a bluetooth enabled wearable transceiver for real-time monitoring of sodium in human perspiration with concentration range of 0.1–100 mM. This sensor can withstand strain caused by bending, stretching and poking (Figure 9b). Gao et al. [12] have designed a fully integrated sensor array for in situ perspiration analysis, which can simultaneously and selectively measure sweat metabolites (e.g., glucose and lactate) and electrolytes (e.g., sodium and potassium ions) as well as the skin temperature for sensor calibration (Figure 9c).

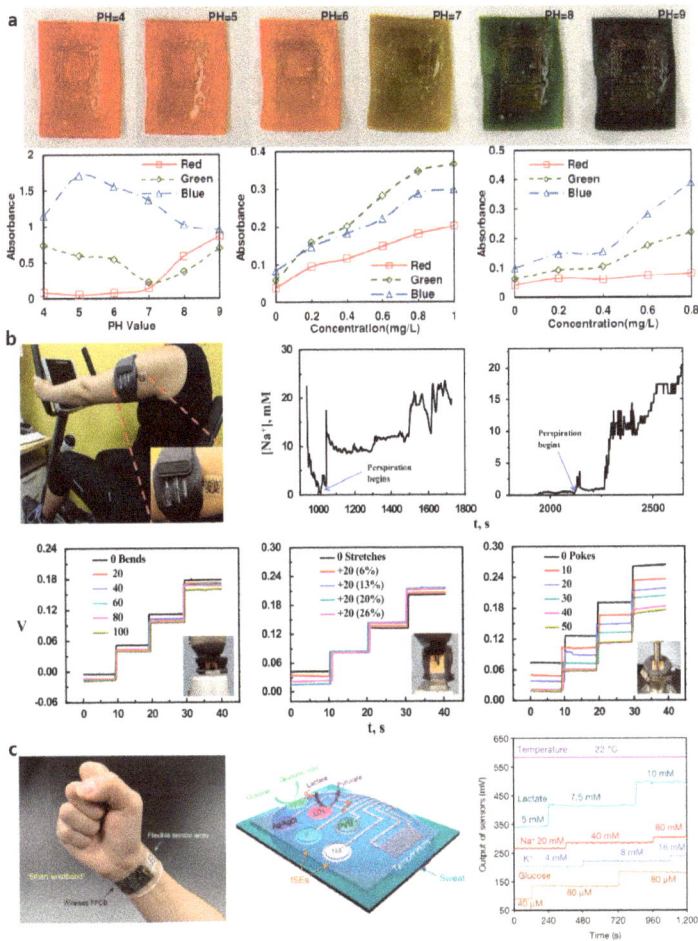

Figure 9. Examples of skin senosrs capable of conducting biomolecule analysis: (**a**) a skin sensor that can monitor biomolecules in sweat based on colorimetry approach (Reprinted with permission from Ref. [13] Copyright 2014 John Wiley and Sons); (**b**) a skin sensor that can monitor sodium in perspiration using electrochemical methods (Reprinted with permission from Ref. [175] Copyright 2014 Elsevier); and (**c**) a integrated system that can analyze multiple compositions in sweat simultaneously and selectively (Reprinted with permission from Ref. [12] Copyright 2016 Nature Publishing Group).

5.6. Other Sensing

Flexible and stretchable skin sensors also have other applications, including oximetry [21,176,177], pressure sensing [178,179] and wound healing monitoring [180,181]. For example, Yokota et al. [177] have developed optoelectronic skins integrated with OLED and organic photodetectors, which can measure the oxygen concentration of blood based on a photoplethysmogram (PPG) approach (Figure 10a). Choong et al. [182] have demonstrated a stretchable resistive pressure sensor within which the conductive electrode is built on the micro-pyramid PDMS arrays grafted with a PEDOT:PSS/PUD composite polymer. The sensor offers a pressure sensitivity of 10.3 kPa^{-1} when stretched by 40% (Figure 10b). Hattori et al. [180] have established an epidermal electronics system that can monitor cutaneous wound healing by recording time-dynamic temperature and thermal conductivity of skin. This system consists of metal traces with fractal and filamentary serpentine (FS) configurations, which can offer stretchability of ~30% (Figure 10c).

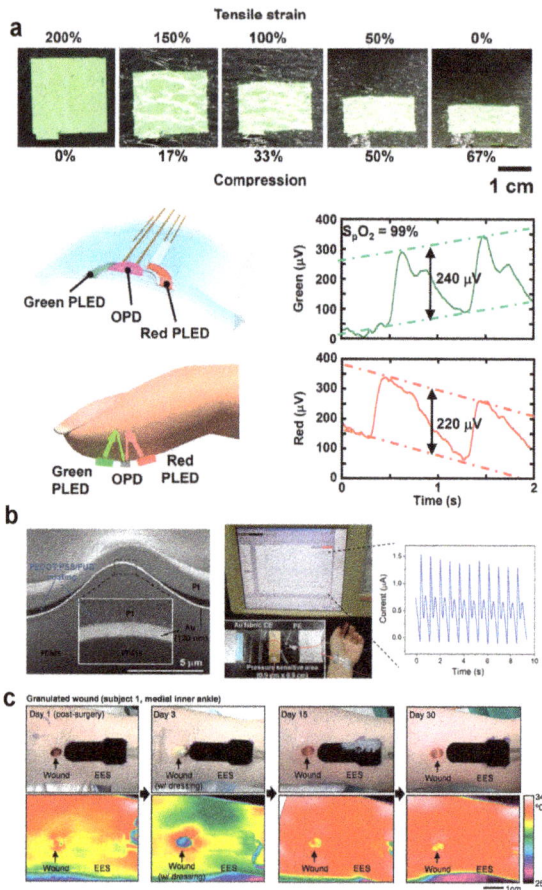

Figure 10. Skin sensors used for other sensing applications: (**a**) a skin sensor that can measure oxygen concentration of blood (Reprinted with permission from Ref [177] Copyright 2016 The Authors); (**b**) a stretchable resistive pressure sensor (Reprinted with permission from Ref [182] Copyright 2014 John Wiley and Sons); and (**c**) an epidermal electronics system that can monitor cutaneous wound healing (Reprinted with permission from Ref. [180] Copyright 2014 John Wiley and Sons).

6. Conclusions and Perspectives

This paper reviews the mechanisms, materials, fabrication techniques, and the representative applications of flexible and stretchable skin sensors. These sensors are constructed by various intrinsic soft materials or stretchable thin film structures with applications in biophysiological signal measurement and activity tracking, offering improved precision and effectiveness of body integration. Flexible and stretchable skin sensors can collect massive amount of data associated with personal biomedical information and life-style. These data can be used to assist more specific diagnosis and effective treatment of disease, and can potentially be used to reveal the underlying connection among biomedical information, environmental effects, and various diseases.

Despite the rapid progress in skin sensors, development of flexible and stretchable skin sensors has encountered several critical issues such as power supplies and system complexity. Firstly, most of the devices mentioned above focus on sensing functions of the stretchable electronic devices. However, power supply, signal conditioning, data communication, and data storage still largely rely on bulky instruments or integrated circuits based on rigid substrates. Some researches tackle the issues of power supplies with stretchable batteries [48,183], piezoelectric generators [184,185], solar cells [186,187], and wireless power harvesting [188], showing promising future in replacing current bulky power sources with components that mechanically and geometrically match stretchable electronic devices. While for signal conditioning and data communication, and data storage, these functions can be realized by integrating multiple commercial-off-the-shelf components connected by flexible and stretchable interconnects. These systems can be best represented by the work of Xu et al. [44] who realized a complex measurement system to detect biopotential, acceleration, temperature and achieve signal processing and wireless data communication with an operational amplifier-based circuit and a voltage control oscillator. The sweat sensing system [12] mentioned in the previous section has also demonstrated the possibility of integrating flexible circuits made of commercial components with stretchable skin sensor for precise analysis of sweat contents. These comprehensive systems will lead to capability to assess multiple biophysiological signals to improve accuracy in diagnosis and treatment. Discrete stretchable electronic components based purely on thin film materials and stretchable structures have been demonstrated as signal amplifiers [189], logic circuits [190,191], oscillators [192], and nonvolatile resistive memory [193,194]. However, a fully stretchable and integrated system has not yet been achieved due to the challenges in the fabrication of different functional components, interconnection of transfer printed components, and low electronic performance as compared to conventional devices based on rigid materials.

Furthermore, some fundamental knowledge of device mechanisms has not yet been well studied mechanically and electrically, and the interaction between the biological tissues and the flexible and stretchable skin sensors has not yet been well understood. For example, the electromagnetic properties of the stretchable structures such as serpentine, wavy, and out-of-plane buckling are largely unexplored, and the negative effects of the biological tissues to electromagnetic signal and optical signal have not yet been addressed to achieve optimized sensor performance. It can be expected that special properties offered by the sensor/skin interaction may be used to achieve more unique functions such as spontaneous actuation and transduction through skin motions, and the skin barrier functions may be overcome to allow analysis of biomolecules in blood and interstitial fluids using skin sensors. With more understanding of the fundamental knowledge of flexible and stretchable skin sensors, more sensing functions and powerful integrated systems may be developed based on the skin sensor platform, allowing revolutionary changes in the formats of continuous, long-term health monitoring devices to improve social health levels.

Acknowledgments: This work is supported by the National Natural Science Foundation of China under Grant No. 61604108 and the Natural Science Foundation of Tianjin under Grant No. 16JCYBJC40600.

Author Contributions: Xian Huang and Yicong Zhao conceived the article, constructed the graphs and wrote the paper. Xian Huang supervised the work.

Conflicts of Interest: The authors declare no conflict of interest.

References

1. Abdali-Mohammadi, F.; Bajalan, V.; Fathi, A. Toward a fault tolerant architecture for vital medical-based wearable computing. *J. Med. Syst.* **2015**, *39*, 149. [CrossRef] [PubMed]
2. Pantelopoulos, A.; Bourbakis, N.G. A survey on wearable sensor-based systems for health monitoring and prognosis. *IEEE Trans. Syst. Man Cybern. C* **2010**, *40*, 1–12. [CrossRef]
3. Yoo, J.; Yan, L.; Lee, S.; Kim, Y. A 5.2 mW self-configured wearable body sensor network controller and a 12 μW wirelessly powered sensor for a continuous health monitoring system. *IEEE J. Solid State Circuits* **2010**, *45*, 178–188. [CrossRef]
4. Evenson, K.R. Systematic review of the validity and reliability of consumer-wearable activity trackers. *Int. J. Behav. Nutr. Phys. Act.* **2015**, *12*, 159. [CrossRef] [PubMed]
5. Yang, C.-C.; Hsu, Y.L. A review of accelerometry-based wearable motion detectors for physical activity monitoring. *Sensors* **2010**, *10*, 7772–7788. [CrossRef] [PubMed]
6. Case, M.A.; Burwick, H.A.; Volpp, K.G.; Patel, M.S. Accuracy of smartphone applications and wearable devices for tracking physical activity data. *JAMA* **2015**, *313*, 625–626. [CrossRef] [PubMed]
7. Patel, S.; Park, H.; Bonato, P.; Chan, L.; Rodgers, M. A review of wearable sensors and systems with application in rehabilitation. *J. Neuroeng. Rehabil.* **2012**, *9*, 21. [CrossRef] [PubMed]
8. Chuah, H.W.; Rauschnabel, P.A.; Krey, N.; Bang, N.; Ramayah, T.; Lade, S. Wearable technologies: The role of usefulness and visibility in smartwatch adoption. *Comput. Hum. Behav.* **2016**, *65*, 276–284. [CrossRef]
9. Castillejo, P.; Martinez, J.F.; Rodriguez-Molina, J.; Cuerva, A. Integration of wearable devices in a wireless sensor network for an E-health application. *IEEE Wirel. Commun.* **2013**, *20*, 38–49. [CrossRef]
10. Ashley, E.A. The precision medicine initiative: A new national effort. *JAMA* **2015**, *313*, 2119–2120. [CrossRef] [PubMed]
11. Shah, S.H.; Arnett, D.; Houser, S.R.; Ginsburg, G.S.; Macrae, C.; Mital, S.; Loscalzo, J.; Hall, J.L. Opportunities for the cardiovascular community in the precision medicine initiative. *Circulation* **2016**, *133*, 226–231. [CrossRef] [PubMed]
12. Gao, W.; Emaminejad, S.; Nyein, H.Y.Y.; Challa, S.; Chen, K.; Peck, A.; Fahad, H.M.; Ota, H.; Shiraki, H.; Kiriya, D. Fully integrated wearable sensor arrays for multiplexed in situ perspiration analysis. *Nature* **2016**, *529*, 509–514. [CrossRef] [PubMed]
13. Huang, X.; Liu, Y.; Chen, K.; Shin, W.J.; Lu, C.J.; Kong, G.W.; Patnaik, D.; Lee, S.H.; Cortes, J.F.; Rogers, J.A. Stretchable, wireless sensors and functional substrates for epidermal characterization of sweat. *Small* **2014**, *10*, 3083–3090. [CrossRef] [PubMed]
14. Huang, X.; Yeo, W.H.; Liu, Y.; Rogers, J.A. Epidermal differential impedance sensor for conformal skin hydration monitoring. *Biointerphases* **2012**, *7*, 52. [CrossRef] [PubMed]
15. Huang, X.; Liu, Y.; Cheng, H.; Shin, W.J.; Fan, J.A.; Liu, Z.; Lu, C.J.; Kong, G.W.; Chen, K.; Patnaik, D. Materials and designs for wireless epidermal sensors of hydration and strain. *Adv. Funct. Mater.* **2014**, *24*, 3846–3854. [CrossRef]
16. Kim, J.; Salvatore, G.A.; Araki, H.; Chiarelli, A.M.; Xie, Z.; Banks, A.; Sheng, X.; Liu, Y.; Lee, J.W.; Jang, K.-I.; et al. Battery-free, stretchable optoelectronic systems for wireless optical characterization of the skin. *Sci. Adv.* **2016**, *2*, e1600418. [CrossRef] [PubMed]
17. Park, J.J.; Hyun, W.J.; Mun, S.C.; Park, Y.T.; Park, O.O. Highly stretchable and wearable graphene strain sensors with controllable sensitivity for human motion monitoring. *ACS Appl. Mater. Interfaces* **2015**, *7*, 6317–6324. [CrossRef] [PubMed]
18. Jeong, J.W.; Kim, M.K.; Cheng, H.; Yeo, W.H.; Huang, X.; Liu, Y.; Zhang, Y.; Huang, Y.; Rogers, J.A. Capacitive epidermal electronics for electrically safe, long-term electrophysiological measurements. *Adv. Healthc. Mater.* **2014**, *3*, 642–648. [CrossRef] [PubMed]
19. Maiti, R.; Gerhardt, L.-C.; Lee, Z.S.; Byers, R.A.; Woods, D.; Sanz-Herrera, J.A.; Franklin, S.E.; Lewis, R.; Matcher, S.J.; Carré, M.J. In vivo measurement of skin surface strain and sub-surface layer deformation induced by natural tissue stretching. *J. Mech. Behav. Biomed. Mater.* **2016**, *62*, 556–569. [CrossRef] [PubMed]
20. Kim, D.-H.; Lu, N.; Ma, R.; Kim, Y.-S.; Kim, R.-H.; Wang, S.; Wu, J.; Won, S.M.; Tao, H.; Islam, A.; et al. Epidermal electronics. *Science* **2011**, *333*, 838–843. [CrossRef] [PubMed]

21. Jang, K.-I.; Han, S.Y.; Xu, S.; Mathewson, K.E.; Zhang, Y.; Jeong, J.-W.; Kim, G.-T.; Webb, R.C.; Lee, J.W.; Dawidczyk, T.J.; et al. Rugged and breathable forms of stretchable electronics with adherent composite substrates for transcutaneous monitoring. *Nat. Commun.* **2014**, *5*, 4779. [CrossRef] [PubMed]

22. Yu, R.J.; Park, H.; Jin, S.W.; Hong, S.Y.; Lee, S.S.; Ha, J.S. Highly stretchable and sensitive strain sensors using fragmentized graphene foam. *Adv. Funct. Mater.* **2015**, *25*, 4228–4236.

23. Rogers, J.A.; Someya, T.; Huang, Y. Materials and mechanics for stretchable electronics. *Science* **2010**, *327*, 1603–1607. [CrossRef] [PubMed]

24. Song, J. Mechanics of stretchable electronics. *Curr. Opin. Solid State Mater. Sci.* **2015**, *19*, 160–170. [CrossRef]

25. Park, S.-I.; Ahn, J.-H.; Feng, X.; Wang, S.; Huang, Y.; Rogers, J.A. Theoretical and experimental studies of bending of inorganic electronic materials on plastic substrates. *Adv. Funct. Mater.* **2008**, *18*, 2673–2684. [CrossRef]

26. Kim, D.H.; Ahn, J.H.; Choi, W.M.; Kim, H.S.; Kim, T.H.; Song, J.; Huang, Y.Y.; Liu, Z.; Lu, C.; Rogers, J.A. Stretchable and foldable silicon integrated circuits. *Science* **2008**, *320*, 507–511. [CrossRef] [PubMed]

27. Huang, X.; Cheng, H.; Chen, K.; Zhang, Y.; Zhang, Y.; Liu, Y.; Zhu, C.; Ouyang, S.C.; Kong, G.W.; Yu, C. Epidermal impedance sensing sheets for precision hydration assessment and spatial mapping. *IEEE Trans. Biomed. Eng.* **2013**, *60*, 2848–2857. [CrossRef] [PubMed]

28. Chen, Y.; Lu, B.; Chen, Y.; Feng, X. Breathable and stretchable temperature sensors inspired by skin. *Sci. Rep.* **2015**, *5*, 11505. [CrossRef] [PubMed]

29. Kim, D.H.; Song, J.; Choi, W.M.; Kim, H.S.; Kim, R.H.; Liu, Z.; Huang, Y.Y.; Hwang, K.C.; Zhang, Y.W.; Rogers, J.A. Materials and noncoplanar mesh designs for integrated circuits with linear elastic responses to extreme mechanical deformations. *Proc. Natl. Acad. Sci. USA* **2008**, *105*, 18675–18680. [CrossRef] [PubMed]

30. Khang, D.Y.; Jiang, H.; Huang, Y.; Rogers, J.A. A stretchable form of single-crystal silicon for high-performance electronics on rubber substrates. *Science* **2006**, *311*, 208–212. [CrossRef] [PubMed]

31. Huang, Y.A.; Dong, W.; Huang, T.; Wang, Y.; Xiao, L.; Su, Y.; Yin, Z. Self-similar design for stretchable wireless LC strain sensors. *Sens. Actuators A Phys.* **2015**, *224*, 36–42. [CrossRef]

32. Lee, J.W.; Xu, R.; Lee, S.; Jang, K.I.; Yang, Y.; Banks, A.; Yu, K.J.; Kim, J.; Xu, S.; Ma, S. Soft, thin skin-mounted power management systems and their use in wireless thermography. *Proc. Natl. Acad. Sci. USA* **2016**, *113*, 201605720. [CrossRef] [PubMed]

33. Xu, S.; Yan, Z.; Jang, K.-I.; Huang, W.; Fu, H.; Kim, J.; Wei, Z.; Flavin, M.; McCracken, J.; Wang, R.; et al. Assembly of micro/nanomaterials into complex, three-dimensional architectures by compressive buckling. *Science* **2015**, *347*, 154–159. [CrossRef] [PubMed]

34. Khang, D.-Y.; Xiao, J.; Kocabas, C.; Maclaren, S.; Banks, T.; Jiang, H.; Huang, Y.Y.; Rogers, J.A. Molecular scale buckling mechanics in individual aligned single-wall carbon nanotubes on elastomeric substrates. *Nano Lett.* **2008**, *8*, 124–130. [CrossRef] [PubMed]

35. Xie, Y.; Liu, Y.; Zhao, Y.; Tsang, Y.; Lau, S.; Huang, H.; Chai, Y. Stretchable all-solid-state supercapacitor with wavy shaped polyaniline/graphene electrode. *J. Mater. Chem. A* **2014**, *2*, 9142–9149. [CrossRef]

36. Lee, G.B.; Sathi, S.G.; Kim, D.Y.; Jeong, K.U.; Nah, C. Wrinkled elastomers for the highly stretchable electrodes with excellent fatigue resistances. *Polym. Test.* **2016**, *53*, 329–337. [CrossRef]

37. Tuukkanen, S.; Hoikkanen, M.; Poikelispää, M.; Honkanen, M.; Vuorinen, T.; Kakkonen, M.; Vuorinen, J.; Lupo, D. Stretching of solution processed carbon nanotube and graphene nanocomposite films on rubber substrates. *Synth. Met.* **2014**, *191*, 28–35. [CrossRef]

38. Feng, X.; Yang, B.D.; Liu, Y.; Wang, Y.; Dagdeviren, C.; Liu, Z.; Carlson, A.; Li, J.; Huang, Y.; Rogers, J.A. Stretchable ferroelectric nanoribbons with wavy configurations on elastomeric substrates. *ACS Nano* **2011**, *5*, 3326–3332. [CrossRef] [PubMed]

39. Ryu, S.Y.; Xiao, J.; Park, W.I.; Son, K.S.; Huang, Y.Y.; Paik, U.; Rogers, J.A. Lateral buckling mechanics in silicon nanowires on elastomeric substrates. *Nano Lett.* **2009**, *9*, 3214–3219. [CrossRef] [PubMed]

40. Choi, W.M.; Song, J.; Khang, D.Y.; Jiang, H.; Huang, Y.Y.; Rogers, J.A. Biaxially stretchable "wavy" silicon nanomembranes. *Nano Lett.* **2007**, *7*, 1655–1663. [CrossRef] [PubMed]

41. Won, Y.; Kim, A.; Yang, W.; Jeong, S.; Moon, J. A highly stretchable, helical copper nanowire conductor exhibiting a stretchability of 700%. *NPG Asia Mater.* **2014**, *6*, e132. [CrossRef]

42. Gonzalez, M.; Axisa, F.; Bossuyt, F.; Hsu, Y.Y.; Vandevelde, B.; Vanfleteren, J. Design and performance of metal conductors for stretchable electronic circuits. *Circuit World* **2009**, *35*, 371–376. [CrossRef]

43. Xu, S.; Zhang, Y.; Cho, J.; Lee, J.; Huang, X.; Jia, L.; Fan, J.A.; Su, Y.; Su, J.; Zhang, H. Stretchable batteries with self-similar serpentine interconnects and integrated wireless recharging systems. *Nat. Commun.* **2013**, *4*, 66–78. [CrossRef] [PubMed]

44. Xu, S.; Zhang, Y.; Jia, L.; Mathewson, K.E.; Jang, K.I.; Kim, J.; Fu, H.; Huang, X.; Chava, P.; Wang, R. Soft microfluidic assemblies of sensors, circuits, and radios for the skin. *Science* **2014**, *344*, 70–74. [CrossRef] [PubMed]

45. Fan, J.A.; Yeo, W.H.; Su, Y.; Hattori, Y.; Lee, W.; Jung, S.Y.; Zhang, Y.; Liu, Z.; Cheng, H.; Falgout, L. Fractal design concepts for stretchable electronics. *Nat. Commun.* **2014**, *5*, 163–180. [CrossRef] [PubMed]

46. Zhu, S.; Huang, Y.; Li, T. Extremely compliant and highly stretchable patterned graphene. *Appl. Phys. Lett.* **2014**, *104*, 173103–173105. [CrossRef]

47. Fan, Z.; Zhang, Y.; Ma, Q.; Zhang, F.; Fu, H.; Hwang, K.C.; Huang, Y. A finite deformation model of planar serpentine interconnects for stretchable electronics. *Int. J. Solids Struct.* **2016**, *91*, 46–54. [CrossRef] [PubMed]

48. Zhang, Y.; Fu, H.; Su, Y.; Xu, S.; Cheng, H.; Fan, J.A.; Hwang, K.C.; Rogers, J.A.; Huang, Y. Mechanics of ultra-stretchable self-similar serpentine interconnects. *Acta Mater.* **2013**, *61*, 7816–7827. [CrossRef]

49. Kim, J.; Lee, M.; Shim, H.J.; Ghaffari, R.; Cho, H.R.; Son, D.; Jung, Y.H.; Soh, M.; Choi, C.; Jung, S. Stretchable silicon nanoribbon electronics for skin prosthesis. *Nat. Commun.* **2014**, *5*, 5747. [CrossRef] [PubMed]

50. Su, Y.; Wang, S.; Huang, Y.A.; Luan, H.; Dong, W.; Fan, J.A.; Yang, Q.; Rogers, J.A.; Huang, Y. Elasticity of fractal inspired interconnects. *Small* **2014**, *11*, 367–373. [CrossRef] [PubMed]

51. Wagner, S.; Bauer, S. Materials for stretchable electronics. *MRS Bull.* **2012**, *37*, 207–213. [CrossRef]

52. Kim, T.A.; Kim, H.S.; Sang, S.L.; Min, P. Single-walled carbon nanotube/silicone rubber composites for compliant electrodes. *Carbon* **2012**, *50*, 444–449. [CrossRef]

53. Choi, K.M.; Rogers, J.A. A photocurable poly (dimethylsiloxane) chemistry designed for soft lithographic molding and printing in the nanometer regime. *J. Am. Chem. Soc.* **2003**, *125*, 4060–4061. [CrossRef] [PubMed]

54. Puig-Lleixà, C.; Jiménez, C.L.; Alonso, J.; Bartrolí, J. Polyurethane–acrylate photocurable polymeric membrane for ion-sensitive field-effect transistor based urea biosensors. *Anal. Chim. Acta* **1999**, *389*, 179–188. [CrossRef]

55. Kokkinis, D.; Schaffner, M.; Studart, A.R. Multimaterial magnetically assisted 3D printing of composite materials. *Nat. Commun.* **2015**, *6*, 8643. [CrossRef] [PubMed]

56. Kramer, R.K.; Boley, J.W.; Stone, H.A.; Weaver, J.C.; Wood, R.J. Effect of microtextured surface topography on the wetting behavior of eutectic gallium-indium alloys. *Langmuir* **2014**, *30*, 533–539. [CrossRef] [PubMed]

57. Dickey, M.D.; Chiechi, R.C.; Larsen, R.J.; Weiss, E.A.; Weitz, D.A.; Whitesides, G.M. Eutectic gallium-indium (eGaIn): A liquid metal alloy for the formation of stable structures in microchannels at room temperature. *Adv. Funct. Mater.* **2008**, *18*, 1097–1104. [CrossRef]

58. Liu, T.; Sen, P.; Kim, C.J. Characterization of nontoxic liquid-metal alloy galinstan for applications in microdevices. *J. Microelectromech. Syst.* **2012**, *21*, 443–450. [CrossRef]

59. Park, Y.-L.; Majidi, C.; Kramer, R.; Bérard, P.; Wood, R.J. Hyperelastic pressure sensing with a liquid-embedded elastomer. *J. Micromech. Microeng.* **2010**, *20*, 125029. [CrossRef]

60. Khan, M.R.; Hayes, G.J.; Zhang, S.; Dickey, M.D.; Lazzi, G. A pressure responsive fluidic microstrip open stub resonator using a liquid metal alloy. *IEEE Microw. Wirel. Compon. Lett.* **2012**, *22*, 577–579. [CrossRef]

61. Cheng, S.; Wu, Z. A microfluidic, reversibly stretchable, large-area wireless strain sensor. *Adv. Funct. Mater.* **2011**, *21*, 2282–2290. [CrossRef]

62. Cheng, S.; Rydberg, A.; Hjort, K.; Wu, Z. Liquid metal stretchable unbalanced loop antenna. *Appl. Phys. Lett.* **2009**, *94*, 144103. [CrossRef]

63. Jalali Mazlouman, S.; Jiang, X.J.; Mahanfar, A.; Menon, C.; Vaughan, R.G. A reconfigurable patch antenna using liquid metal embedded in a silicone substrate. *IEEE Trans. Antennas Propag.* **2011**, *59*, 4406–4412. [CrossRef]

64. Kim, H.-J.; Son, C.; Ziaie, B. A multiaxial stretchable interconnect using liquid-alloy-filled elastomeric microchannels. *Appl. Phys. Lett.* **2008**, *92*, 011904. [CrossRef]

65. Zhu, S.; So, J.-H.; Mays, R.; Desai, S.; Barnes, W.R.; Pourdeyhimi, B.; Dickey, M.D. Ultrastretchable fibers with metallic conductivity using a liquid metal alloy core. *Adv. Funct. Mater.* **2013**, *23*, 2308–2314. [CrossRef]

66. Vosgueritchian, M.; Lipomi, D.J.; Bao, Z. Highly conductive and transparent pedot: PSS films with a fluorosurfactant for stretchable and flexible transparent electrodes. *Adv. Funct. Mater.* **2012**, *22*, 421–428. [CrossRef]

67. Hyun, D.C.; Park, M.; Park, C.; Kim, B.; Xia, Y.; Hur, J.H.; Kim, J.M.; Park, J.J.; Jeong, U. Ordered zigzag stripes of polymer gel/metal nanoparticle composites for highly stretchable conductive electrodes. *Adv. Mater.* **2011**, *23*, 2946–2950. [CrossRef] [PubMed]

68. Joseph, N.; Janardhanan, C.; Sebastian, M.T. Electromagnetic interference shielding properties of butyl rubber-single walled carbon nanotube composites. *Compos. Sci. Technol.* **2014**, *101*, 139–144. [CrossRef]

69. Park, M.; Im, J.; Shin, M.; Min, Y.; Park, J.; Cho, H.; Park, S.; Shim, M.B.; Jeon, S.; Chung, D.Y.; et al. Highly stretchable electric circuits from a composite material of silver nanoparticles and elastomeric fibres. *Nat. Nanotechnol.* **2012**, *7*, 803–809. [CrossRef] [PubMed]

70. Shang, S.; Zeng, W.; Tao, X. High stretchable MWNTs/polyurethane conductive nanocomposites. *J. Mater. Chem.* **2011**, *21*, 7274–7280. [CrossRef]

71. Niu, Z.; Dong, H.; Zhu, B.; Li, J.; Hng, H.H.; Zhou, W.; Chen, X.; Xie, S. Highly stretchable, integrated supercapacitors based on single-walled carbon nanotube films with continuous reticulate architecture. *Adv. Mater.* **2013**, *25*, 1058–1064. [CrossRef] [PubMed]

72. Yamada, T.; Hayamizu, Y.; Yamamoto, Y.; Yomogida, Y.; Izadi-Najafabadi, A.; Futaba, D.N.; Hata, K. A stretchable carbon nanotube strain sensor for human-motion detection. *Nat. Nano* **2011**, *6*, 296–301. [CrossRef] [PubMed]

73. Lee, K.; Park, J.; Lee, M.-S.; Kim, J.; Hyun, B.G.; Kang, D.J.; Na, K.; Lee, C.Y.; Bien, F.; Park, J.-U. In-situ synthesis of carbon nanotube–graphite electronic devices and their integrations onto surfaces of live plants and insects. *Nano Lett.* **2014**, *14*, 2647–2654. [CrossRef] [PubMed]

74. Xu, F.; Lu, W.; Zhu, Y. Controlled 3D buckling of silicon nanowires for stretchable electronics. *ACS Nano* **2010**, *5*, 672–678. [CrossRef] [PubMed]

75. Xu, F.; Zhu, Y. Highly conductive and stretchable silver nanowire conductors. *Adv. Mater.* **2012**, *24*, 5117–5122. [CrossRef] [PubMed]

76. Kim, M.; Park, J.; Ji, S.; Shin, S.-H.; Kim, S.-Y.; Kim, Y.-C.; Kim, J.-Y.; Park, J.-U. Fully-integrated, bezel-less transistor arrays using reversibly foldable interconnects and stretchable origami substrates. *Nanoscale* **2016**, *8*, 9504–9510. [CrossRef] [PubMed]

77. Lee, M.-S.; Lee, K.; Kim, S.-Y.; Lee, H.; Park, J.; Choi, K.-H.; Kim, H.-K.; Kim, D.-G.; Lee, D.-Y.; Nam, S. High-performance, transparent, and stretchable electrodes using graphene–metal nanowire hybrid structures. *Nano Lett.* **2013**, *13*, 2814–2821. [CrossRef] [PubMed]

78. Akinwande, D.; Petrone, N.; Hone, J. Two-dimensional flexible nanoelectronics. *Nat. Commun.* **2014**, *5*, 5678. [CrossRef] [PubMed]

79. Roch, A.; Greifzu, M.; Talens, E.R.; Stepien, L.; Roch, T.; Hege, J.; Van Nong, N.; Schmiel, T.; Dani, I.; Leyens, C.; et al. Ambient effects on the electrical conductivity of carbon nanotubes. *Carbon* **2015**, *95*, 347–353. [CrossRef]

80. Bolotin, K.I.; Sikes, K.J.; Jiang, Z.; Klima, M.; Fudenberg, G.; Hone, J.; Kim, P.; Stormer, H.L. Ultrahigh electron mobility in suspended graphene. *Solid State Commun.* **2008**, *146*, 351–355. [CrossRef]

81. Lee, C.; Wei, X.; Kysar, J.W.; Hone, J. Measurement of the elastic properties and intrinsic strength of monolayer graphene. *Science* **2008**, *321*, 385–388. [CrossRef] [PubMed]

82. Lu, C.C.; Lin, Y.C.; Yeh, C.H.; Huang, J.C.; Chiu, P.W. High mobility flexible graphene field-effect transistors with self-healing gate dielectrics. *ACS Nano* **2012**, *6*, 4469–4474. [CrossRef] [PubMed]

83. Park, J.U.; Nam, S.W.; Lee, M.S.; Lieber, C.M. Synthesis of monolithic graphene–graphite integrated electronics. *Nat. Mater.* **2011**, *11*, 120–125. [CrossRef] [PubMed]

84. Sire, C.; Ardiaca, F.; Lepilliet, S.; Seo, J.W.T.; Hersam, M.C.; Dambrine, G.; Happy, H.; Derycke, V. Flexible gigahertz transistors derived from solution-based single-layer graphene. *Nano Lett.* **2012**, *12*, 1184. [CrossRef] [PubMed]

85. Lee, H.; Choi, T.K.; Lee, Y.B.; Cho, H.R.; Ghaffari, R.; Wang, L.; Choi, H.J.; Chung, T.D.; Lu, N.; Hyeon, T.; et al. A graphene-based electrochemical device with thermoresponsive microneedles for diabetes monitoring and therapy. *Nat. Nano* **2016**, *11*, 566–572. [CrossRef] [PubMed]

86. Kim, J.; Lee, M.-S.; Jeon, S.; Kim, M.; Kim, S.; Kim, K.; Bien, F.; Hong, S.Y.; Park, J.-U. Highly transparent and stretchable field-effect transistor sensors using graphene–nanowire hybrid nanostructures. *Adv. Mater.* **2015**, *27*, 3292–3297. [CrossRef] [PubMed]

87. Roh, E.; Hwang, B.U.; Kim, D.; Kim, B.Y.; Lee, N.E. Stretchable, transparent, ultrasensitive, and patchable strain sensor for human-machine interfaces comprising a nanohybrid of carbon nanotubes and conductive elastomers. *ACS Nano* **2015**, *9*, 6252–6261. [CrossRef] [PubMed]

88. Jariwala, D.; Sangwan, V.K.; Lauhon, L.J.; Marks, T.J.; Hersam, M.C. Emerging device applications for semiconducting two-dimensional transition metal dichalcogenides. *ACS Nano* **2014**, *8*, 1102–1120. [CrossRef] [PubMed]

89. Jang, H.; Park, Y.J.; Chen, X.; Das, T.; Kim, M.S.; Ahn, J.H. Graphene-based flexible and stretchable electronics. *Adv. Mater.* **2016**, *28*. [CrossRef] [PubMed]

90. Wang, C.; Takei, K.; Takahashi, T.; Javey, A. Carbon nanotube electronics–moving forward. *Chem. Soc. Rev.* **2013**, *42*, 2592–2609. [CrossRef] [PubMed]

91. Yeo, W.H.; Kim, Y.S.; Lee, J.; Ameen, A.; Shi, L.; Li, M.; Wang, S.; Ma, R.; Jin, S.H.; Kang, Z.; et al. Multifunctional epidermal electronics printed directly onto the skin. *Adv. Mater.* **2013**, *25*, 2773–2778. [CrossRef] [PubMed]

92. Karnaushenko, D.D.; Karnaushenko, D.; Makarov, D.; Schmidt, O.G. Compact helical antenna for smart implant applications. *NPG Asia Mater.* **2015**, *7*, e188. [CrossRef]

93. Sun, Y.; Kumar, V.; Adesida, I.; Rogers, J.A. Buckled and wavy ribbons of gaas for high-performance electronics on elastomeric substrates. *Adv. Mater.* **2006**, *18*, 2857–2862. [CrossRef]

94. Park, K.; Lee, D.-K.; Kim, B.-S.; Jeon, H.; Lee, N.-E.; Whang, D.; Lee, H.-J.; Kim, Y.J.; Ahn, J.-H. Stretchable, transparent zinc oxide thin film transistors. *Adv. Funct. Mater.* **2010**, *20*, 3577–3582. [CrossRef]

95. Wallentin, J.; Anttu, N.; Asoli, D.; Huffman, M.; Aberg, I.; Magnusson, M.H.; Siefer, G.; Fuss-Kailuweit, P.; Dimroth, F.; Witzigmann, B.; et al. InP nanowire array solar cells achieving 13.8% efficiency by exceeding the ray optics limit. *Science* **2013**, *339*, 1057–1060. [CrossRef] [PubMed]

96. Kim, R.-H.; Tao, H.; Kim, T.; Zhang, Y.; Kim, S.; Panilaitis, B.; Yang, M.; Kim, D.H.; Jung, Y.H.; Kim, B.H. Materials and designs for wirelessly powered implantable light-emitting systems. *Small* **2012**, *8*, 2812–2818. [CrossRef] [PubMed]

97. Savagatrup, S.; Printz, A.D.; Wu, H.; Rajan, K.M.; Sawyer, E.J.; Zaretski, A.V.; Bettinger, C.J.; Lipomi, D.J. Viability of stretchable poly(3-heptylthiophene) (P3HpT) for organic solar cells and field-effect transistors. *Synth. Met.* **2015**, *203*, 208–214. [CrossRef]

98. Mikhnenko, O.V.; Blom, P.W.M.; Nguyen, T.Q. Exciton diffusion in organic semiconductors. *Energy Environ. Sci.* **2015**, *8*, 1867–1888. [CrossRef]

99. Gwinner, M.C.; Pietro, R.D.; Vaynzof, Y.; Greenberg, K.J.; Ho, P.K.H.; Friend, R.H.; Sirringhaus, H. Doping of organic semiconductors using molybdenum trioxide: A quantitative time-dependent electrical and spectroscopic study. *Adv. Funct. Mater.* **2011**, *21*, 1432–1441. [CrossRef]

100. Kim, D.-H.; Rogers, J.A. Stretchable electronics: Materials strategies and devices. *Adv. Mater.* **2008**, *20*, 4887–4892. [CrossRef]

101. Sun, Y.; Choi, W.M.; Jiang, H.; Huang, Y.Y.; Rogers, J.A. Controlled buckling of semiconductor nanoribbons for stretchable electronics. *Nat. Nanotechnol.* **2006**, *1*, 201–207. [CrossRef] [PubMed]

102. Rojas, J.P.; Hussain, M.M. Flexible semi-transparent silicon (100) fabric with high-k/metal gate devices. *Phys. Status Solidi* **2013**, *7*, 187–191. [CrossRef]

103. Ghoneim, M.T.; Kutbee, A.; Nasseri, F.G.; Bersuker, G.; Hussain, M.M. Mechanical anomaly impact on metal-oxide-semiconductor capacitors on flexible silicon fabric. *Appl. Phys. Lett.* **2014**, *104*, 234104. [CrossRef]

104. Ghoneim, M.T.; Zidan, M.A.; Salama, K.N.; Hussain, M.M. Towards neuromorphic electronics: Memristors on foldable silicon fabric. *Microelectron. J.* **2014**, *45*, 1392–1395. [CrossRef]

105. Ghoneim, M.T.; Hussain, M.M. Study of harsh environment operation of flexible ferroelectric memory integrated with PZT and silicon fabric. *Appl. Phys. Lett.* **2015**, *107*, 495–502. [CrossRef]

106. Rojas, J.P.; Torres Sevilla, G.A.; Ghoneim, M.T.; Inayat, S.B.; Ahmed, S.M.; Hussain, A.M.; Hussain, M.M. Transformational silicon electronics. *ACS Nano* **2014**, *8*, 1468–1474. [CrossRef] [PubMed]

107. Ahmed, S.M.; Hussain, A.M.; Rojas, J.P.; Hussain, M.M. Solid state MEMS devices on flexible and semi-transparent silicon (100) platform. In Proceedings of the 2014 IEEE 27th International Conference on Micro Electro Mechanical Systems (MEMS), San Francisco, CA, USA, 26–30 January 2014; pp. 548–551.

108. Minemawari, H.; Yamada, T.; Matsui, H.; Tsutsumi, J.; Haas, S.; Chiba, R.; Kumai, R.; Hasegawa, T. Inkjet printing of single-crystal films. *Nature* **2011**, *475*, 364–367. [CrossRef] [PubMed]

109. Someya, T.S.T. Stretchable organic integrated circuits for large-area electronic skin surfaces. *MRS Bull.* **2012**, *37*, 236–245.

110. Faddoul, R.; Reverdy-Bruas, N.; Blayo, A. Formulation and screen printing of water based conductive flake silver pastes onto green ceramic tapes for electronic applications. *Mater. Sci. Eng. B* **2012**, *177*, 1053–1066. [CrossRef]

111. Kuo, H.P.; Yang, C.F.; Huang, A.N.; Wu, C.T.; Pan, W.C. Preparation of the working electrode of dye-sensitized solar cells: Effects of screen printing parameters. *J. Taiwan Inst. Chem. E* **2014**, *45*, 2340–2345. [CrossRef]

112. Jabbour, G.E.; Radspinner, R.; Peyghambarian, N. Screen printing for the fabrication of organic light-emitting devices. *IEEE J. Sel. Top. Quantum Electron.* **2001**, *7*, 769–773. [CrossRef]

113. Ren, X.; Pei, K.; Peng, B.; Zhang, Z.; Wang, Z.; Wang, X.; Chan, K. Low operating power and flexible active matrix organic transistor temperature sensors array. *Adv. Mater.* **2016**, *28*, 4832–4838. [CrossRef] [PubMed]

114. Rogers, J.A.; Bao, Z.; Raju, V.R. Nonphotolithographic fabrication of organic transistors with micron feature sizes. *Appl. Phys. Lett.* **1998**, *72*, 2716–2718. [CrossRef]

115. Xia, C.; Chen, F.; Liu, M. Reduced-temperature solid oxide fuel cells fabricated by screen-printing. *Electrochem. Solid State Lett.* **2001**, *4*, A52–A54. [CrossRef]

116. Kang, K.Y.; Lee, Y.G.; Dong, O.S.; Kim, J.C.; Kim, K.M. Performance improvements of pouch-type flexible thin-film lithium-ion batteries by modifying sequential screen-printing process. *Electrochim. Acta* **2014**, *138*, 294–301. [CrossRef]

117. Tehrani, Z.; Korochkina, T.; Govindarajan, S.; Thomas, D.J.; O'Mahony, J.; Kettle, J.; Claypole, T.C.; Gethin, D.T. Ultra-thin flexible screen printed rechargeable polymer battery for wearable electronic applications. *Org. Electron.* **2015**, *26*, 386–394. [CrossRef]

118. Krebs, F.C. Polymer solar cell modules prepared using roll-to-roll methods: Knife-over-edge coating, slot-die coating and screen printing. *Sol. Energy Mater. Sol. Cell* **2009**, *93*, 465–475. [CrossRef]

119. Shaheen, S.E.; Radspinner, R.; Peyghambarian, N.; Jabbour, G.E. Fabrication of bulk heterojunction plastic solar cells by screen printing. *Appl. Phys. Lett.* **2001**, *79*, 2996–2998. [CrossRef]

120. Song, L.; Myers, A.C.; Adams, J.J.; Zhu, Y. Stretchable and reversibly deformable radio frequency antennas based on silver nanowires. *ACS Appl. Mater. Interfaces* **2014**, *6*, 4248–4253. [CrossRef] [PubMed]

121. Yu, Y.; Yan, C.; Zheng, Z. Polymer-assisted metal deposition (PAMD): A full-solution strategy for flexible, stretchable, compressible, and wearable metal conductors. *Adv. Mater.* **2014**, *26*, 5508–5516. [CrossRef] [PubMed]

122. Guo, R.; Yu, Y.; Xie, Z.; Liu, X.; Zhou, X.; Gao, Y.; Liu, Z.; Zhou, F.; Yang, Y.; Zheng, Z. Matrix-assisted catalytic printing for the fabrication of multiscale, flexible, foldable, and stretchable metal conductors. *Adv. Mater.* **2014**, *25*, 3343–3350. [CrossRef] [PubMed]

123. Liang, J.; Tong, K.; Pei, Q. A water-based silver-nanowire screen-print ink for the fabrication of stretchable conductors and wearable thin-film transistors. *Adv. Mater.* **2016**, *28*, 5986–5996. [CrossRef] [PubMed]

124. Tong, K.; Liang, J.; Pei, Q. (Invited) intrinsically-stretchable, transparent thin film transistors. *ECS Trans.* **2016**, *75*, 205–212. [CrossRef]

125. Bao, Z.; Feng, Y.; Dodabalapur, A.; And, V.R.R.; Lovinger, A.J. High-performance plastic transistors fabricated by printing techniques. *Chem. Mater.* **1997**, *9*, 1299–1301. [CrossRef]

126. Lee, Y.; Choi, J.; Lee, K.J.; Stott, N.E.; Kim, D. Large-scale synthesis of copper nanoparticles by chemically controlled reduction for applications of inkjet-printed electronics. *Nanotechnology* **2008**, *19*, 415604. [CrossRef] [PubMed]

127. Li, D.; Sutton, D.; Burgess, A.; Graham, D.; Calvert, P.D. Conductive copper and nickel lines via reactive inkjet printing. *J. Mater. Chem.* **2009**, *19*, 3719–3724. [CrossRef]

128. Kic, P.; Liška, R. Formation of air-stable copper–silver core–shell nanoparticles for inkjet printing. *J. Mater. Chem.* **2009**, *19*, 3057–3062.

129. Ko, S.H.; Pan, H.; Grigoropoulos, C.P.; Luscombe, C.K.; Fréchet, J.M.J.; Poulikakos, D. All-inkjet-printed flexible electronics fabrication on a polymer substrate by low-temperature high-resolution selective laser sintering of metal nanoparticles. *Nanotechnology* **2007**, *18*, 345202. [CrossRef]

130. Tekin, E.; Smith, P.J.; Schubert, U.S. Inkjet printing as a deposition and patterning tool for polymers and inorganic particles. *Soft Matter* **2008**, *4*, 703–713. [CrossRef]

131. Crowley, K.; Morrin, A.; Hernandez, A.; O'Malley, E.; Whitten, P.G.; Wallace, G.G.; Smyth, M.R.; Killard, A.J. Fabrication of an ammonia gas sensor using inkjet-printed polyaniline nanoparticles. *Talanta* **2008**, *77*, 710–717. [CrossRef]

132. Kawase, T.; Sirringhaus, H.; Friend, R.H.; Shimoda, T. Inkjet printed via-hole interconnections and resistors for all-polymer transistor circuits. *Adv. Mater.* **2001**, *13*, 1601–1605. [CrossRef]

133. Jung, S.; Sou, A.; Gili, E.; Sirringhaus, H. Inkjet-printed resistors with a wide resistance range for printed read-only memory applications. *Org. Electron.* **2013**, *14*, 699–702. [CrossRef]

134. Kang, B.J.; Chang, K.L.; Oh, J.H. All-inkjet-printed electrical components and circuit fabrication on a plastic substrate. *Microelectron. Eng.* **2012**, *97*, 251–254. [CrossRef]

135. Cook, B.S.; Cooper, J.R.; Tentzeris, M.M. Multi-layer RF capacitors on flexible substrates utilizing inkjet printed dielectric polymers. *IEEE Microw. Wirel. Compon. Lett.* **2013**, *23*, 353–355. [CrossRef]

136. Riggs, B.; Elupula, R.; Grayson, S.; Chrisey, D. Photonic curing of aromatic thiol-ene click dielectric capacitors via inkjet printing. *J. Mater. Chem. A* **2014**, *2*, 17380–17386. [CrossRef]

137. Graddage, N.; Chu, T.Y.; Ding, H.; Py, C.; Dadvand, A.; Tao, Y. Inkjet printed thin and uniform dielectrics for capacitors and organic thin film transistors enabled by the coffee ring effect. *Org. Electron.* **2016**, *29*, 114–119. [CrossRef]

138. Lee, H.; Cook, B.S.; Murali, K.P.; Raj, M. Inkjet printed high-Q RF inductors on paper substrate with ferromagnetic nanomaterial. *IEEE Microw. Wirel. Compon. Lett.* **2016**, *26*, 1–3. [CrossRef]

139. Maccurdy, R.; Katzschmann, R.; Kim, Y.; Rus, D. Synthesis of ZnO nanoparticles to fabricate a mask-free thin-film transistor by inkjet printing. *J. Nanotechnol.* **2012**, *2012*.

140. Kawase, T.; Shimoda, T.; Newsome, C.; Sirringhaus, H.; Friend, R.H. Inkjet printing of polymer thin film transistors. *Thin Solid Films* **2003**, *438–439*, 279–287. [CrossRef]

141. Yan, H.; Chen, Z.; Zheng, Y.; Newman, C.; Quinn, J.R.; Dötz, F.; Kastler, M.; Facchetti, A. A high-mobility electron-transporting polymer for printed transistors. *Nature* **2009**, *457*, 679–686. [CrossRef] [PubMed]

142. O'Toole, M.; Shepherd, R.; Wallace, G.G.; Diamond, D. Inkjet printed led based pH chemical sensor for gas sensing. *Anal. Chim. Acta* **2009**, *652*, 308–314. [CrossRef] [PubMed]

143. Verma, A.; Zink, D.M.; Fléchon, C.; Carballo, J.L.; Flügge, H.; Navarro, J.M.; Baumann, T.; Volz, D. Efficient, inkjet-printed TADF-OLEDS with an ultra-soluble NHetPHOS complex. *Appl. Phys. A* **2016**, *122*, 1–5. [CrossRef]

144. Haverinen, H.M.; Myllyla, R.A.; Jabbour, G.E. Inkjet printed RGB quantum dot-hybrid LED. *J. Disp. Technol.* **2010**, *6*, 87–89. [CrossRef]

145. Park, S.I.; Xiong, Y.; Kim, R.H.; Elvikis, P.; Meitl, M.; Kim, D.H.; Wu, J.; Yoon, J.; Yu, C.J.; Liu, Z. Printed assemblies of inorganic light-emitting diodes for deformable and semitransparent displays. *Science* **2010**, *325*, 977–981. [CrossRef] [PubMed]

146. Ahn, J.H.; Kim, H.S.; Lee, K.J.; Jeon, S.; Kang, S.J.; Sun, Y.; Nuzzo, R.G.; Rogers, J.A. Heterogeneous three-dimensional electronics by use of printed semiconductor nanomaterials. *Science* **2006**, *314*, 1754–1757. [CrossRef] [PubMed]

147. Lee, K.J.; Ahn, H.; Motala, M.J.; Nuzzo, R.G.; Menard, E.; Rogers, J.A. Fabrication of microstructured silicon (μs-Si) from a bulk Si wafer and its use in the printing of high-performance thin-film transistors on plastic substrates. *J. Micromech. Microeng.* **2010**, *20*, 75018–75025. [CrossRef]

148. Cao, Q.; Kim, H.S.; Pimparkar, N.; Kulkarni, J.P.; Wang, C.; Shim, M.; Roy, K.; Alam, M.A.; Rogers, J.A. Medium-scale carbon nanotube thin-film integrated circuits on flexible plastic substrates. *Nature* **2008**, *454*, 495–500. [CrossRef] [PubMed]

149. Jeong, J.W.; Yeo, W.H.; Akhtar, A.; Norton, J.J.; Kwack, Y.J.; Li, S.; Jung, S.Y.; Su, Y.; Lee, W.; Xia, J. Materials and optimized designs for human-machine interfaces via epidermal electronics. *Adv. Mater.* **2013**, *25*, 6839–6846. [CrossRef] [PubMed]

150. Amjadi, M.; Yoon, Y.J.; Park, I. Ultra-stretchable and skin-mountable strain sensors using carbon nanotubes-ecoflex nanocomposites. *Nanotechnology* **2015**, *26*, 9–21. [CrossRef] [PubMed]

151. Majidi, C.; Kramer, R.; Wood, R.J. A non-differential elastomer curvature sensor for softer-than-skin electronics. *Smart Mater. Struct.* **2011**, *20*, 1487–1490. [CrossRef]

152. Zhang, J.; Liu, J.; Zhuang, R.; Mäder, E.; Heinrich, G.; Gao, S. Single MWNT-glass fiber as strain sensor and switch. *Adv. Mater.* **2011**, *23*, 3392–3397. [CrossRef] [PubMed]

153. Kang, I.; Schulz, M.J.; Kim, J.H.; Shanov, V.; Shi, D. A carbon nanotube strain sensor for structural health monitoring. *Smart Mater. Struct.* **2006**, *15*, 737–748. [CrossRef]

154. Muth, J.T.; Vogt, D.M.; Truby, R.L.; Mengüç, Y.; Kolesky, D.B.; Wood, R.J.; Lewis, J.A. Embedded 3D printing of strain sensors within highly stretchable elastomers. *Adv. Mater.* **2014**, *26*, 6307–6312. [CrossRef] [PubMed]

155. Yan, C.; Wang, J.; Kang, W.; Cui, M.; Wang, X.; Foo, C.Y.; Chee, K.J.; Lee, P.S. Highly stretchable piezoresistive graphene-nanocellulose nanopaper for strain sensors. *Adv. Mater.* **2014**, *26*, 2022–2027. [CrossRef] [PubMed]

156. Mcevoy, M.A.; Correll, N. Materials science. Materials that couple sensing, actuation, computation, and communication. *Science* **2015**, *347*. [CrossRef] [PubMed]

157. Majidi, C. Soft robotics: A perspective—Current trends and prospects for the future. *Soft Robot.* **2014**, *1*, 5–11. [CrossRef]

158. Kang, D.; Pikhitsa, P.V.; Choi, Y.W.; Lee, C.; Shin, S.S.; Piao, L.; Park, B.; Suh, K.Y.; Kim, T.I.; Choi, M. Ultrasensitive mechanical crack-based sensor inspired by the spider sensory system. *Nature* **2014**, *516*, 222–226. [CrossRef] [PubMed]

159. Webb, R.C.; Bonifas, A.P.; Behnaz, A.; Zhang, Y.; Yu, K.J.; Cheng, H.; Shi, M.; Bian, Z.; Liu, Z.; Kim, Y.S. Ultrathin conformal devices for precise and continuous thermal characterization of human skin. *Nat. Mater.* **2013**, *12*, 938–944. [CrossRef] [PubMed]

160. Trung, T.Q.; Ramasundaram, S.; Hwang, B.U.; Lee, N.E. An all-elastomeric transparent and stretchable temperature sensor for body-attachable wearable electronics. *Adv. Mater.* **2016**, *28*, 502–509. [CrossRef] [PubMed]

161. Sator, P.G.; Schmidt, J.B.; Hönigsmann, H. Comparison of epidermal hydration and skin surface lipids in healthy individuals and in patients with atopic dermatitis. *J. Am. Acad. Dermatol.* **2003**, *48*, 352–358. [CrossRef] [PubMed]

162. Kim, S.D.; Huh, C.H.; Seo, K.I.; Suh, D.H.; Youn, J.I. Evaluation of skin surface hydration in korean psoriasis patients: A possible factor influencing psoriasis. *Clin. Exp. Dermatol.* **2002**, *27*, 147–152. [CrossRef] [PubMed]

163. Berents, T.L.; Carlsen, K.C.L.; Mowinckel, P.; Skjerven, H.O.; Kvenshagen, B.; Rolfsjord, L.B.; Bradley, M.; Lieden, A.; Carlsen, K.; Gaustad, P. Skin barrier function and staphylococcus aureus colonization in vestibulum nasi and fauces in healthy infants and infants with eczema: A population-based cohort study. *PLoS ONE* **2015**, *10*, e0130145. [CrossRef] [PubMed]

164. Morton, C.A.; Lafferty, M.; Hau, C.; Henderson, I.; Jones, M.; Lowe, J.G. Pruritus and skin hydration during dialysis. *Nephrol. Dial. Transplant.* **1996**, *11*, 2031–2036. [CrossRef] [PubMed]

165. Cravello, B.; Ferri, A. Relationships between skin properties and environmental parameters. *Skin Res. Technol.* **2008**, *14*, 180–186. [CrossRef] [PubMed]

166. Woo, C.J.; Hyo, K.S.; Hun, H.C.; Chan, P.K.; Woong, Y.S. The influences of skin visco-elasticity, hydration level and aging on the formation of wrinkles: A comprehensive and objective approach. *Skin Res. Technol.* **2013**, *19*, e349–e355.

167. Sator, P.G.; Schmidt, J.B.; Rabe, T.; Zouboulis, C.C. Skin aging and sex hormones in women—Clinical perspectives for intervention by hormone replacement therapy. *Exp. Dermatol.* **2004**, *13* (Suppl. 4), 36–40. [CrossRef] [PubMed]

168. Parrilla, M.; Ferré, J.; Guinovart, T.; Andrade, F.J. Wearable potentiometric sensors based on commercial carbon fibres for monitoring sodium in sweat. *Electroanal* **2016**, *28*, 1267–1275. [CrossRef]

169. Parrilla, M.; Cánovas, R.; Jeerapan, I.; Andrade, F.J.; Wang, J. A textile-based stretchable multi-ion potentiometric sensor. *Adv. Healthc. Mater.* **2016**, *5*, 996–1001. [CrossRef] [PubMed]

170. Cazalé, A.; Sant, W.; Ginot, F.; Launay, J.C.; Savourey, G.; Revol-Cavalier, F.; Lagarde, J.M.; Heinry, D.; Launay, J.; Temple-Boyer, P. Physiological stress monitoring using sodium ion potentiometric microsensors for sweat analysis. *Sens. Actuators B Chem.* **2015**, *225*, 1211–1219. [CrossRef]

171. Guinovart, T.; Bandodkar, A.J.; Windmiller, J.R.; Andrade, F.J.; Wang, J. A potentiometric tattoo sensor for monitoring ammonium in sweat. *Analyst* **2013**, *138*, 7031–7038. [CrossRef] [PubMed]

172. Bandodkar, A.J.; Jeerapan, I.; You, J.M.; Nuñezflores, R.; Wang, J. Highly stretchable fully-printed cnt-based electrochemical sensors and biofuel cells: Combining intrinsic and design-induced stretchability. *Nano Lett.* **2016**, *16*, 721–727. [CrossRef] [PubMed]

173. Kudo, H.; Sawada, T.; Kazawa, E.; Yoshida, H.; Iwasaki, Y.; Mitsubayashi, K. A flexible and wearable glucose sensor based on functional polymers with soft-mems techniques. *Biosens. Bioelectron.* **2006**, *22*, 558–562. [CrossRef] [PubMed]
174. Labroo, P.; Cui, Y. Flexible graphene bio-nanosensor for lactate. *Biosens. Bioelectron.* **2013**, *41*, 852–856. [CrossRef] [PubMed]
175. Bandodkar, A.J.; Molinnus, D.; Mirza, O.; Guinovart, T.; Windmiller, J.R.; Valdés-Ramírez, G.; Andrade, F.J.; Schöning, M.J.; Wang, J. Epidermal tattoo potentiometric sodium sensors with wireless signal transduction for continuous non-invasive sweat monitoring. *Biosens. Bioelectron.* **2013**, *54*, 603–609. [CrossRef] [PubMed]
176. Lochner, C.M.; Khan, Y.; Pierre, A.; Arias, A.C. All-organic optoelectronic sensor for pulse oximetry. *Nat. Commun.* **2014**, *5*, 5745. [CrossRef] [PubMed]
177. Yokota, T.; Zalar, P.; Kaltenbrunner, M.; Jinno, H.; Matsuhisa, N.; Kitanosako, H.; Tachibana, Y.; Yukita, W.; Koizumi, M.; Someya, T. Ultraflexible organic photonic skin. *Sci. Adv.* **2016**, *2*, e1501856. [CrossRef] [PubMed]
178. Gong, S.; Schwalb, W.; Wang, Y.; Chen, Y.; Tang, Y.; Si, J.; Shirinzadeh, B.; Cheng, W. A wearable and highly sensitive pressure sensor with ultrathin gold nanowires. *Nat. Commun.* **2013**, *5*, 163–180. [CrossRef] [PubMed]
179. Vandeparre, H.; Watson, D.; Lacour, S.P. Extremely robust and conformable capacitive pressure sensors based on flexible polyurethane foams and stretchable metallization. *Appl. Phys. Lett.* **2013**, *103*, 204103. [CrossRef]
180. Hattori, Y.; Falgout, L.; Lee, W.; Jung, S.Y.; Poon, E.; Lee, J.W.; Na, I.; Geisler, A.; Sadhwani, D.; Zhang, Y. Multifunctional skin-like electronics for quantitative, clinical monitoring of cutaneous wound healing. *Adv. Healthc. Mater.* **2014**, *3*, 1597–1607. [CrossRef] [PubMed]
181. Lin, S.; Yuk, H.; Zhang, T.; Parada, G.A.; Koo, H.; Yu, C.; Zhao, X. Stretchable hydrogel electronics and devices. *Adv. Mater.* **2015**, *53*, 735–741. [CrossRef] [PubMed]
182. Choong, C.L.; Shim, M.B.; Lee, B.S.; Jeon, S.; Ko, D.S.; Kang, T.H.; Bae, J.; Lee, S.H.; Byun, K.E.; Im, J. Highly stretchable resistive pressure sensors using a conductive elastomeric composite on a micropyramid array. *Adv. Mater.* **2014**, *26*, 3451–3458. [CrossRef] [PubMed]
183. Gaikwad, A.M.; Zamarayeva, A.M.; Rousseau, J.; Chu, H.; Derin, I.; Steingart, D.A. Highly stretchable alkaline batteries based on an embedded conductive fabric. *Adv. Mater.* **2012**, *24*, 5071–5076. [CrossRef] [PubMed]
184. Lee, J.-H.; Lee, K.Y.; Gupta, M.K.; Kim, T.Y.; Lee, D.-Y.; Oh, J.; Ryu, C.; Yoo, W.J.; Kang, C.-Y.; Yoon, S.-J.; et al. Highly stretchable piezoelectric-pyroelectric hybrid nanogenerator. *Adv. Mater.* **2014**, *26*, 765–769. [CrossRef] [PubMed]
185. Duan, Y.; Huang, Y.; Yin, Z.; Bu, N.; Dong, W. Non-wrinkled, highly stretchable piezoelectric devices by electrohydrodynamic direct-writing. *Nanoscale* **2014**, *6*, 3289–3295. [CrossRef] [PubMed]
186. Lipomi, D.J.; Tee, B.C.K.; Vosgueritchian, M.; Bao, Z. Stretchable organic solar cells. *Adv. Mater.* **2011**, *23*, 1771–1775. [CrossRef] [PubMed]
187. Li, G.; Zhu, R.; Yang, Y. Polymer solar cells. *Nat. Photon.* **2012**, *6*, 153–161. [CrossRef]
188. Huang, X.; Liu, Y.; Kong, G.W.; Seo, J.H.; Ma, Y.; Jang, K.-I.; Fan, J.A.; Mao, S.; Chen, Q.; Li, D.; et al. Epidermal radio frequency electronics for wireless power transfer. *Microsyst. Nanoeng.* **2016**, *2*, 16052. [CrossRef]
189. Münzenrieder, N.; Cantarella, G.; Vogt, C.; Petti, L.; Büthe, L.; Salvatore, G.A.; Fang, Y.; Andri, R.; Lam, Y.; Libanori, R.; et al. Stretchable and conformable oxide thin-film electronics. *Adv. Electron. Mater.* **2015**, *1*, 1400038. [CrossRef]
190. Zhang, X.; Zhao, J.; Dou, J.; Tange, M.; Xu, W.; Mo, L.; Xie, J.; Xu, W.; Ma, C.; Okazaki, T.; et al. Flexible CMOS-like circuits based on printed p-type and n-type carbon nanotube thin-film transistors. *Small* **2016**, *12*, 5066–5073. [CrossRef] [PubMed]
191. Xu, W.; Liu, Z.; Zhao, J.; Xu, W.; Gu, W.; Zhang, X.; Qian, L.; Cui, Z. Flexible logic circuits based on top-gate thin film transistors with printed semiconductor carbon nanotubes and top electrodes. *Nanoscale* **2014**, *6*, 14891–14897. [CrossRef] [PubMed]
192. Kim, B.; Park, J.; Geier, M.L.; Hersam, M.C.; Dodabalapur, A. Voltage-controlled ring oscillators based on inkjet printed carbon nanotubes and zinc tin oxide. *ACS Appl. Mater. Interfaces* **2015**, *7*, 12009–12014. [CrossRef] [PubMed]

193. Cheng, C.H.; Yeh, F.S.; Chin, A. Low-power high-performance non-volatile memory on a flexible substrate with excellent endurance. *Adv. Mater.* **2011**, *23*, 902–905. [CrossRef] [PubMed]
194. Ji, Y.; Cho, B.; Song, S.; Kim, T.W.; Choe, M.; Kahng, Y.H.; Lee, T. Stable switching characteristics of organic nonvolatile memory on a bent flexible substrate. *Adv. Mater.* **2010**, *22*, 3071–3075. [CrossRef] [PubMed]

micromachines

MDPI

Review

Towards Flexible Transparent Electrodes Based on Carbon and Metallic Materials

Minghui Luo [1,2], Yanhua Liu [1,2,*], Wenbin Huang [1,2], Wen Qiao [1,2], Yun Zhou [1,2], Yan Ye [1,2] and Lin-Sen Chen [1,2,*]

[1] College of Physics, Optoelectronics and Energy & Collaborative Innovation Center of Suzhou Nano Science and Technology, Soochow University, Suzhou 215006, China;
lmhsuda@yeah.net (M.L.); wbhuang@suda.edu.cn (W.H.); wqiao@suda.edu.cn (W.Q.);
zyun@suda.edu.cn (Y.Z.); yanye@suda.edu.cn (Y.Y.)
[2] Key Laboratory of Advanced Optical Manufacturing Technologies of Jiangsu Province & Key Laboratory of Modern Optical Technologies of Education Ministry of China, Soochow University, Suzhou 215006, China
* Correspondence: yhliu@suda.edu.cn (Y.L.); lschen@suda.edu.cn (L.-S.C.);
Tel.: +86-512-6787-3745 (Y.L.); +86-512-6286-8882 (L.-S.C.)

Academic Editors: Seung Hwan Ko, Daeho Lee and Zhigang Wu
Received: 1 November 2016; Accepted: 27 December 2016; Published: 4 January 2017

Abstract: Flexible transparent electrodes (FTEs) with high stability and scalability are in high demand for the extremely widespread applications in flexible optoelectronic devices. Traditionally, thin films of indium thin oxide (ITO) served the role of FTEs, but film brittleness and scarcity of materials limit its further application. This review provides a summary of recent advances in emerging transparent electrodes and related flexible devices (e.g., touch panels, organic light-emitting diodes, sensors, supercapacitors, and solar cells). Mainly focusing on the FTEs based on carbon nanomaterials (e.g., carbon nanotubes and graphene) and metal materials (e.g., metal grid and metal nanowires), we discuss the fabrication techniques, the performance improvement, and the representative applications of these highly transparent and flexible electrodes. Finally, the challenges and prospects of flexible transparent electrodes will be summarized.

Keywords: flexible transparent electrodes; flexible electronics; metal nanowire; metal grid; optoelectronics devices

1. Introduction

Flexible transparent electrodes (FTEs) are essential components for numerous flexible optoelectronic devices due to their excellent capacity for transparency and flexibility, including organic light-emitting diodes [1–6], solar cells [7–12], touch panels [13,14], and wearable devices [15]. Conventionally, transparent conductive oxides (TCO) such as the indium tin oxide films (ITO), fluorine doped tin oxide (FTO) [16], ZnO:Al (AZO) [17], and ZnO:Ga (GZO) [18,19] have governed the domain of optoelectronic devices for several decades. However, some innate drawbacks of the TCO films such as brittleness due to its ceramic nature [20] and high cost for the scarcity of materials such as indium limit the widespread use in flexible devices, where stretching, twisting, or bending are usually requested. Recently, potential alternative materials to TCO have been widely explored, including ultra-thin metallic film [21,22], carbon-based nanomaterials (e.g., carbon nanotubes (CNTs) [23,24], and graphene [25–28]), conducting polymer [29,30], and metallic materials (e.g., metal grid [7,31,32] and metal nanowires [33–37]). To replace conventional TCO films, new types of FTEs should have low processing costs and mechanical flexibility while maintaining a low sheet resistance (R_s) and high optical transparency (T). As for FTEs based on conducting polymer, although the flexibility is improved, low conductivity and T limit their optoelectronic performance, and stability in ambient atmosphere is not good enough. Commonly, CNTs with percolating networks cannot process a low R_s and a high T simultaneously. The low conductivity of the CNTs remains the main limiting factor of the overall conductivity. For example, to achieve an R_s of the CNTs based FTE less than 10 $\Omega\square^{-1}$, the T will decrease drastically because the required thickness of CNTs exceeds 100 nm [38]. There remain challenges to improve the electrical conductivity of CNT-based FTEs.

Monolayer graphene only absorbs 2.3% of visible light and can sustain 4% strain with negligible cracking [39,40]. The theoretical R_s of the graphene is as low as 30 $\Omega\square^{-1}$ [41]. However, the R_s of synthesized graphene usually exceeds several hundred $\Omega\square^{-1}$ using different synthesis methods (e.g., epitaxial grown graphene on silicon carbide, chemical vapor (CVD) deposited on Cu catalysts), due to lower quality graphene with polycrystalline structures and plenty of defects [42,43].

Metallic based FTEs constructed from random networks of nanowires or regular metal grids regarded as another potential alternative to TCO, due to their better R_s-T performance than other alternatives. The metal nanothough networks can obtain a FTE with R_s of 2 $\Omega\square^{-1}$ at a transmittance of 90% [44]. In addition, the fabrication of metal structures coincides with printing and roll-to-roll technology, which reduces the cost for mass-production of FTEs significantly. However, the stability of the metal-based FTEs needs to be investigated further.

In this review, we provide a summary of recent advances in emerging FTEs and related flexible optoelectronic devices, mainly focusing on works reported in the past three years. For early work on FTEs, readers can refer to the review by Hecht et al. [45] and Ellmer et al. [46]. Carbon-based nanomaterials and metallic nanomaterials are promising to replace the dominance of the TCO films due to their superior performance, which will be the focus of our review. Then, we discuss the fabrication techniques, the performance improvement, and the representative applications of these FTEs. The challenges and prospects of the FTEs will eventually be summarized.

2. Currently Emerging Materials

Among those emerging alternatives, carbon-based materials and metallic materials are considered promising candidates for next-generation FTEs, due to their high mechanical flexibility paralleling good optical transparency and electrical conductivity. The material properties, combined with low material costs and fabrication techniques, make these emerging materials very attractive for FTEs.

2.1. One-Deminsional CNT-Based Nanomaterials

Carbon nanotubes (CNTs) have been evaluated and verified as one of the future FTEs relying on their remarkable characteristics. In the past several years, transparent, conductive, and flexible CNT-based FTEs have been investigated widely, involving many applications (e.g., OLED and supercapacitor). They can be fabricated using different methods [47–49], including wet and dry processing, which could be further exploited by combination with the roll-to-roll process. Both single-walled CNTs (SWCNTs) and multiwalled CNT (MWCNT)-based FTEs have been fabricated via solution approaches [24,50,51]. In detail, the SWCNT powders made from the CVD process are centrifuged to remove large conglomerations, before blending into an SWCNT solution. After that, the FTEs can be fabricated by various methods with the solution, including spray coating, dip-coating, and infiltration. Wu et al. have reported a simple process for the fabrication of FTEs using pure SWCNTs [52]. Low sheet resistance of 30 $\Omega\square^{-1}$ was obtained with 90% transmittance. However, owing to the SWCNTs tendency of twining during the CVD process, only a small part of them can be efficiently used, which increases the production costs.

To explore cost-effective method, Feng et al. employed a straightforward roll-to-roll process to fabricate flexible and stretchable MWCNT films as FTEs, where the CNTs possess excellent performance with low sheet resistance (208 $\Omega\square^{-1}$) and high transmittance (90%) [47]. In the work, CNT arrays are obtained by batch growth. Lin et al. reported polyaniline composite films incorporated with aligned MWCNTs, which were fabricated using an easy electrodeposition process [48]. Additionally, Recep et al. reported FTEs based on SWCNTs (Figure 1a), which exhibited an optical transmittance of 82% with 0.02 mg SWCNTs (Figure 1b). The supercapacitors fabricated by the SWCNT FTEs showed good capacity retention (94%) upon cycling over 500 times (Figure 1c) [49].

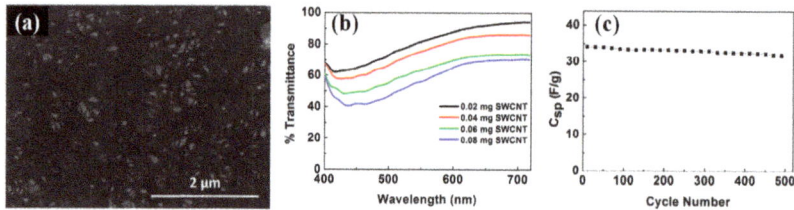

Figure 1. (a) SEM image of a SWCNT thin film with a sheet resistance of 75 $\Omega\square^{-1}$; (b) Optical transmittance characteristics of the full devices with respect to total SWCNT weight in both electrodes; (c) Cycle performance of flexible supercapacitors at a current density of 1.25 A/g. Reproduced with permission from Reference [49]. Copyright 2014, American Chemical Society.

2.2. Two-Dimensional Graphene-Based Nanomaterials

Graphene, a two-dimensional carbon allotrope, becomes a popular material for transparent electrodes due to the excellent performance [25,53,54]. Generally, the R_s of synthesized graphene usually exceed several hundred $\Omega\square^{-1}$ [42,43]. To improve the photoelectric performance, combing graphene with other conductors is an effective method.

Liu et al. prepared the graphene/ITO flexible hybrid transparent electrode, which showed excellent mechanical and optoelectronic characteristics [55]. Figure 2a shows the optical image of the graphene/ITO hybrid electrode. The SEM image shows the surface of ITO bridges on left and right sides as well as the graphene film upon them. Figure 2b is the transmittance spectra of the samples on the PET substrate. Besides, combination of graphene with metal materials can significantly improve the optoelectronic performance. The metal materials and the graphene offer extra conductive paths for each other. Qiu et al. reported the combination of the metal grid with graphene oxide films via a facile, green, and room-temperature method [56]. The excellent photoelectric properties with the sheet resistance of 18 $\Omega\square^{-1}$ and the transmittance of 80% can be obtained. Hong et al. demonstrated omnidirectionally stretchable and transparent graphene electrodes with good mechanical durability and performance reliability (Figure 2c) [57]. The multilayered graphene based FTE exhibits a transmittance of 87.1% at 550 nm (Figure 2d). Figure 2e,f show that sheet resistances of the graphene/PDMS film systematically monitored under various bending and stretching conditions, respectively. The textured graphene/PDMS film sustains its electrical properties when the film is folded with a bending radius (r) as small as r = 0.5 mm or stretched up to 30% of the tensile strain. Deng et al. developed a full roll-to-roll production of FTEs based on metal NWs (AgNWs and CuNWs) encapsulated by a monolayer graphene film, as shown in Figure 3a–c [26]. The R_s of the encapsulated NWs film show a 60%–90% decrease relative to the pure one (Figure 3d), which is due to the graphene conduction channel. Meanwhile, the encapsulated film processes lower R_s than pristine graphene because metal NWs provide additional conduction channels.

In addition to the graphene-based hybrid FTEs, doped graphene-based FTEs also exhibit good performance [41,58]. Bae et al. demonstrated roll-to-roll production and wet-chemical doping of predominantly monolayer 30-inch graphene films grown by CVD [41]. The doped graphene based FTEs have sheet resistances as low as ~125 $\Omega\square^{-1}$ with 97.4% optical transmittance. Park et al. found that AuCl$_3$ doping on graphene could improve the conductivity and shift the work function of the graphene FTEs, which results in improved power conversion efficiency of the OPV devices [58].

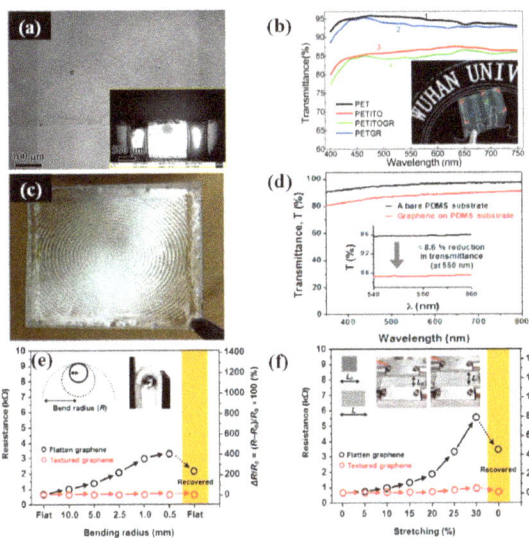

Figure 2. (**a**) Optical image of graphene/ITO hybrid electrode; (**b**) Transmittance spectra of the samples on PET substrate. (**a**,**b**) Reproduced with permission from Ref. [55]. Copyright 2016, Springer Nature; (**c**) Photograph of a graphene/PDMS free-standing film; (**d**) Transmission spectra of a flat PDMS substrate (black) and multilayered graphene on a flat PDMS substrate (red); (**e**) bending radius and (**f**) tensile strains (insets: actual test images for a graphene/PDMS film); (**c**–**f**) Reproduced with permission from Reference [57]. Copyright 2016, American Chemical Society.

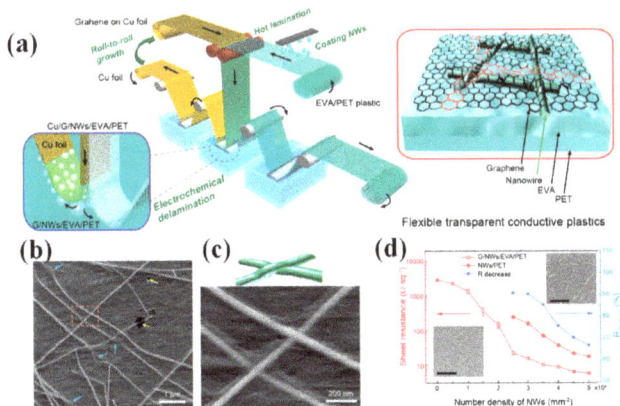

Figure 3. (**a**) Schematic diagram of the fabrication process includes coating of metal nanowires on polymer substrate (EVA/PET), hot-press lamination with graphene/Cu foil, delamination of graphene and Cu foil by electrochemical bubbling method, and the reuse of Cu foil to grow graphene by a continuous chemical vapor deposition system. The detailed structural schematic of the hybrid film labeled in the red cycle shows that nanowires are partly embedded into the EVA substrate and fully encapsulated by monolayer graphene film; (**b**) SEM image of the monolayer graphene; (**c**) Enlarged side-view SEM image of the hybrid film of graphene and AgNWs; (**d**) Sheet resistance versus number density of AgNWs for pure AgNW film and graphene/AgNW hybrid film. Reproduced with permission from Reference [26]. Copyright 2015, American Chemical Society.

2.3. Metallic Materials

As mentioned above, the progress in carbon-based FTEs is exciting, but the R_s is still not low enough. For the realization of the next generation of flexible and large-area electronics, further improvement in R_s-T performance is required. Metal-based FTEs have gained prominence in both academic and industrial use in flexible electronics. Metal-based FTEs can be constructed from random distribution metallic NWs works [33–37] to regular metal grids [7,31,32], which have demonstrated great potential in optical transparency, electrical conductivity, and mechanical flexibility.

2.3.1. Random Distribution Metallic NWs as FTEs

Most metals have excellent electrical conductivity due to their high free-electron density. To utilize metal materials in FTEs, high transparency should be satisfied. Among common metallic nanomaterials, silver nanowires (AgNWs) are commerciogenic due to the excellent electrical conductivity and high transmittance in a broad wavelength range [33–37].

The fabrication of AgNW FTEs has been explored by many researchers in various fabrication processes [59], which can achieve a low R_s (8–50 $\Omega\square^{-1}$) with a high transmittance (80%–98%). However, metal NW-based FTEs suffer from degradation of the R_s-T performance because of high junction resistances, which have been systematical investigated by several groups [60,61]. Mutiso et al. demonstrated that a higher NW aspect ratio could decrease the R_s while keeping the T a constant [61]. To obtain $R_s \leq 10\ \Omega\square^{-1}$ at T = 90%, the aspect ratio should exceed 800 at a given junction resistance of 2 KΩ. Commonly, the strategies to reduce junction resistance involve bulk heating [62], plasmonic treatment [63], and chemical modifications [64], which tend to damage the plastic substrate and are not suitable for flexible applications. Therefore, other additional processes are introduced to produce hybrid structure electrodes. For example, Xiong et al. exhibited electroless-welding of an AgNW network coated with conductive ion gel that significantly reduces junction resistance between AgNWs, while its effect on the T of the FTEs (R_s of 8.4 $\Omega\square^{-1}$ at T = 86%) is negligible [65].

In addition, the material of AgNWs exhibits the limitation of high surface roughness, which leads to low carrier transport. Currently, AgNW-based composite materials have become promising materials as FTEs, exhibiting excellent optical and electric properties [34,66]. Lee et al. reported a percolating network of AgNWs with densities above the percolation threshold integrated into graphene as hybrid FTEs [67]. The hybrid structure can reduce R_s down to 33 $\Omega\square^{-1}$ with a transmittance of 94%. Other research groups demonstrated the combination of an AgNW network with graphene or graphene oxide could enhance electrical conductivity and decrease surface roughness [68–70]. Xu et al. mixed Ag NWs with graphene oxide and obtained Ag-doped graphene fibers, whose electrical conductivity increased from 4.1×10^4 S/m to 9.3×10^4 S/m, exhibiting a 330% enhancement factor [71]. Meanwhile, chemical and thermal stabilities also can be further improved. It is obvious that the composite materials enhance electrical properties of FTEs by decreasing junction resistance or providing new conductive pathways.

Not merely AgNWs, copper NWs (CuNWs) and gold NWs (AuNWs) are promising candidates for FTEs, due to the similarly excellent conductivity (Figure 4a–c) [34,72,73]. Cui et al. have reported a new synthetic approach for obtaining ultrathin high-quality CuNWs (Figure 4b), where oleylamine is used as a coordinating ligand [72]. Maurer et al. have demonstrated that ultrathin AuNWs can be synthesized at room temperature [73]. Triisopropylsilane (TIPS, 2 mL) is added into the solution after the gold salt was completely dissolved. The AuNWs will format standing for two days until the color turns from yellow into dark-red (Figure 4c). A high transparency of 79% can be achieved and maintained over 80 stretching cycles.

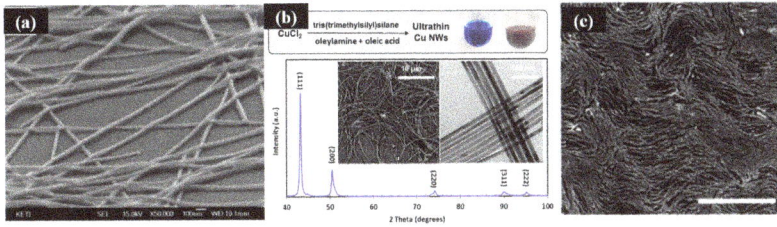

Figure 4. (**a**) SEM image of AgNWs. Reproduced with permission from Reference [34]. Copyright 2016, American Chemical Society; (**b**) Synthesis of ultrathin copper nanowires. Reproduced with permission from Reference [72]. Copyright 2015, American Chemical Society; (**c**) SEM image of monolayer AuNWs on PDMS substrate. Reproduced with permission from Reference [73]. Copyright 2016, American Chemical Society.

2.3.2. Metal Grids

The randomly dispersed network in the whole scale cannot be used immediately as prepared. Subsequently, patterning processes are necessarily adopted depending on different device architectures. In particular, metal grid FTEs offer advantages over other NW FTEs due to the nature of artificial design. For example, the electrical and optical properties can be managed by modulating the grid pitch, width, and thickness.

Uniform metal grids as FTEs fabricated by many processes, including nano-patterning techniques such as photolithography, crackle, electroless plating, nano-transfer printing, and electrospinning [74–78]. Almost all of these processes involve deposition of a metal film onto a template to form inter-connected network excluding junction resistance. In contrast to the limited material choices for NWs, a variety of metallic materials can be exploited in different applications. Cui et al. employed electrospun and metal deposited technique to generate nanotrough networks. Correspondingly, the width and thickness of the networks are 420 nm and 80–100 nm, respectively. The new kind FTEs exhibited both superior optoelectronic performance (R_s of 2 $\Omega\square^{-1}$ at 90% transmission) (Figure 5a) and remarkable mechanical flexibility (Figure 5b,c) [44]. The R_s-T performance achieved is much better than that of TCO, carbon-based materials, or solution processed mental NWs networks.

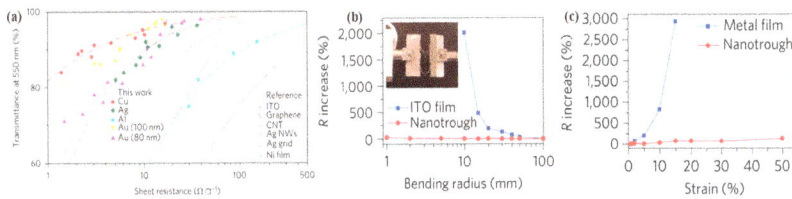

Figure 5. (**a**) Sheet resistance versus optical transmission (at 550 nm) for copper, gold, silver, and aluminum nanotrough networks, described by percolation theory. The performances of device-grade ITO, CNTs, graphene, silver nanowires (NWs), silver grid, and nickel thin films are shown for comparison; (**b**) R_s versus bending radius for bendable transparent electrodes consisting of gold nanotrough networks or ITO films on 178-μm-thick PET substrates; (**c**) R_s versus uniaxial strain for a stretchable transparent electrode consisting of gold nanotrough networks on 0.5-mm-thick PDMS substrate. Reproduced with permission from Ref. [44]. Copyright 2013, Nature Publishing Group.

In another work, Han et al. proposed the cracked TiO_2 gel film as a template to make Ag networks [75]. The Ag networks with diameters of 1–2 μm, and widths of 4–100 μm, exhibited good electro-optical properties. The transmittance ranges from 82% to 45%, and correspondingly the R_s ranges from 4.2 $\Omega\square^{-1}$ to 0.5 $\Omega\square^{-1}$. To further improve the R_s-T performance of the NWs based FTEs,

Hsu et al. introduced a mesoscale metal-wire concept in conjunction with NWs [78]. The mesoscale metal-wire networks show the extraordinary R_s-T performance with the R_s being 0.36 and the T being 92%. However, fabricating metal grid FTEs usually employs the physical deposition of metal materials involving thermal evaporation or sputtering, which requires expensive vacuum-based processing. Therefore, the fabrication process is not simple and cost-effective.

Recently, electrohydrodynamic (EHD) jet printing employed to fabricate metal grid FTEs [31,79]. Lee et al. fabricated FTEs by thermal pressing of metal lines, which was provided by EHD jet printing [31]. The excellent properties have been demonstrated with the sheet resistance of 0.5 $\Omega\square^{-1}$ at the transmittance of 80%. Seong et al. have reported a method of fabricated FTEs by EHD jet printing. Sheet resistance of 1.49 $\Omega\square^{-1}$ can be achieved by printing the Ag mesh on the convex glass [79]. Another route utilized to fabricate FTE by using inject printing [80]. Mohl et al. obtained grid meshes by inject printing and subsequent chemical copper plating [32]. The printed metal grids from reactive ink plated with copper can create high performance FTEs. The achieved R_s is 10 $\Omega\square^{-1}$ and transmittance is 80%.

A novel fabrication method via hybrid printing technique has been demonstrated for FTEs with embedded metal grids [81]. Cui et al. developed high-resolution metal mesh as FTEs by nanoimprinting technology [14]. The silver nanoparticle inks embedded into the metal mesh, where the metal mesh was fabricated by the roll-to-roll progress. Low-cost fabrication is the most important advantage of this new approach. Moreover, high performance could be obtained with the low sheet resistance of 0.69 $\Omega\square^{-1}$ at 88% transparency. Most important is that the fabrication cost can be further decreased by exploiting their compatibility with printing technologies, attributing to the efficient use of material, a simple fabrication process, and easy scalability to large scale. In addition, Kiruthika et al. employed the roll and spray coating methods to fabricate the FTEs by a simple solution process using crackle lithography [82]. A transmittance of 78% and sheet resistance of ~20 $\Omega\square^{-1}$ can be obtained (Figure 6a). Figure 6b shows the optical micrographs of Ag meshes with different widths on PET substrate. Figure 6c–f exhibits the mechanical stability of the metallized Ag in the crackle network. The mesh was subjected to 500 bending cycles with 1.5 mm radius, the change in the resistance remained within 5%.

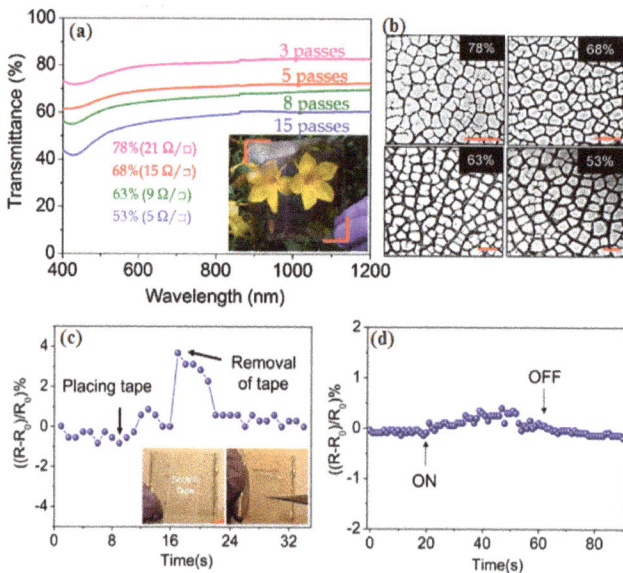

Figure 6. *Cont.*

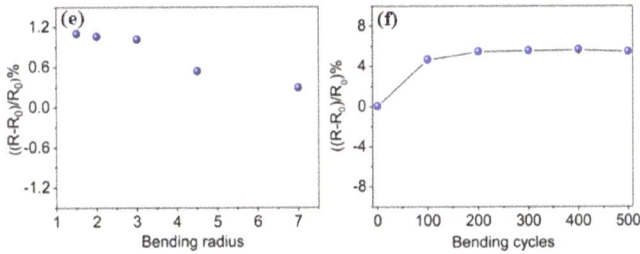

Figure 6. (**a**) Specular optical transmittance of Ag mesh prepared from crackles of different widths. The sheet resistances of the samples are mentioned in parentheses. Photograph of the prepared TCE with 78% transmittance at 550 nm is shown as an inset. (**b**) Optical micrographs of Ag mesh of different widths. Scale bar is 1 mm. Relative variations in the resistance of the Ag mesh during (**c**) the scotch tape adhesion test, (**d**) sonication test, (**e**) bending to different radii, and (**f**) 500 bending cycles with a radius of 1.5 mm. The photographs in the inset in (**a**) show the scotch tape pasted over the Ag mesh and while peeling off. Reproduced with permission from Reference [82]. Copyright 2015, American Chemical Society.

Finally, some typical values of sheet resistance and transmission of the novel FTEs up to date are summarized in Figure 7. Great progress has made in FTEs towards lower sheet resistance and high transmittance.

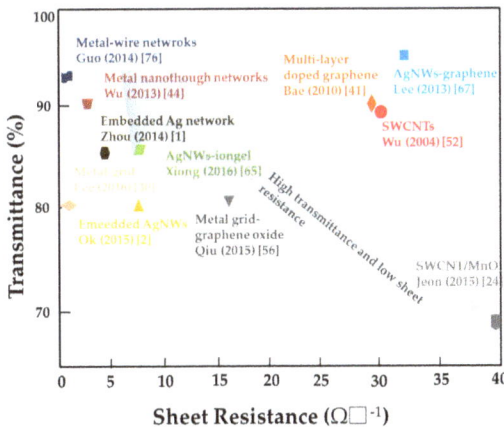

Figure 7. Sheet resistance/Transmission values of FTEs reported in recent works.

3. Applications

Based on the excellent mechanical compliance, high transmittance, and electric conductivity, the FTEs can be utilized in many essential applications, including supercapacitor, OLED, solar cell, and touch panel [45,46]. Different performances are needed for different applications, due to the requirement of technical indexes [60]. To satisfy the requirement of touch screens, the R_s of FTE should range from 100 $\Omega\square^{-1}$ to 1000 $\Omega\square^{-1}$, a T exceeds 85% combined with low haze. The FTEs must have the R_s ~10 $\Omega\square^{-1}$ for T > 90% to provide solar cells and OLEDs. In addition, for use of FTEs in flexible optoelectronics, several requirements must be met, including the reduced haze value for the optical transparency, and ultra-smooth surface of the electrode to avoid the disconnection problem. Finally, a simple, low-cost, and large-scale process in fabricating FTEs is a necessity for commercial applications.

To function as an FTE in OLEDs, the metal-based electrodes must alleviate the leakage current by planarizing its surface. Zhou et al. investigated embedded Ag-grid FTEs as the anode of OLED, which exhibits a power efficiency of 106 lm·W^{-1} at 1000 cd·m^{-2} (Figure 8) [1].

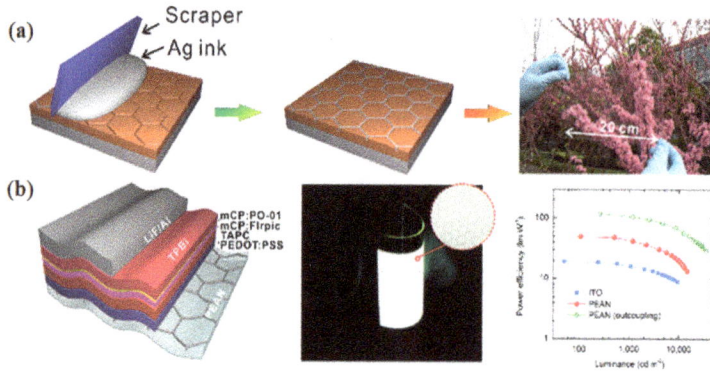

Figure 8. (**a**) Schematic illustration of the fabrication process of an embedded Ag network on PET (PEAN) and its optical image; (**b**) Device structure and performance of flexible white OLED using PEAN anode. Reproduced with permission from Reference [1]. Copyright 2014, American Chemical Society.

Liu et al. reported a composite electrode developed from poly (3,4-ethylenedioxythiophene): poly (styrenesulfonate) (PEDOT: PSS) and AgNWs with a template stripping method [3]. The efficiency of the resultant OLEDs improved by 25% compared to that of traditional PEDOT: PSS FTEs, attributing to the reduction of surface roughness and the improvement of electric conductivity. Ok and co-worker developed an ultra-thin and smooth FTE by embedding AgNWs in a colorless polymide (cPI) [2]. The AgNWs-cPI composite electrode exhibited a T > 80%, a low R$_s$ of 8 Ω□$^{-1}$, and ultra-smooth surfaces comparable to glass (4.1 nm), as shown in Figure 9a. The OLEDs fabricated from such composite electrodes showed a stable performance with a luminance reduction of <3% after 10 repeated bendings at a radius of 30 μm (Figure 9b).

For applications in display devices and touch screens, high visibility is needed. Haze property is a vital value to the display industry. Ag networks that are fabricated by cracked template and AgNWs solution processed methods [75,83] can survive multiple finger touchings. The low R$_s$ (<10 Ω□$^{-1}$) and the large area fabrication are rapidly improving in the large format touch panels market. Currently, iV-touch can offer large projected capacitive touch panels from 21.5″ to 55″ with stunning touch performance [84].

Similar to the case of OLEDs, the application of NW networks and metal grids as the TCEs employed in functional organic solar cells. In order to achieve the required R$_s$ (<50 Ω□$^{-1}$) and T (>90%), constructing hybrid structured FTEs is a promising technology. Singh et al. demonstrated solution processed solar cells employing AgNW film binding with a 40 nm thin overlayer of sputtered ZnO as FTEs [85]. The ZnO overlayer is used to increase adhesion between AgNW and ZnO/buffer layer and to interconnect the AgNWs junctions. More recently, Li et al. integrated embedded Ag grids and conducting polymer hybrid electrodes into a perovskite-based photovoltaic cell and demonstrated an ultrathin flexible device delivering a power conversion efficiency of 14.0% [7]. Wu et al. demonstrated AgNW grids with multi-length scaled structures as FTEs for an organic solar cell [86]. A power conversion efficiency of 9% was achieved for the organic solar cell devices.

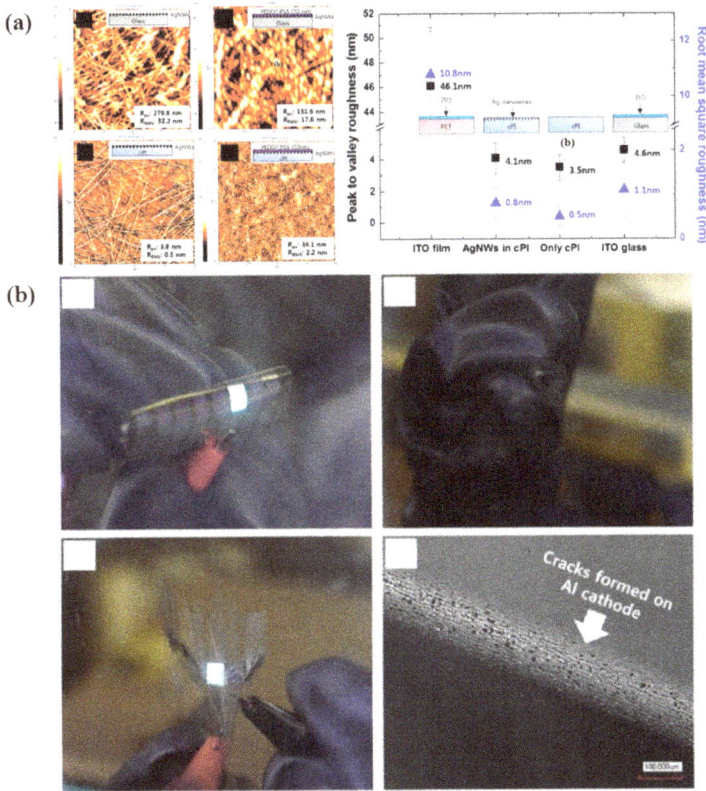

Figure 9. (**a**) Comparison of morphologies by AFM analyses and a diagram comparing the roughness of various samples; (**b**) Bending stability of the flexible OLEDs based on AgNWs-cPI composite electrodes. Reproduced with permission from Reference [2]. Copyright 2015, Macmillan Publishers Limited.

Transparent supercapacitors have been proposed in the past decade, which function both as a current collector to transport the electrons, and an active material to store electrochemical energy. For the application in supercapacitors, Lin et al. has synthesized polyaniline composite film incorporated with aligned MWCNTs through an easy electrodeposition process [48]. The conductive films are sufficiently used to fabricate transparent, flexible, and efficient supercapacitors with a maximum specific capacitance of 233 F/g at a current density of 1 A/g. Cai et al. demonstrated that the electrochemical stability of Ag grid FTEs can be enhanced by coating one layer of PEDOT:PSS [87]. The film sustained an optical modulation and a specific capacitance of 87.7% and 67.2% at 10 A·g^{-1}, respectively. Cheng et al. reported hybrid electrodes composed of PEDOT:PSS and Ag-grids prepared via inject printing [88]. The hybrid structures not only compensated for the shortcomings of the single materials but also fully combined their advantages. The comparatively high Ct (capacitance of the supercapacitor) and Csc (capacitance of the electrode) of the supercapacitor based on PEDOT: PSS (three layers)/Ag grid electrodes were 1.13 mF·cm^{-2} and 4.52 mF·cm^{-2}, respectively. In addition, Lee et al. have introduced a highly flexible and transparent supercapacitor based on electrochemically stable Ag-Au core-shell (AACS) nanowire percolation network electrode (Figure 10) [89]. Figure 10c confirms that the AACS NW networks possess superior optical transparency, exceeding 85% in the entire range of the visible wavelength. Figure 10d shows strain-dependent electrical resistance of Ag–Au core–shell NW electrode on an elastic PDMS substrate.

The various applications make carbon-based and metal-based FTE one of the most promising materials in optoelectronic devices. Considering the progress and the ongoing efforts on FTEs, there is no doubt that performance of these FTE-based flexible devices (OLED, OSC, supercapacitor, touch screen) can be further improved, and the new applications can served to an amazing degree.

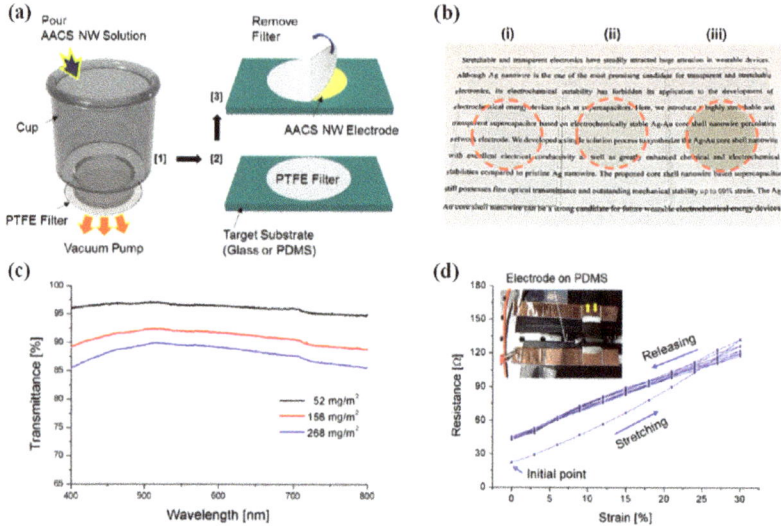

Figure 10. Fabrication and characterization of Ag–Au core–shell NW-based electrodes. (**a**) Schematic illustration of the electrode fabrication through vacuum filtration and transfer method; (**b**) Digital image of as-prepared transparent Ag–Au core–shell NW network electrodes at various sheet resistances; (**c**) Optical transmittance of the transparent electrodes at various areal NW densities; (**d**) Strain-dependent electrical resistance of Ag–Au core–shell NW electrode on an elastic PDMS substrate. Reproduced with permission from Reference [89]. Copyright 2016, American Chemical Society.

4. Conclusions and Challenges

In conclusion, a comprehensive overview on recent developments and achievements in carbon-based and metal-based FTEs and related flexible optoelectronic devices is provided in this review. Carbon-based materials and metallic materials are promising to dominate the FTEs, especially hybrid FTEs exhibiting superior properties. Although outstanding performances—including low sheet resistance, high transmittance, and good mechanical properties—have been demonstrated, there are still many restrictions hindering large-scale fabrication. In addition, more exploration in the microcosmic aspect and the fundamental theories are needed. The present FTEs already play a significant role in many optical and electric applications, which urges more studies to improve the performance (stretchability, flexibility, transmittance, stability, and electric conductivity) from the material synthesis to device fabrication.

Acknowledgments: This work is supported by the National Natural Science Foundation of China (NSFC) (61405133, 61505131, 91323303), and by the Specialized Research Fund for the Doctoral Program of Higher Education (No. 20133201120027), and by the Natural Science Foundation of Jiangsu Province (No. BK20140348, BK20150309), and by the China Postdoctoral Science Foundation (Grant No. 2015M571816), and by a project funded by the Priority Academic Program Development (PAPD) of Jiangsu Higher Education Institutions.

Author Contributions: Minghui Luo gathered information and wrote the first draft of this review. Yanhua Liu supervised and amended this review. Wenbin Huang, Wen Qiao, Yun Zhou, and Yan Ye were auxiliary for perfecting this work. Lin-Sen Chen supervised and supported the work. All authors reviewed the manuscript.

Conflicts of Interest: The authors declare no conflict of interest.

References

1. Zhou, L.; Xiang, H.Y.; Shen, S.; Li, Y.Q.; Chen, J.D.; Xie, H.J.; Goldthorpe, I.A.; Chen, L.S.; Lee, S.T.; Tang, J.X. High-performance flexible organic light-emitting diodes using embedded silver network transparent electrodes. *ACS Nano* **2014**, *8*, 12796–12805. [CrossRef] [PubMed]

2. Ok, K.H.; Kim, J.; Park, S.R.; Kim, Y.; Lee, C.J.; Hong, S.J.; Kwak, M.G.; Kim, N.; Han, C.J.; Kim, J.W. Ultra-thin and smooth transparent electrode for flexible and leakage-free organic light-emitting diodes. *Sci. Rep.* **2015**, *5*, 9464. [CrossRef] [PubMed]

3. Liu, Y.S.; Feng, J.; Ou, X.L.; Cui, H.F.; Xu, M.; Sun, H.B. Ultrasmooth, highly conductive and transparent PEDOT:PSS/silver nanowire composite electrode for flexible organic light-emitting devices. *Org. Electron.* **2016**, *31*, 247–252. [CrossRef]

4. Lee, H.; Lee, D.; Ahn, Y.; Lee, E.W.; Park, L.S.; Lee, Y. Highly efficient and low voltage silver nanowire-based OLEDs employing a n-type hole injection layer. *Nanoscale* **2014**, *6*, 8565–8570. [CrossRef] [PubMed]

5. Yun, S.O.; Hwang, Y.; Park, J.; Jeong, Y.; Kim, S.H.; Noh, B.I.; Jung, H.S.; Jang, H.S.; Hyun, Y.; Choa, S.H. Sticker-Type Alq 3-Based OLEDs Based on Printable Ultrathin Substrates in Periodically Anchored and Suspended Configurations. *Adv. Mater.* **2013**, *25*, 5626–5631. [CrossRef] [PubMed]

6. Meng, H.; Luo, J.; Wang, W.; Shi, Z.; Niu, Q.; Dai, L.; Qin, G. Top-Emission Organic Light-Emitting Diode with a Novel Copper/Graphene Composite Anode. *Adv. Funct. Mater.* **2013**, *23*, 3324–3328. [CrossRef]

7. Li, Y.; Meng, L.; Yang, Y.; Xu, G.; Hong, Z.; Chen, Q.; You, J.; Li, G.; Yang, Y.; Li, Y. High-efficiency robust perovskite solar cells on ultrathin flexible substrates. *Nat. Commun.* **2016**, *7*. [CrossRef] [PubMed]

8. Kim, B.J.; Dong, H.K.; Lee, Y.Y.; Shin, H.W.; Sang, G.; Han, G.S.; Sug, J.; Mahmood, K.; Ahn, T.K.; Joo, Y.C. Highly efficient and bending durable perovskite solar cells: Toward a wearable power source. *Energy Environ. Sci.* **2014**, *8*, 916–921. [CrossRef]

9. Shin, S.S.; Yang, W.S.; Noh, J.H.; Suk, J.H.; Jeon, N.J.; Park, J.H.; Ju, S.K.; Seong, W.M.; Sang, I.S. High-performance flexible perovskite solar cells exploiting Zn_2SnO_4 prepared in solution below 100 °C. *Nat. Commun.* **2015**, *6*. [CrossRef] [PubMed]

10. Kranz, L.; Gretener, C.; Perrenoud, J.; Schmitt, R.; Pianezzi, F.; La, M.F.; Blösch, P.; Cheah, E.; Chirilă, A.; Fella, C.M. Doping of polycrystalline CdTe for high-efficiency solar cells on flexible metal foil. *Nat. Commun.* **2013**, *4*, 1431–1442. [CrossRef] [PubMed]

11. Li, Z.; Kulkarni, S.A.; Boix, P.P.; Shi, E.; Cao, A.; Fu, K.; Batabyal, S.K.; Zhang, J.; Xiong, Q.; Wong, L.H. Laminated Carbon Nanotube Networks for Metal Electrode-Free Efficient Perovskite Solar Cells. *ACS Nano* **2014**, *8*, 6797–6804. [CrossRef] [PubMed]

12. Han, J.; Yuan, S.; Liu, L.; Qiu, X.; Gong, H.; Yang, X.; Li, C.; Hao, Y.; Cao, B. Fully indium-free flexible Ag nanowires/ZnO:F composite transparent conductive electrodes with high haze. *J. Mater. Chem. A* **2015**, *3*, 5375–5384. [CrossRef]

13. Lee, J.; Lee, P.; Lee, H.B.; Hong, S.; Lee, I.; Yeo, J.; Lee, S.S.; Kim, T.S.; Lee, D.; Ko, S.H. Room-Temperature Nanosoldering of a Very Long Metal Nanowire Network by Conducting-Polymer-Assisted Joining for a Flexible Touch-Panel Application. *Adv. Funct. Mater.* **2013**, *23*, 4171–4176. [CrossRef]

14. Cui, Z.; Gao, Y. 27.5L: Late-News Paper: Hybrid Printing of High Resolution Metal Mesh as A Transparent Conductor for Touch Panels and OLED Displays. *SID Symp. Dig. Tech. Pap.* **2015**, *46*, 398–400. [CrossRef]

15. Wu, C.; Kim, T.W.; Li, F.; Guo, T. Wearable Electricity Generators Fabricated Utilizing Transparent Electronic Textiles Based on Polyester/Ag Nanowires/Graphene Core-Shell Nanocomposites. *ACS Nano* **2016**, *10*, 6449–6457. [CrossRef] [PubMed]

16. Baek, W.H.; Choi, M.; Yoon, T.S.; Lee, H.H.; Kim, Y.S. Use of fluorine-doped tin oxide instead of indium tin oxide in highly efficient air-fabricated inverted polymer solar cells. *Appl. Phys. Lett.* **2010**, *96*, 133506. [CrossRef]

17. Pei, Z.L.; Zhang, X.B.; Zhang, G.P.; Gong, J.; Sun, C.; Huang, R.F.; Wen, L.S. Transparent conductive ZnO:Al thin films deposited on flexible substrates prepared by direct current magnetron sputtering. *Thin Solid Films* **2005**, *68*, 20–23. [CrossRef]

18. Assuncao, V.; Fortunato, E.; Marques, A.; Aguas, H.; Ferreira, I.; Costa, M.; Martins, R. Influence of the deposition pressure on the properties of transparent and conductive ZnO:Ga thin-film produced by r.f. sputtering at room temperature. *Thin Solid Films* **2003**, *427*, 401–405. [CrossRef]

19. Abduev, A.; Akmedov, A.; Asvarov, A.; Chiolerio, A. A revised growth model for transparent conducting Ga doped ZnO films: Improving crystallinity by means of buffer layers. *Plasma Process. Polym.* **2015**, *12*, 725–733. [CrossRef]

20. Gutruf, P.; Shah, C.M.; Walia, S.; Nili, H.; Zoolfakar, A.S.; Karnutsch, C.; Kalantarzadeh, K.; Sriram, S.; Bhaskaran, M. Transparent functional oxide stretchable electronics: Micro-tectonics enabled high strain electrodes. *NPG Asia Mater.* **2013**, *5*, 759–760. [CrossRef]

21. Zhang, C.; Zhao, D.; Gu, D.; Kim, H.; Ling, T.; Wu, Y.K.R.; Guo, L.J. An Ultrathin, Smooth, and Low-Loss Al-Doped Ag Film and Its Application as a Transparent Electrode in Organic Photovoltaics. *Adv. Mater.* **2014**, *26*, 5696–5701. [CrossRef] [PubMed]

22. Vj, L.; Kobayashi, N.P.; Islam, M.S.; Wu, W.; Chaturvedi, P.; Fang, N.X.; Wang, S.Y.; Williams, R.S. Ultrasmooth silver thin films deposited with a germanium nucleation layer. *Nano Lett.* **2009**, *9*, 178–182.

23. Wang, X.; Li, Z.; Xu, W.; Kulkarni, S.A.; Batabyal, S.K.; Zhang, S.; Cao, A.; Wong, L.H. TiO$_2$ nanotube arrays based flexible perovskite solar cells with transparent carbon nanotube electrode. *Nano Energy* **2015**, *11*, 728–735. [CrossRef]

24. Jeon, I.; Cui, K.; Chiba, T.; Anisimov, A.; Nasibulin, A.G.; Kauppinen, E.I.; Maruyama, S.; Matsuo, Y. Direct and Dry Deposited Single-Walled Carbon Nanotube Films Doped with MoO(x) as Electron-Blocking Transparent Electrodes for Flexible Organic Solar Cells. *J. Am. Chem. Soc.* **2015**, *137*, 7982–7985. [CrossRef] [PubMed]

25. Chen, Y.Z.; Medina, H.; Tsai, H.W.; Wang, Y.C.; Yen, Y.T.; Manikandan, A.; Chueh, Y.L. Low Temperature Growth of Graphene on Glass by Carbon-Enclosed Chemical Vapor Deposition Process and Its Application as Transparent Electrode. *Chem. Mater.* **2015**, *27*, 1636–1655. [CrossRef]

26. Deng, B.; Hsu, P.C.; Chen, G.; Chandrashekar, B.N.; Liao, L.; Ayitimuda, Z.; Wu, J.; Guo, Y.; Lin, L.; Zhou, Y. Roll-to-Roll Encapsulation of Metal Nanowires between Graphene and Plastic Substrate for High-Performance Flexible Transparent Electrodes. *Nano Lett.* **2015**, *15*, 4206–4213. [CrossRef] [PubMed]

27. Li, N.; Yang, G.; Sun, Y.; Song, H.; Cui, H.; Yang, G.; Wang, C. Free-Standing and Transparent Graphene Membrane of Polyhedron Box-Shaped Basic Building Units Directly Grown Using a NaCl Template for Flexible Transparent and Stretchable Solid-State Supercapacitors. *Nano Lett.* **2015**, *15*, 3139–3203. [CrossRef] [PubMed]

28. Liu, Z.; Parvez, K.; Li, R.; Dong, R.; Feng, X.; Müllen, K. Transparent Conductive Electrodes from Graphene/PEDOT:PSS Hybrid Inks for Ultrathin Organic Photodetectors. *Adv. Mater.* **2015**, *27*, 669–675. [CrossRef] [PubMed]

29. Lipomi, D.J.; Lee, J.A.; Vosgueritchian, M.; Tee, C.K.; Bolander, J.A.; Bao, Z. Electronic Properties of Transparent Conductive Films of PEDOT:PSS on Stretchable Substrates. *Chem. Mater.* **2012**, *24*, 373–382. [CrossRef]

30. Jung, S.; Lim, K.; Kang, J.W.; Kim, J.K.; Oh, S.I.; Eun, K.; Kim, D.G.; Choa, S.H. Electromechanical properties of indium-tin-oxide/poly(3,4-ethylenedioxythiophene): Poly(styrenesulfonate) hybrid electrodes for flexible transparent electrodes. *Thin Solid Films* **2014**, *550*, 435–443. [CrossRef]

31. Lee, Y.; Jin, W.Y.; Cho, K.Y.; Kang, J.W.; Kim, J. Thermal pressing of a metal-grid transparent electrode into a plastic substrate for flexible electronic devices. *J. Mater. Chem. C* **2016**, *4*, 4577–4583. [CrossRef]

32. Mohl, M.; Dombovari, A.; Vajtai, R.; Ajayan, P.M.; Kordas, K. Self-assembled large scale metal alloy grid patterns as flexible transparent conductive layers. *Sci. Rep.* **2015**, *5*, 13710. [CrossRef] [PubMed]

33. Kim, D.H.; Yu, K.C.; Kim, Y.; Kim, J.W. Highly Stretchable and Mechanically Stable Transparent Electrode Based on Composite of Silver Nanowires and Polyurethane-Urea. *ACS Appl. Mater. Interfaces* **2015**, *7*, 15214–15222. [CrossRef] [PubMed]

34. Sciacca, B.; Groep, J.V.D.; Polman, A.; Garnett, E.C. Nanowires: Solution-Grown Silver Nanowire Ordered Arrays as Transparent Electrodes. *Adv. Mater.* **2016**, *28*, 976. [CrossRef]

35. Guo, F.; Azimi, H.; Hou, Y.; Przybilla, T.; Hu, M.; Bronnbauer, C.; Langner, S.; Spiecker, E.; Forberich, K.; Brabec, C.J. High-performance semitransparent perovskite solar cells with solution-processed silver nanowires as top electrodes. *Nanoscale* **2015**, *7*, 1642–1649. [CrossRef] [PubMed]

36. José, A.L.; Fe, M.M.; Gómez, D.; Luisa, M.A.; Bristow, N.; Paul, K.J.; Menéndez, A.; Ruiz, B. Rapid synthesis of ultra-long silver nanowires for tailor-made transparent conductive electrodes: Proof of concept in organic solar cells. *Nanotechnology* **2015**, *26*, 26.

37. Li, B.; Ye, S.; Stewart, I.E.; Alvarez, S.; Wiley, B.J. Synthesis and Purification of Silver Nanowires To Make Conducting Films with a Transmittance of 99. *Nano Lett.* **2015**, *15*, 6722–6726. [CrossRef] [PubMed]

38. Kumar, A.; Zhou, C. The race to replace tin-doped indium oxide: Which material will win? *ACS Nano* **2010**, *4*, 11–14. [CrossRef] [PubMed]

39. Nair, R.R.; Blake, P.; Grigorenko, A.N.; Novoselov, K.S.; Booth, T.J.; Stauber, T.; Peres, N.M.; Geim, A.K. Fine structure constant defines visual transparency of graphene. *Science* **2008**, *320*, 1308. [CrossRef] [PubMed]

40. Chen, J.H.; Jang, C.; Xiao, S.; Ishigami, M.; Fuhrer, M.S. Intrinsic and extrinsic performance limits of graphene devices on SiO2. *Nat. Nanotechnol.* **2008**, *3*, 206–209. [CrossRef] [PubMed]

41. Mcdonald, C.; Salter, D.M.; Chetty, U.; Anderson, T.J. Roll-to-roll production of 30-inch graphene films for transparent electrodes. *Nat. Nanotechnol.* **2010**, *5*, 574–578.

42. Park, J.U.; Nam, S.W.; Lee, M.S.; Lieber, C.M. Synthesis of monolithic graphene–graphite integrated electronics. *Nat. Mater.* **2011**, *11*, 120–125. [CrossRef] [PubMed]

43. Li, X.; Cai, W.; An, J.; Kim, S.; Nah, J.; Yang, D.; Piner, R.; Velamakanni, A.; Jung, I.; Tutuc, E. Large-area synthesis of high-quality and uniform graphene films on copper foils. *Science* **2009**, *324*, 1312–1314. [CrossRef] [PubMed]

44. Wu, H.; Kong, D.; Ruan, Z.; Hsu, P.C.; Wang, S.; Yu, Z.; Carney, T.J.; Hu, L.; Fan, S.; Cui, Y. A transparent electrode based on a metal nanotrough network. *Nat. Nanotechnol.* **2013**, *8*, 421–425. [CrossRef] [PubMed]

45. Hecht, D.S.; Hu, L.; Irvin, G. Emerging transparent electrodes based on thin films of carbon nanotubes, graphene, and metallic nanostructures. *Adv. Mater.* **2011**, *23*, 1482–1513. [CrossRef] [PubMed]

46. Ellmer, K. Past achievements and future challenges in the development of optically transparent electrodes. *Nat. Photonics* **2012**, *6*, 809–817. [CrossRef]

47. Feng, C.; Liu, K.; Wu, J.S.; Liu, L.; Cheng, J.S.; Zhang, Y.; Sun, Y.; Li, Q.; Fan, S.; Jiang, K. Flexible, Stretchable, Transparent Conducting Films Made from Superaligned Carbon Nanotubes. *Adv. Funct. Mater.* **2010**, *20*, 885–891. [CrossRef]

48. Lin, H.; Li, L.; Ren, J.; Cai, Z.; Qiu, L.; Yang, Z.; Peng, H. Conducting polymer composite film incorporated with aligned carbon nanotubes for transparent, flexible and efficient supercapacitor. *Sci. Rep.* **2013**, *3*, 1353. [CrossRef] [PubMed]

49. Yuksel, R.; Sarioba, Z.; Cirpan, A.; Hiralal, P.; Unalan, H.E. Transparent and Flexible Supercapacitors with Single Walled Carbon Nanotube Thin Film Electrodes. *ACS Appl. Mater. Interfaces* **2014**, *6*, 15434–15439. [CrossRef] [PubMed]

50. Ren, L.; Pint, C.L.; Arikawa, T.; Takeya, K.; Kawayama, I.; Tonouchi, M.; Hauge, R.H.; Kono, J. Broadband Terahertz Polarizers with Ideal Performance Based on Aligned Carbon Nanotube Stacks. *Nano Lett.* **2012**, *12*, 787–790. [CrossRef] [PubMed]

51. Kim, B.J.; Park, J.S.; Hwang, Y.J.; Park, J.S. Characteristics of silver meshes coated with carbon nanotubes via spray-coating and electrophoretic deposition for touch screen panels. *Thin Solid Films* **2015**, *596*, 68–71. [CrossRef]

52. Wu, Z.; Chen, Z.; Du, X.; Logan, J.M.; Sippel, J.; Nikolou, M.; Kamaras, K.; Reynolds, J.R.; Tanner, D.B.; Hebard, A.F. Transparent, conductive carbon nanotube films. *Science* **2004**, *305*, 1273–1276. [CrossRef] [PubMed]

53. Eda, G.; Fanchini, G.; Chhowalla, M. Large-area ultrathin films of reduced graphene oxide as a transparent and flexible electronic material. *Nat. Nanotechnol.* **2008**, *3*, 270–274. [CrossRef] [PubMed]

54. Hu, W.; Peng, C.; Luo, W.; Lv, M.; Li, X.; Li, D.; Huang, Q.; Fan, C. Graphene-based antibacterial paper. *ACS Nano* **2010**, *4*, 4317–4323. [CrossRef] [PubMed]

55. Liu, J.; Yi, Y.; Zhou, Y.; Cai, H. Highly Stretchable and Flexible Graphene/ITO Hybrid Transparent Electrode. *Nanoscale Res. Lett.* **2016**, *11*, 1–7. [CrossRef] [PubMed]

56. Qiu, T.; Luo, B.; Liang, M.; Ning, J.; Wang, B.; Li, X.; Zhi, L. Hydrogen reduced graphene oxide/metal grid hybrid film: Towards high performance transparent conductive electrode for flexible electrochromic devices. *Carbon* **2015**, *81*, 232–238. [CrossRef]

57. Hong, J.Y.; Kim, W.; Choi, D.; Kong, J.; Park, H.S. Omni-Directionally Stretchable and Transparent Graphene Electrodes. *ACS Nano* **2016**, *10*, 9446–9455. [CrossRef] [PubMed]

58. Park, H.; Rowehl, J.A.; Kim, K.K.; Bulovic, V.; Kong, J. Doped graphene electrodes for organic solar cells. *Nanotechnology* **2010**, *21*, 505204. [CrossRef] [PubMed]

59. Groep, J.V.D.; Spinelli, P.; Polman, A. Transparent Conducting Silver Nanowire Networks. *Nano Lett.* **2012**, *12*, 3138–3144. [CrossRef] [PubMed]

60. Bergin, S.M.; Chen, Y.H.; Rathmell, A.R.; Charbonneau, P.; Li, Z.Y.; Wiley, B.J. The effect of nanowire length and diameter on the properties of transparent, conducting nanowire films. *Nanoscale* **2012**, *4*, 1996–2004. [CrossRef] [PubMed]

61. Mutiso, R.M.; Sherrott, M.C.; Rathmell, A.R.; Wiley, B.J.; Winey, K.I. Integrating simulations and experiments to predict sheet resistance and optical transmittance in nanowire films for transparent conductors. *ACS Nano* **2013**, *7*, 7654–7663. [CrossRef] [PubMed]

62. Jiu, J.; Nogi, M.; Sugahara, T.; Tokuno, T.; Araki, T.; Komoda, N.; Suganuma, K.; Uchida, H.; Shinozaki, K. Strongly adhesive and flexible transparent silver nanowire conductive films fabricated with a high-intensity pulsed light technique. *J. Mater. Chem.* **2012**, *22*, 23561–23567. [CrossRef]

63. Garnett, E.C.; Cai, W.; Cha, J.J.; Mahmood, F.; Connor, S.T.; Christoforo, M.G.; Cui, Y.; Mcgehee, M.D.; Brongersma, M.L. Self-limited plasmonic welding of silver nanowire junctions. *Nat. Mater.* **2012**, *11*, 241–249. [CrossRef] [PubMed] ·

64. Chung, C.H.; Song, T.B.; Bob, B.; Zhu, R.; Yang, Y. Solution-processed flexible transparent conductors composed of silver nanowire networks embedded in indium tin oxide nanoparticle matrices. *Nano Res.* **2012**, *5*, 805–814. [CrossRef]

65. Xiong, W.; Liu, H.; Chen, Y.; Zheng, M.; Zhao, Y.; Kong, X.; Wang, Y.; Zhang, X.; Kong, X.; Wang, P. Highly Conductive, Air-Stable Silver Nanowire@Iongel Composite Films toward Flexible Transparent Electrodes. *Adv. Mater.* **2016**, *28*. [CrossRef] [PubMed]

66. Menamparambath, M.M.; Ajmal, C.M.; Kim, K.H.; Yang, D.; Roh, J.; Park, H.C.; Chan, K.; Choi, J.Y.; Baik, S. Silver nanowires decorated with silver nanoparticles for low-haze flexible transparent conductive films. *Sci. Rep.* **2015**, *5*, 16371. [CrossRef] [PubMed]

67. Lee, M.S.; Lee, K.; Kim, S.Y.; Lee, H.; Park, J.; Choi, K.H.; Kim, H.K.; Kim, D.G.; Lee, D.Y.; Nam, S. High-performance, transparent, and stretchable electrodes using graphene-metal nanowire hybrid structures. *Nano Lett.* **2013**, *13*, 2814–2821. [CrossRef] [PubMed]

68. Liang, J.; Li, L.; Tong, K.; Ren, Z.; Hu, W.; Niu, X.; Chen, Y.; Pei, Q. Silver Nanowire Percolation Network Soldered with Graphene Oxide at Room Temperature and Its Application for Fully Stretchable Polymer Light-Emitting Diodes. *ACS Nano* **2014**, *8*, 1590–1600. [CrossRef] [PubMed]

69. Moon, I.K.; Kim, J.I.; Lee, H.; Hur, K.; Kim, W.C.; Lee, H. 2D Graphene Oxide Nanosheets as an Adhesive Over-Coating Layer for Flexible Transparent Conductive Electrodes. *Sci. Rep.* **2012**, *3*, 1112. [CrossRef]

70. Liu, Y.; Chang, Q.; Huang, L. Transparent, flexible conducting graphene hybrid films with a subpercolating network of silver nanowires. *J. Mater. Chem. C* **2013**, *1*, 2970–2974. [CrossRef]

71. Xu, Z.; Liu, Z.; Sun, H.; Gao, C. Highly Electrically Conductive Ag-Doped Graphene Fibers as Stretchable Conductors. *Adv. Mater.* **2013**, *25*, 3249–3253. [CrossRef] [PubMed]

72. Cui, F.; Yu, Y.; Dou, L.; Sun, J.; Yang, Q.; Schildknecht, C.; Schierle-Arndt, K.; Yang, P. Synthesis of Ultrathin Copper Nanowires Using Tris(trimethylsilyl)silane for High-Performance and Low-Haze Transparent Conductors. *Nano Lett.* **2015**, *15*, 7610–7615. [CrossRef] [PubMed]

73. Maurer, J.H.; Gonzalez-Garcia, L.; Reiser, B.; Kanelidis, I.; Kraus, T. Templated Self-Assembly of Ultrathin Gold Nanowires by Nanoimprinting for Transparent Flexible Electronics. *Nano Lett.* **2016**, *16*, 2921–2925. [CrossRef] [PubMed]

74. An, B.W.; Gwak, E.J.; Kim, K.; Kim, Y.C.; Jang, J.; Kim, J.Y.; Park, J.U. Stretchable, Transparent Electrodes as Wearable Heaters Using Nanotrough Networks of Metallic Glasses with Superior Mechanical Properties and Thermal Stability. *Nano Lett.* **2015**, *16*, 471–478. [CrossRef] [PubMed]

75. Han, B.; Pei, K.; Huang, Y.; Zhang, X.; Rong, Q.; Lin, Q.; Guo, Y.; Sun, T.; Guo, C.; Carnahan, D. Uniform Self-Forming Metallic Network as a High-Performance Transparent Conductive Electrode. *Adv. Mater.* **2014**, *26*, 873–877. [CrossRef] [PubMed]

76. Guo, C.F.; Sun, T.; Liu, Q.; Suo, Z.; Ren, Z. Highly stretchable and transparent nanomesh electrodes made by grain boundary lithography. *Nat. Commun.* **2014**, *5*, 155–164. [CrossRef] [PubMed]

77. Khan, A.; Lee, S.; Jang, T.; Xiong, Z.; Zhang, C.; Tang, J.; Guo, L.J.; Li, W.D. High-Performance Flexible Transparent Electrode with an Embedded Metal Mesh Fabricated by Cost-Effective Solution Process. *Small* **2016**, *12*, 3021–3030. [CrossRef] [PubMed]

78. Hsu, P.C.; Wang, S.; Wu, H.; Narasimhan, V.K.; Kong, D.; Ryoung, L.H.; Cui, Y. Performance enhancement of metal nanowire transparent conducting electrodes by mesoscale metal wires. *Nat. Commun.* **2013**, *4*, 2522. [CrossRef] [PubMed]

79. Seong, B.; Yoo, H.; Dat Nguyen, V.; Jang, Y.; Ryu, C.; Byun, D. Metal-mesh based transparent electrode on a 3-D curved surface by electrohydrodynamic jet printing. *J. Micromech. Microeng.* **2014**, *24*, 9. [CrossRef]

80. Bao, B.; Jiang, J.; Li, F.; Zhang, P.; Chen, S.; Yang, Q.; Wang, S.; Su, B.; Jiang, L.; Song, Y. Fabrication of Patterned Concave Microstructures by Inkjet Imprinting. *Adv. Funct. Mater.* **2015**, *25*, 3286–3294. [CrossRef]

81. Liu, Y.; Shen, S.; Hu, J.; Chen, L. Embedded Ag mesh electrodes for polymer dispersed liquid crystal devices on flexible substrate. *Opt. Express* **2016**, *24*, 25774–25784. [CrossRef] [PubMed]

82. Kiruthika, S.; Gupta, R.; Anand, A.; Kumar, A.; Kulkarni, G.U. Fabrication of Oxidation Resistant Metal Wire Network Based Transparent Electrodes by a Spray-Roll Coating Process. *ACS Appl. Mater. Interfaces* **2015**, *7*, 27215–27222. [CrossRef] [PubMed]

83. Lee, J.; Lee, P.; Lee, H.; Lee, D.; Lee, S.S.; Ko, S.H. Very long Ag nanowire synthesis and its application in a highly transparent, conductive and flexible metal electrode touch panel. *Nanoscale* **2012**, *4*, 6408–6414. [CrossRef] [PubMed]

84. SVG Optronics. Available online: http://www.svgoptronics.com/en/product.asp?id=29 (accessed on 29 December 2016).

85. Singh, M.; Rana, T.R.; Kim, S.Y.; Kim, K.; Yun, J.H.; Kim, J.H. Silver Nanowires Binding with Sputtered ZnO to Fabricate Highly Conductive and Thermally Stable Transparent Electrode for Solar Cell Applications. *ACS Appl. Mater. Interfaces* **2016**, *8*, 12764–12771. [CrossRef] [PubMed]

86. Wu, J.; Que, X.; Hu, Q.; Luo, D.; Liu, T.; Liu, F.; Russell, T.P.; Zhu, R.; Gong, Q. Organic Solar Cells: Multi-Length Scaled Silver Nanowire Grid for Application in Efficient Organic Solar Cells. *Adv. Funct. Mater.* **2016**, *26*, 4822–4828. [CrossRef]

87. Cai, G.; Darmawan, P.; Cui, M.; Wang, J.; Chen, J.; Magdassi, S.; Lee, P.S. Highly Stable Transparent Conductive Silver Grid/PEDOT:PSS Electrodes for Integrated Bifunctional Flexible Electrochromic Supercapacitors. *Adv. Energy Mater.* **2015**, *6*. [CrossRef]

88. Cheng, T.; Zhang, Y.; Yi, J.; Yang, L.; Zhang, J.D.; Lai, W.Y.; Huang, W. Inkjet Printed Flexible, Transparent and Aesthetic Energy Storage Devices based on PEDOT:PSS/Ag Grids Electrodes. *J. Mater. Chem. A* **2016**, *4*, 13754–13763. [CrossRef]

89. Lee, H.; Hong, S.; Lee, J.; Suh, Y.D.; Kwon, J.; Moon, H.; Kim, H.; Yeo, J.; Ko, S.H. Highly Stretchable and Transparent Supercapacitor by Ag–Au Core–Shell Nanowire Network with High Electrochemical Stability. *ACS Appl. Mater. Interfaces* **2016**, *8*, 15449–15458. [CrossRef] [PubMed]

micromachines

MDPI

Review

Materials, Mechanics, and Patterning Techniques for Elastomer-Based Stretchable Conductors

Xiaowei Yu, Bikram K. Mahajan, Wan Shou and Heng Pan *

Department of Mechanical and Aerospace Engineering, Missouri University of Science and Technology,
Rolla, MO 65401, USA; xy5w8@mst.edu (X.Y.); bkm6g7@mst.edu (B.K.M.); ws9n5@mst.edu (W.S.)
* Correspondence: hp5c7@mst.edu; Tel.: +1-573-341-4896

Academic Editors: Seung Hwan Ko, Daeho Lee and Zhigang Wu
Received: 29 October 2016; Accepted: 20 December 2016; Published: 27 December 2016

Abstract: Stretchable electronics represent a new generation of electronics that utilize soft, deformable elastomers as the substrate or matrix instead of the traditional rigid printed circuit boards. As the most essential component of stretchable electronics, the conductors should meet the requirements for both high conductivity and the capability to maintain conductive under large deformations such as bending, twisting, stretching, and compressing. This review summarizes recent progresses in various aspects of this fascinating and challenging area, including materials for supporting elastomers and electrical conductors, unique designs and stretching mechanics, and the subtractive and additive patterning techniques. The applications are discussed along with functional devices based on these conductors. Finally, the review is concluded with the current limitations, challenges, and future directions of stretchable conductors.

Keywords: stretchable conductors; elastomers; patterning techniques; direct printing; transfer printing

1. Introduction

Since the pioneering work on buckling phenomena in metallic film/elastomer composite [1], researchers started to pay attentions to the field of "stretchable electronics". Stretchable electronics represent a new generation of electronics that utilize soft, deformable elastomers as the substrate or matrix instead of rigid printed circuit boards. Unlike "flexible electronics" which utilize thin plastic substrates to endow the devices with conformity to bending and twisting [2,3], "stretchable electronics" are fascinating and challenging because of their capability to stretch and compress over a large scale, in addition to conformability [4–11].

Materials have been playing an important role in the development of stretchable conductors in the past two decades. Due to their high stretchability, elastomeric materials are commonly chosen as the supporting substrate or matrix for stretchable conductors and integrated device systems. Along with the traditional metallic conductors [1,12–14], several research groups have focused on developing conductive nanomaterials, including silver nanowires (AgNWs) [15], carbon nanotubes (CNTs) [16], etc. A substantial amount of work has been done to combine the rigid, brittle conductors with the flexible and soft elastomers, and overcoming the mismatches between their elastic behaviors. Unique designs and stretching mechanics have been proposed to harmonize the mismatches and integrate materials with widely different properties as one unique system. Patterning the conductors is one of the key techniques for the successful fabrication of stretchable electronic devices. The common patterning techniques for stretchable conductors include lithography [17,18], screen/stencil printing [19–21], direct printing [14,22–31], and transfer printing [32–34]. The patterning methods used in wafer-based electronics have also been adopted in stretchable electronics. Besides that, direct printing is emerging as an alternative to the conventional subtractive patterning method, with the recent development of additive manufacturing.

Despite their short history and limited scalability in manufacturing, it is quite clear that stretchable electronic devices would have a huge impact on future consumer electronics [4]. Because of their soft and conformable nature, stretchable electronics have shown great potential in biomedical engineering, e.g., epidermal electronic devices [18,35] and implantable devices [36,37]. Many studies have demonstrated the application of stretchable conductors in various devices and integrated systems, with functions such as sensing, display, and energy storage and conversion.

In this review, we focus on the basic building block of stretchable electronics, the elastomer-based stretchable conductors. The materials for elastomeric substrates and electrical conductors, and the mechanics of various types of stretchable conductors, are summarized in Sections 2 and 3. Section 4 reviews the patterning techniques used in stretchable conductors, followed by a brief discussion on their applications in Section 5. Finally, in Section 6, current limitations, challenges, and the future directions of stretchable conductors are discussed.

2. Materials

It has been recognized that there are two basic approaches to making stretchable conductors: to exploit structures that are stretchable and materials that are stretchable [4,8,10,38,39]. The first approach involves designing commonly used conductors into particular structures that are stretchable. The latter employs a simple design, but utilizes materials that are stretchable, such as carbon nanotubes, metallic nanowires, liquid metals, and ionic liquids. Some studies combine these two approaches to achieve enhanced stretchability [40–47].

In this section, the materials for elastomeric substrate and electrical conductors are summarized. The preparation, modification, and performance of the stretchable conductive materials are discussed. Along with these materials, the traditional metallic conductors are discussed, with an emphasis on the recently developed additive manufacturing methods.

2.1. Elastomers as the Substrate or Supporting Matrix

Elastomers are the fundamental supporting materials for stretchable conductors. They serve either as the matrix for conductive fillers and networks [48–50], or the substrate for the conductive films, tracks, and functional devices [13,18,51]. Elastomers not only provide the stretchability when the whole system is under strain, but also protect the devices and interconnects from large-scale deformation [12,52–54] and corrosive environments [55]. In epidermal electronics, elastomers facilitate the conformal contact of human skin with the sensing electrodes, and prevent the unnecessary contact with other electrical components [18,35,56]. Besides that, elastomers can also be used as functional dielectric materials [20,57–59].

The most important characteristic of the elastomers is their elasticity. Elastomers can easily sustain repetitive strain (typically larger than 100%), for thousands of stretch/release cycles. Desirable properties also include optical transparency, which facilitates optical applications in optoelectronics, photodetectors, light-emitting devices, solar cells, etc. [8,10]. They also have good biocompatibility, which makes the elastomers suitable for biomedical applications [60,61]. Elastomers typically have a high thermal expansion coefficient and have an inherent tendency of swelling in common solvents [62], which can result in device failures [63,64]. However, thermal expansion and induced deformation can be exploited to make 3D buckling structures [1], which is one of the most common approaches to fabricating stretchable conductors. Swelling can also be cleverly utilized to introduce wrinkling patterns on the substrate surface [65,66].

The most extensively used elastomer is polydimethylsiloxane (PDMS), a commercially available silicone-based elastomer [55]. PDMS is an ideal choice for a stretchable substrate/matrix as it is low-cost, transparent, biocompatible, easy to process, and permeable to air [67]. It is also known to have surface hydrophobicity, which can be modified to hydrophilicity by O_2 plasma [68,69], UV/ozone [24,70,71], and chemical treatment [68]. However, after exposure to air for a certain time, the surface reverts back to its hydrophobic nature. Such property has been utilized to modify the surface stiffness (Young's modulus) of PDMS [66,70,72], and to control the interfacial adhesion of PDMS with conductive materials [24,69]. Besides acting as the substrate/matrix, PDMS has been widely used in soft lithography [73], as an intermediate substrate (stamp) to transfer functional materials from the donor substrate to the target substrate.

Along with PDMS, some other elastomers are also employed as the substrate/matrix for various purposes. For example, Ecoflex [40,74,75] and polyurethane (PU) [49,76,77] are used in systems that require stretchability beyond the range of PDMS. Poly(isobutylene-co-isoprene) (IIR) is suitable for air- or moisture-sensitive applications [55] as it has a good gas diffusion barrier. Some rubbers, like polyolefin (POE) and poly (styrene-block-butadiene-block-styrene) (SBS), can be made into elastic fiber mats, taking advantage of their spinnability. These fiber mats have better stretchability than bulk elastomer blocks due to the network structure [78,79]. SBS fiber mat substrate is shown in Figure 1 as an example. Other reported elastomers include Dragon Skin [80], Solaris [81], acrylic elastomer [82–85], and nitrile butadiene rubber (NBR) [86]. Their elastic properties are summarized in Table 1.

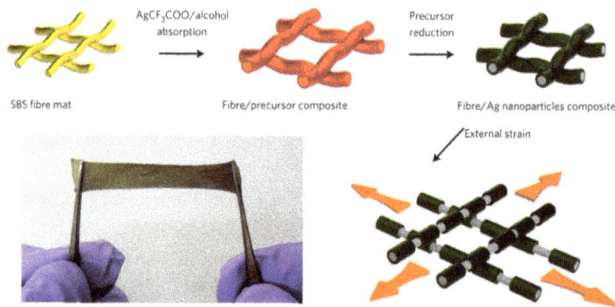

Figure 1. The schematic of fabrication process and the photograph of an SBS (styrene-block-butadiene-block-styrene) fiber mat-based stretchable conductor. Reprinted by permission from Macmillan Publishers Ltd.: *Nat. Nanotechnol.* [79], Copyright 2012.

Table 1. Elastomers for stretchable conductors and their elastic properties.

Material	Tensile Strength (MPa)	Maximum Strain (%)	Young's Modulus (MPa)	Reference
PDMS	6.25	120–160	2.05 (strain < 40%)	[52,55]
Ecoflex® 00-30	1.38	900	0.07 (strain = 100%)	[87]
Polyurethane	7.32	760	7.82 (initial)	[76]
IIR	3.51	170	0.41 (strain < 40%)	[55]
POE fiber mat	-	>600	-	[78]
SBS fiber mat	-	>530	0.47 (low strain)	[79]
Dragon Skin® 30	3.45	364	0.59 (strain = 100%)	[88]
Solaris	1.24	290	0.17 (strain = 100%)	[89]
Acrylic elastomer (VHB 4910)	0.69	-	-	[90]
NBR	10	250	-	[91]

2.2. Electrical Conductors

The conductive materials for stretchable conductors have more diversity compared to the elastomeric materials. They vary from inorganic to organic materials, solids to liquids, and bulk films to nano-scale percolation systems. In this section, the commonly used electrical conductors are introduced, including bulk metal films, metallic nanowires, carbon-based nanomaterials, conductive polymers, liquid metals, and ionic liquids. Bulk metal films have almost the same performance as the regular metallic conductors, with the exception that they are deformable owing to their ultra-small thickness (typically smaller than 1 μm) [12]. They have been extensively studied [1,12,13,92,93] and incorporated into various stretchable devices [4,35,94–96]. Recently, additive manufacturing has brought the bulk metal films and tracks to the research scope as materials for maskless, printable conductors [97–99]. Comparatively, conductive nanomaterials are relatively new, but their development has opened up a whole new area for stretchable conductors. Conductive nanomaterials include carbon-based nanomaterials and metallic nanowires. They can be incorporated into elastomeric materials by direct blending, or by fine architectures. Compared to metallic materials, polymeric materials have better intrinsic flexibility. The discovery of conductive polymers has brought another promising option for stretchable conductors. Liquid conductors, including liquid metals and ionic liquids, are advantageous in mobility in comparison to solid conductors. They can easily fit into the shape defined by the channel and have repeatable conductivity without forming any crack.

2.2.1. Bulk Metal Films and Tracks

Bulk metal films were studied extensively at the initial stages of development of stretchable conductors. It was found by Whitesides and his colleagues that disordered and ordered buckling (Figure 2a,b) occurred as the thin gold film evaporated on PDMS due to thermal contraction [1,72]. Researchers utilized this phenomenon to make stretchable gold stripes, which can be stretched up to 22%, as seen in Figure 2c [12]. Shortly after that, the lithographic method was incorporated and sinuous patterned gold tracks encapsulated in PDMS were fabricated, with a stretchability of 54% [13]. Gold was initially used because it is the most ductile metal. Later, emerging serpentine geometry designs were proposed to reduce strain in the metal [100–103]. Copper turned out to be the preferred material for serpentine-shape conductors due to lower cost and comparable conductivity.

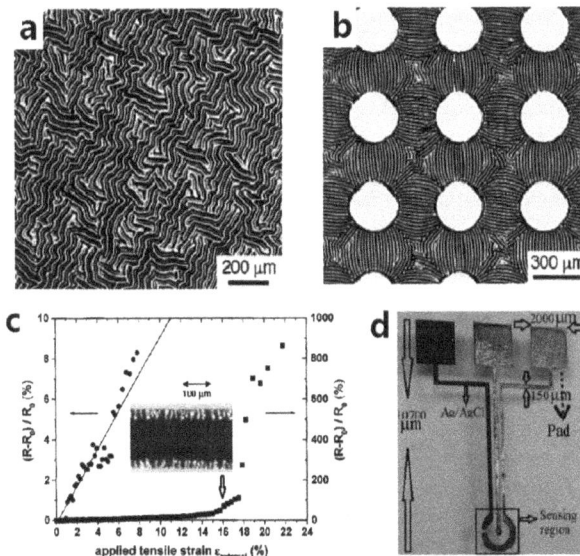

Figure 2. (a) Disordered buckling of Au film on flat, unconstraint PDMS (polydimethylsiloxane); (b) Ordered buckling of Au film on PDMS with flat circles (150 μm in radius) elevated by 10–20 μm; (a,b) are reprinted by permission from Macmillan Publishers Ltd.: *Nature* [1], Copyright 1998; (c) Normalized change in resistance of the Au stripe on PDMS under tensile strain. The left curve represents the linear behavior when the strain is below 8%. The inset is a photograph of the stripe stretched to 16.4%. Adapted from [12], with permission of AIP Publishing; (d) Ink-jet-printed Ag three-electrode electrochemical sensor. Reproduced from [25] with permission of The Royal Society of Chemistry.

With the development of additive manufacturing, direct writing techniques are showing up as an alternative to the traditional subtractive patterning methods. In the direct printing method, nanoparticle (NP) or metal-organic decomposition (MOD) inks are used, which are deposited from a printing head or nozzle to the substrate [98]. NP inks are usually a suspension of metallic nanoparticles dispersed in solvent. Silver NP inks are primarily used due to their high conductivity, good solvent-stability, air-inertness, and accessibility. MOD inks contain metal precursor and need additional steps (i.e., reduction) to generate metallic materials in-situ. In addition, sintering is required for both types of inks to evaporate solvent, decompose organic additives, and solidify the metallic particles. Figure 2d shows an optical image of an electrochemical sensor fabricated by ink-jet printing silver NP ink. The details of printing methods are introduced in Section 4.3.

2.2.2. Metallic Nanowires

Despite the high conductivity of bulk metal films and tracks, their utilization in stretchable electronics has to rely on special designs based on stretching mechanics due to their rigid and brittle nature. On the other hand, metallic nanowires are typically arranged into networks (shown in Figure 3a) where the nanowires with high aspect ratio are connected to each other and form the conductive paths. The intrinsic stretchability of the networks makes metallic nanowires an ideal candidate for stretchable conductors.

Silver nanowires (AgNWs) [20,40,59,82,104], copper nanowires (CuNWs) [85,105], and gold nanowires (AuNWs) [106,107] were reported serving as the conductors in stretchable electronics. AuNWs are costly, while CuNWs are easily oxidized in air [105]. Comparatively, AgNWs turned out to be the most popular choice because of their high conductivity and the well-established synthesis methods [108].

AgNWs were largely studied for transparent conductors [109,110] and were not incorporated into stretchable conductors until about five years ago [15,40,104]. AgNWs networks are usually fabricated by solution processes such as drop casting [20,59,82,104,111], vacuum filtration [40], and spray deposition [112,113]. Because elastomers are at risk to be attacked by the solvent and heat, AgNWs are typically solution-processed on glass or silicon, and then transferred to elastomeric substrate after evaporating the solvent. The performance of the AgNW-based stretchable conductors is related to various factors, such as the aspect ratio of AgNWs [40], the loading density of AgNWs [82,113], and interfacial adhesion between AgNWs and the substrates [82,112]. A short AgNW percolation network fabricated on 50% pre-stretched Ecoflex failed at 60% strain, while a long AgNW percolation network was able to retain conductivity with a strain of more than 80% [40]. Figure 3b shows the conductivity of an AgNW/PDMS conductor as a function of the areal density of AgNW, in which the conductivity increases with the loading density until the percolation threshold is reached [113]. The bonding between polyacrylate substrate and AgNWs was enhanced by introducing acrylic acid (AA) into the polymerization reaction. The sheet resistance of the AgNW/polyacrylate composite without AA was found to be 26% higher than that with AA [112].

Despite the high conductivity of the nanowires, the significant junction resistance makes the actual resistance of the whole nanowire network higher than expected [6]. The junction resistance can be effectively reduced by soldering. Thermal annealing, an effective method to get the junctions fused, was reported to reduce the sheet resistance of an AgNW network by one third [114]. Soldering with secondary functional materials or networks [74,115,116] would further improve the performance of the AgNW-based conductors under large tensile deformation. There are reports where AgNW networks were soldered by silver nanoparticles [116] and graphene oxides [74]. The conductor made by the former approach exhibited no obvious change in electrical conductivity, with strain up to 120%, and that by the latter was stretched to greater than 100% strain without losing electrical conductivity.

2.2.3. Carbon-Based Nanomaterials

Carbon-based nanomaterials are a burning topic in materials science due to their unusual structures, which endow these materials with excellent mechanical and electrical properties. The application of the carbon-based nanomaterials in stretchable conductors is dated back from 1982 [117], where carbon black particles were dispersed in a silicon rubber matrix to make conductive elastomers. With the progress in materials research, carbon nanotubes [77] and graphene [118] were subsequently introduced to stretchable conductors. However, carbon-based nanomaterials offer lower electrical conductivity compared to metallic materials [8]. The simplest way of increasing the conductivity is to increase the loading density of the carbon-based nanomaterials, at the cost of sacrificing some of the mechanical properties and the transparency. High loading density would also make it difficult to obtain a uniform dispersion in solvents. Hence, there is always a trade-off between electrical performance and stretchability, transparency, and ease of processing. However, researchers have found clever ways to bypass this challenge and obtain good conductivity as well as good stretchability.

Dispersing carbon black [50,117,119] or graphite [120] in the elastomer matrix is a simple but effective method to fabricate stretchable conductors. Such composites have relatively low conductivity, which is desirable for micro-heaters [50]. Single-wall carbon nanotubes (SWCNTs) were reported to be made into composite dispersion and aerogel. The elastic conductor made from the dispersion of SWCNT, ionic liquid, and fluorinated copolymer retained good conductivity up to 134% strain [121]. The resistivity of SWCNT aerogels was found to increase only slightly (14%) at a strain of 100% [122]. Multi-wall carbon nanotubes (MWCNTs) can be dispersed in elastomer using a similar method to carbon blacks. The MWCNT/elastomer composites were reported to have percolation concentration ranging from 0.2–5 wt % [48].

Besides direct dispersion, many designs and processing techniques to assemble CNTs into stretchable conductors have been proposed. The vertically grown MWCNTs were embedded into polyurethane, retaining their conductivity when stretched up to 300% strain [76], and into PDMS for a load-bearing antenna [123]. CNT ribbons drawn from vertically grown CNT forest [124] were embedded into PDMS and stretched to 120% strain. Figure 3c,d show the SEM images of the vertically aligned MWCNT forest and the CNT ribbons, respectively. The CNT ribbons were fabricated into "out-of-plane" and "lateral" buckling architectures, which further enhanced the stretchability [42,43]. Cross-stacked super-aligned CNT film possessed a high intrinsic tensile strain of more than 35%, which was further improved by embedding the film into PDMS [125]. In addition, gap-island networks made by the overlapping SWCNT films [41], spring-like CNT ropes [126], hierarchical reticulate SWCNT architecture [44], etc. were demonstrated for stretchable, transparent, and highly conductive carbon-based conductors.

Graphene is a two-dimensional material with hexagonal honeycomb architecture and boasts of very high conductivity, stretchability, and transparency [46,47]. Successful synthesis of graphene in a large scale by chemical vapor deposition facilitated its application in stretchable conductors [118,127]. Graphene film transferred to a pre-stretched PDMS substrate was able to sustain isotopic strain up to 12% without change in resistance, as seen in Figure 3e [118]. Graphene was further patterned into structures such as meshes [47], ribbons [46,128], and serpentine tracks [27,129] for better stretchability by using pre-designed metallic templates or photolithography.

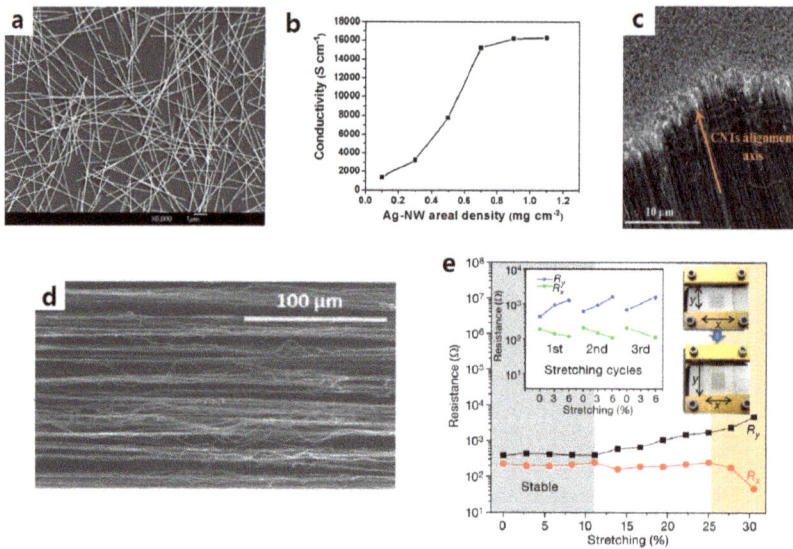

Figure 3. (**a**) SEM image of an AgNW (silver nanowire) network on glass [82]. © IOP Publishing. Reproduced with permission. All rights reserved; (**b**) Conductivity of an AgNW/PDMS conductor as a function of the areal density of AgNW. Reused from [113], under the terms of CC-BY license; (**c**) SEM image of vertically aligned CNT (carbon nanotube) forests. Reproduced from [130] with permission of The Royal Society of Chemistry; (**d**) SEM image of CNT ribbons drawn from a vertically aligned CNT forest. Reproduced from [131] with permission of The Royal Society of Chemistry; (**e**) Resistance of a graphene film transferred to a pre-stretched PDMS substrate isotopically stretched by 12%. The left inset shows the case in which the graphene film is transferred to an unstretched PDMS substrate. The right insets are photographs of the graphene film under vertical and horizontal strain. Reprinted by permission from Macmillan Publishers Ltd.: *Nature* [118], Copyright 2009.

2.2.4. Conductive Polymers

Compared to metals, polymers have better intrinsic flexibility. The discovery of conductive polymers has brought another promising option for stretchable conductors. Conductive polymers like polyaniline (PAni) [132], polypyrrole [133,134], poly(3,4-ethylenediox-ythiophene) (PEDOT) [49], and poly(3,4-ethylenediox-ythiophene):poly(styrene sulfonate) (PEDOT:PSS) [27,51,71] have been used in stretchable conductors despite some limitations such as environmental instability [135] and low stretchability [49,136]. PEDOT:PSS receives most of the attention among the conductive polymers because of its good conductivity and ease of processing. However, the breaking strain of a pristine PEDOT:PSS thin film is below 10% [136,137]. The stretchability of PEDOT:PSS film was improved by adding dimethylsulfoxide (DMSO) and Zonyl fluorosurfactant with the maximum strain of more than 20% [137]. The Zonyl modified PEDOT:PSS film on PDMS was found to form a 3D wavy buckling structure which promotes stretchability. The resistance of the film increased by a factor of only two at 50% strain [71]. Recently, a strain-insensitive PEDOT:PSS/acrylamide organogel was reported, which was stretched to more than 350% strain while it retained invariant resistance up to 50% strain [138]. Besides that, elastomeric conducting polyaniline networks with maximum elongation up to 90% ± 10% were fabricated by synthesizing polyaniline via molecular templates [139].

2.2.5. Liquid Metals and Ionic Liquids

The most common liquid metal used in stretchable conductors is eutectic gallium indium alloy (EGaIn, with 75.5% Ga and 24.5% In). The advantages of EGaIn include mobility, self-healability [140], and processability at room temperature [141]. Typically, it is injected into microchannels fabricated by molding elastomers [53]. Some other methods, such as microcontact printing, stencil printing [142], and photolithography [143,144], are also used to fabricate stretchable conductors based on liquid metals.

Ionic liquids are molten salts that have conductivity comparable to many organic electrolyte solutions at room-temperature and are stable up to 300–400 °C [145]. Ma et al. reported the integration of ionic liquids with various supporting materials, such as cotton fabric, rubber film, rubber band, and sponge, which exhibited exceptional performance with high conductivity at strain greater than 600% [78].

3. Mechanics of the Stretchable Conductive Structure

As introduced in last section, elastomers serve as the stretchable substrate or matrix for stretchable conductors, while conductive materials provide the electrical functionalities. However, they have huge differences in elastic properties. Substantial effort has been made to combine the rigid, brittle conductors with the flexible and soft elastomers. From the aspect of the mechanics, many fancy structures and clever designs were proposed to harmonize the mismatches and integrate materials with widely different properties as a whole system.

3.1. Blending

Blending is a pure mechanical process which is simple and cost-effective. Several material, like carbon blacks [50,117,119], silver particles/flakes [19,50,86,146,147], MWCNTs [48,77], and PEDOTs [49] were reported as blending into elastomers. Such composites can be analyzed by the percolation theory, where the electrical conductivity is determined by the concentration of the fillers [86]:

$$\sigma = \sigma_0 (V_f - V_c)^s \qquad (1)$$

where σ is the electrical conductivity, σ_0 is the conductivity of the conductive filler, V_f is the volumetric fraction of the filler, V_c is the volumetric fraction at the percolation threshold, and s is the fitting exponent. The percolation threshold is the key to control the electrical conductivity of composites. Table 2 gives the percolation thresholds of the conductive fillers in elastomer matrix (in weight percent) for comparison. As seen in the table, the percolation threshold has a strong dependence on the size and shape of the materials. Compared to the particle fillers, MWCNT (as a 1D nanomaterial), has lower percolation threshold due to its high aspect ratio, and thus is the ideal candidate for transparent stretchable conductors. The electrical properties of these conductive composites are also dependent on dispersion techniques [48] and properties of matrix [77,119].

Table 2. Conductive fillers for stretchable conductors based on blending.

Material	Size	Percolation Threshold (wt %)	Reference
Silver particles	1–2 μm	83	[50]
Carbon black	40–100 nm	10	[50]
MWCNT	Diameter: 60–100 nm Length: 5–15 μm	2	[48]
PEDOT	-	23	[49]

The blended conductive composites can be further patterned or used as conductive inks. Photo-patternable conductive PDMS was fabricated by blending in both conductive fillers and photosensitizers [146]. A composite ink of soluble silver salt and adhesive rubber was directly filled into ballpoint pens to write on different substrate to form adhesive conductive tracks [147]. Ag/PDMS composite ink was stencil printed and screen printed onto PDMS substrate to fabricate soft printed circuit boards (PCBs) [19].

3.2. Conductors in Microchannels

Filling the PDMS microchannels with liquid metals is another straightforward method of fabricating stretchable conductors. The PDMS microchannels are typically fabricated by molding top and bottom layers from the SU-8 master and adhering these two layers by oxygen plasma treat, heat, and pressure [144,148]. In the initial stages, solder was melted and filled into PDMS channels to fabricate stretchable conductors [149]. This process needed heating (up to 180 °C) and cooling steps, which was hence incompatible with heat-sensitive materials [141]. Filling liquid metals into the microchannels circumvents the heating during fabrication. The liquid nature endows them with excellent stretchability and self-healability [140]. Conductors fabricated by filling diamond-shape PDMS microchannels with EGaIn were stretched biaxially to 100% strain [144] while retaining their characteristics. Wetting of PDMS with the liquid metal was an issue which was solved effectively by putting on a metal layer [144] or applying silane adhesion promoter [149]. This strategy was also exploited to strengthen the brittle areas, such as the sharp corners of the diamond-shaped metallic conductive tracks [150].

3.3. Stretchable Network on/Embedded in Elastomers

As discussed in Section 2, the nanomaterials with high aspect ratio are typically arranged into networks possessing stretchability. Such network films are ready to be fabricated onto the elastomeric substrate without applying any other mechanics. Some nanomaterials can be directly solution- processed onto the modified elastomeric substrate [112]. For example, the self-assembly of functionalized SWCNTs onto PDMS substrate was reported [151]. Besides the direct fabrication of network films onto elastomer, transferring a solid film to the elastomeric substrate by conformal contact (due to the good adhesion between PDMS substrate and the conductive network film), or using the transfer media [27,41], provides alternative routes to obtain conductive network films on the elastomer surface [42,43]. In addition to these methods, vacuum suction is also an effective method to transfer conductive network film from membrane filter to elastomeric substrate [40,75,122].

The nanomaterial-based conductive networks can be embedded into the elastomer as well. The conductive networks are fabricated on a donor substrate. The conductive composites are obtained by coating the liquid elastomer precursor onto the donor substrate (thus the precursor infiltrates into the networks), curing and peeling off [82,104]. This method is also suitable for fabricating conductive networks with relatively large thickness, such as the vertically aligned CNT forests [45,76,123]. Figure 4a gives a schematic of such fabrication process of an AgNW/PDMS stretchable conductor.

3.4. Geometry Design

Designing the geometry of conductive tracks to reduce the strain in the metals, is an intelligent strategy in the fabrication of metallic film-based stretchable conductors. The realization of the geometry largely relies upon the patterning techniques, which are discussed in Section 4.

Here, we categorize those geometries which endow stretchability to the conductors into three types: polygonous, serpentine, and fractal designs. The polygonous designs make metallic tracks into connected polygonous networks, such as diamond [150,152] and hexagon (honeycomb) [153]. Serpentine designs include zigzag [100], half circle (U-shape) [101,154], sinuous [6,13], and horseshoe [100,101,155]. Among them, the horseshoe design provided the best stretchability [101]. Parameters such as the linewidth of metal, thickness of metal, and height/spacing ratio of geometry in the serpentine design play an important role in the performance of these conductive tracks. It was proven experimentally [13] and with the finite element method (FEM) analyses [101] that subdividing a wide conductive line into several thinner lines largely improved the stretchability and alleviated the stretching-induced stress. Analysis of postbuckling of the serpentine interconnects revealed that the elastic stretchability of the serpentine interconnects increased with decreased thickness [154,156].

The height/spacing ratio of geometry refers to the ratio of height (perpendicular to the stretching direction) to spacing width (parallel to the stretching direction). Increased height/spacing aspect ratio benefits the stretchcability when it is under a certain range [102,154]. In Figure 4b, the simulated uniaxial stretchability of the horseshoe patterns, with varied height/spacing ratio (arc length), is given [102]. As shown in the figure, the stretchability keeps increasing until the arc angle reaches 235°. Fractal designs can be perceived as the assembly of serpentine designs, which enable unusual mechanics with implications in stretchable device design [102]. In Figure 4c, the stretchability of a freestanding U-shape serpentine copper interconnect in symmetric buckling mode is demonstrated both experimentally and with FEM analyses. Some other representative geometry designs are shown in Figure 4d–g.

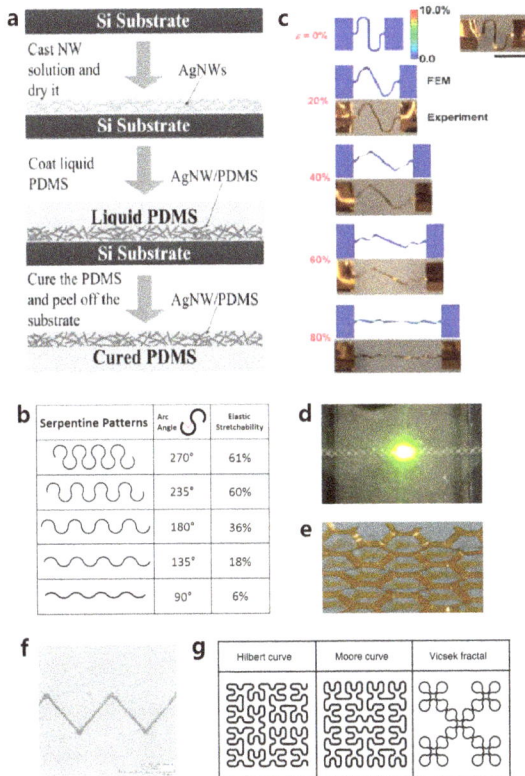

Figure 4. (**a**) Schematic of the fabrication process of an AgNW/PDMS stretchable conductor. Reprinted with permission from [21]. Copyright 2014 American Chemical Society; (**b**) Simulated uniaxial elastic stretchability for horseshoe patterns as a function of height/spacing ratio (arc angle). Reprinted by permission from Macmillan Publishers Ltd.: *Nat. Commun.* [102], Copyright 2014; (**c**) Experiments and FEM analyses on the symmetric buckling behavior of the serpentine interconnect with strain from 0% to 80% (scale bar: 1 mm). Reproduced from [156] with permission of The Royal Society of Chemistry; (**d**) Diamond-shaped stretchable interconnects based on liquid metal. Reprinted from [144], with the permission of AIP Publishing; (**e**) Honeycomb-patterned polyimide substrate at stretched state. Reprinted with permission from [153]. Copyright 2011 American Chemical Society; (**f**) Zigzag Cu conductive tracks on PDMS [100]. Reproduced with permission from © IOP Publishing. All rights reserved; (**g**) Three representative fractal patterns. Reprinted by permission from Macmillan Publishers Ltd.: *Nat. Commun.* [102], Copyright 2014.

3.5. Buckling

The buckling phenomenon was first found in the metal thin film on PDMS. Due to the large thermal expansion of PDMS, the subsequent cooling processes caused compressive stress in the metal film, which was calculated using Equation (2).

$$\sigma_0 = \frac{E_m (\alpha_p - \alpha_m)(T_D - T)}{(1 - \nu_m)} \tag{2}$$

where σ_0 was the compressive stress in the metal film. E, α and ν were referred to as Young's modulus, thermal expansion coefficient, and Poisson's ratio, respectively. The subscripts m and p referred to metal film and PDMS, respectively. T and T_D were the real-time temperature and the deposition temperature [1]. The compressive stress developed in the metal film led to buckling. It is worth mentioning that an interfacial adhesion layer of titanium or chromium was critical in making metal film compliant to PDMS. The buckling of metal film on a flat, unconstrained PDMS was however, uncontrollable. Ordered buckling was obtained by molding the PDMS surface with certain patterns or by partial modification of the surface [72].

The idea to utilize the pre-stretched elastomer to get buckled metal film further propelled the application of buckling strategy in stretchable conductors. In this approach, metal film was evaporated on 15% pre-stretched PDMS with an adhesion layer and patterned by standard lift-off process. Upon releasing the PDMS from the fixture, the metal film stayed compliant on the surface of PDMS and buckled. The as-prepared metal film was stretched, with stable resistance up to 25% [92]. Analytical models were proposed to predict the wavelength and amplitude of the buckling [39,70,157,158]. Interested readers can refer to a review on mechanical buckling [159], where the buckling phenomenon was studied in cases of compliant/small strain, compliant/large strain, and delaminating.

With the advents of more and more electrical conductive materials in stretchable conductors, the buckling strategy was explored in stretchable conductors based on CNTs [42,57], graphene [128], AgNWs [113], conductive polymers [133], and a combination of other interesting mechanics [160]. Figure 5b shows the SEM image of the buckling of an elastomer-infiltrated vertically aligned carbon nanotube (VACNT) film. This film was reported as having a very small resistance change (ratio of less than 6%) when stretched to the level of pre-strain (100%).

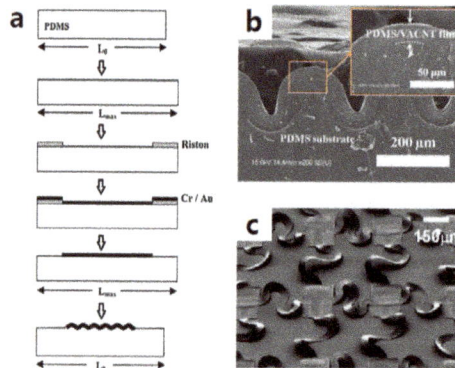

Figure 5. (a) The schematic of the fabrication of Au interconnects on PDMS substrate with buckling strategy. © 2004 IEEE. Reprinted, with permission, from [92]; **(b)** Cross-sectional SEM image of a VACNT/PDMS conductive film. The inset shows the magnified SEM image of the composite part. Reprinted with permission from [45]. Copyright 2014 American Chemical Society; **(c)** SEM image of an array of stretchable complementary metal-oxide-semiconductor (CMOS) inverters with non-coplanar bridges [160]. Copyright 2008 National Academy of Science.

Buckling was also reported without the pre-strain of substrate. The buckling of AgNW/PDMS film emerged upon the first stretch/release cycle. Such films could maintain stable conductivity of 5285 S·cm^{-1} at the strain range of 0%–50% [104]. A similar strategy was employed to fabricate laterally buckled CNT ribbons on PDMS [43].

3.6. Out-of-Plane Design

According to some studies of stretchable conductors, the conductive materials were supposed to have good adhesion to the substrate [10,161]. However, it was also pointed out that, for serpentine metallic lines on elastomer, the lines would deform out of plane of the substrate to alleviate the strain [101]. The strong bonding with the substrate would constrain such deformation and thus increase the plastic strain and in-plane shear strain [155].

An innovative approach to fabricate stretchable conductors is to make them free of substrate constraints. Figure 5c shows an array of stretchable complementary metal-oxide-semiconductor (CMOS) inverters with non-coplanar bridges. The islands with devices are well-attached to the substrate, while the interconnecting bridges are lifted off from the substrate to accommodate deformation [160]. Similar structures which contained anchored square islands and S-shaped suspensions were also reported [162].

3.7. Designs in Substrate

Generally, electrical conductors of a stretchable device are prone to failure. That is why most of the mechanics of strechability are focused on the electrical conductors. However, designing the substrate is also a good strategy to broaden the mechanics of stretchable electronics. One way to enhance stretchability of the substrate is to use two substrates with different Young's modulus so that the substrate with a higher modulus can sustain lesser strain [52,53]. For example, in a stretchable device array, stiff PDMS islands were used for device assembly, while soft Ecoflex substrate was designed to sustain the deformation. PDMS relief, consisting of raised islands and recessed trenches, was used as the substrate for rigid GaAs photovoltaics. It was found that stretching the substrate to an overall strain of 20% induced 123% strain in the trenches and only 0.4% strain at the islands [54]. Fabricating the substrate into a wavy shape would largely improve the stretchability of the whole system [24,125,160,163]. Besides that, a novel shape-memory shrinkable polymeric substrate utilized in the fabrication of highly stretchable gold films was reported [164].

4. Patterning Techniques

Patterning is one of the key techniques in the fabrication of both wafer-based electronics and stretchable conductors. The common patterning techniques for stretchable conductors are lithography, screen/stencil printing, direct printing, and transfer printing. The techniques from conventional wafer-based electronics, such as photolithography and lift-off, are subsequently adopted in stretchable electronics [17]. Besides, stencil printing and screen printing are employed to pattern features that do not require high complexity and superfine resolution [19–21]. With the development of additive manufacturing, direct printing emerges as an alternative to the conventional subtractive patterning method [14,22–31]. Transfer printing generally means to transfer functional materials from one substrate to another. With a mold-patterned elastomer stamp, those materials can be selectively transferred from donor substrate to the target substrate [32–34].

4.1. Lithography

Subtractive patterning techniques, such as photolithography and lift-off, are well-developed and dominatingly used in wafer-based electronics. However, in elastomer-based stretchable conductors, the elastomers are mostly incompatible with the chemicals and high energy beams used in these processes. Despite that, the patterning of devices still relies on lithography, where high resolution and complexity is required. In addition to the photo-defining and etching steps, an extra transfer step is needed. The patterning of the conductive materials is initially processed on regular substrates such as silicon wafer or glass and then transferred to elastomeric substrate. The transfer step can be done by directly coating the liquid elastomer precursor, curing, and peeling off (similar to the process shown in Figure 4a) [165]. It is also achievable by picking up and releasing the conducive materials with the help of an intermediate substrate [102,166,167].

4.2. Stencil/Screen Printing

Compared to the lithographic approach, direct deposition of conductive materials through a mask, stencil, or screen onto the substrate is relatively simple and cost-effective. It is suitable for applications that do not require high complexity and superfine resolution. Figure 6a is an SEM image of stencil-printed Ag/PDMS composite with the highest resolution achieved (linewidth of 150 μm with spacing of 100 μm).

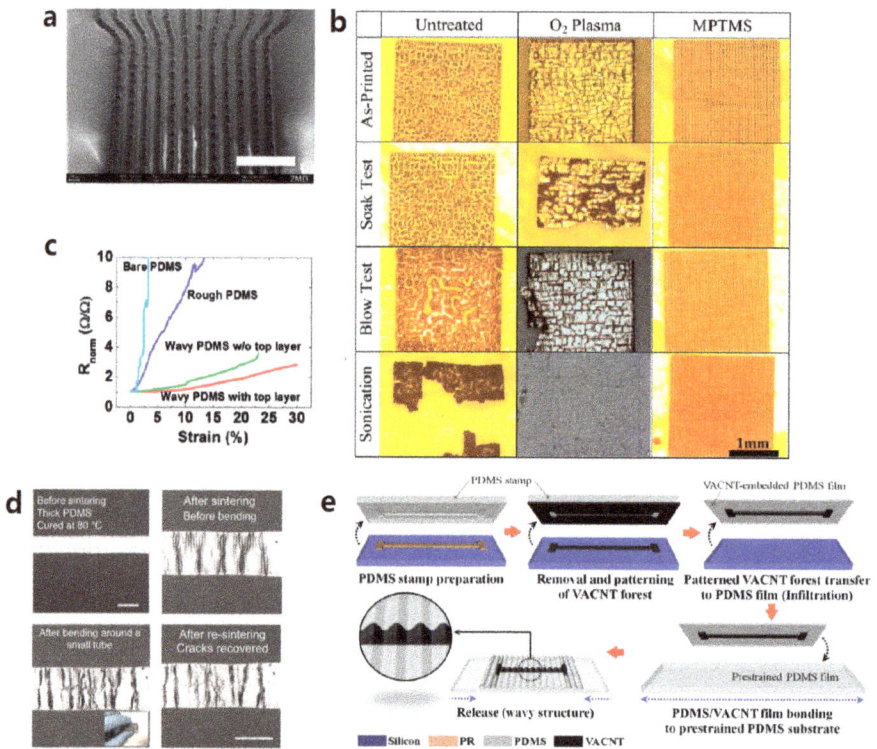

Figure 6. (**a**) SEM image of stencil printed Ag/PDMS tracks with the highest resolution achieved (scale bar: 1 mm). Reprinted by permission from Macmillan Publishers Ltd.: *Sci. Rep.* [19], Copyright 2014; (**b**) Photographs from adhesion tests of untreated, O_2-plasma treated, and adhesion-promoter (MPTMS)-treated 2 mm square ink-jet-printed Ag patterns on PDMS under destructive tests. © 2014 IEEE. Reprinted, with permission, from [69]; (**c**) Changes in resistance as a function of strain of the ink-jet-printed silver tracks on PDMS substrate with varied adhesion situations. Reprinted from [24], with the permission of AIP Publishing; (**d**) Morphologies of ink-jet-printed silver track on PDMS before and after sintering, after bending, and after re-sintering (scale bar: 100 μm) [26]. © IOP Publishing. Reproduced with permission. All rights reserved; (**e**) The schematic of fabrication process of PDMS/VACNT-film-based wavy-configured stretchable conductors. Reprinted with permission from [45]. Copyright 2014 American Chemical Society.

This approach is versatilely utilized with various conductive materials. Gold strip on PDMS was made by evaporating gold on PDMS laminated with patterned Dupont Riston photoresist film [92]. Liquid metals embedded into elastomer were fabricated using masked deposition [168]. A strain sensor array was fabricated by cross-folding screen printing AgNW stripes on PDMS [20]. Ag/PDMS composite was both stencil printed and screen printed to obtain soft PCBs [19]. PEDOT:PSS ink was stencil printed on Ecoflex to fabricate electrochemical sensors [51].

4.3. Direct Printing

Owing to the development of additive manufacturing, conductive inks can be directly printed on various substrates without physical contact. Various printing techniques, including laser-aided direct writing [22], ink-jet printing [23–26], and aerosol-jet printing [14,27–30] have been developed for printed electronics. The maskless processes are well suited for rapid prototyping. Also, they offer a cost-effective way to achieve large-area electronics production with minimal materials waste and without lengthy subtractive processes. Both planar and 3D architectures are achievable by laser-aided direct writing [169]. Ink-jet printing is a well-established technique for graphical printing and has been extensively used for printed electronics [170]. In this technique, the ink drop is driven by thermal, piezoelectric, or electrostatic actuation and delivered through a printing orifice to the substrate at a demanded area [171]. Aerosol-jet printing is a relatively new technique. Instead of printing the ink drop by drop, it prints the aerosol mist generated by ultrasonic or pneumatic atomizer. Aerosol-jet print has the capability of generating smaller feature size [30], as compared to the ink-jet printing and is compatible with 3D non-planar substrates [170]. The broad applications of such direct printing methods have been demonstrated by printing various conductive materials, such as metallic NP inks [24,69], CNTs [28,30,172], conductive polymers [27,173,174], and dielectric materials, such as ion gel [27,174] and polyimide [28], on not only regular flexible plastic substrate but also elastomeric substrates.

One of the issues in the direct printing technique is that of wetting and adhesion between elastomers and conductive inks. PDMS is known to have surface hydrophobicity, which can be modified to hydrophilicity by O_2 plasma [68,69], UV/ozone [24,70,71], and chemical treatment [68]. The surface treatment of PDMS imparts good wetting of PDMS with various inks and promotes the adhesion between PDMS and conductive materials. However, in some cases, such adhesion might not be strong enough and require further deposition of adhesion layer or blending of an adhesion promoter in the initial stages. Super-thin (5 nm) titanium or chromium adhesion layer was largely used to bond gold thin film with PDMS [1,12,92,162,175]. Poly-dopamine adhesion layer was employed to significantly modify the surface property of PDMS and enhance the adhesion between PDMS and spray-deposited AgNWs [112]. Zonyl, as a non-ionic fluorosurfactant, can be mixed in both elastomer precursors and conductive inks, which enhances the interfacial adhesion between elastomers and various conductive materials [7,51,71,176]. Silane adhesion promoter MPTMS (3-mercaptopropyl trimethoxysilane) is widely used in PDMS-based stretchable conductors. It was used as the adhesion promoter between PDMS and molten liquid solder [149], evaporated gold film [165,177], and printed metallic inks as well [25,69]. In Figure 6b, the adhesion tests of the ink-jet- printed silver patterns on PDMS with blank treatment, O_2 plasma treatment, and MPTMS treatment are compared. Only the MPTMS-treated sample survived the three destructive tests. [69]. The influence of the adhesion on the stretchability of ink-jet-printed silver tracks is demonstrated in Figure 6c. The PDMS rough surface obtained from UV/ozone treatment provided better adhesion, thus making the tracks more stretchable. Besides that, a top PDMS layer encapsulated the silver tracks, making them compliant to the wavy PDMS and hence further enhancing the stretchability [24].

Besides the adhesion issue, many other problems, such as the swelling [62], thermal hardening, and large thermal expansion of the elastomers, still exist in the direct printing of conductive inks [26], resulting in poor performance of printed stretchable conductors. Cracking [26,31] and low stretchability are some of the resulting issues found in printed stretchable conductors [14]. As an example, Figure 6d shows the morphologies of ink-jet-printed silver track on PDMS before and after sintering, after bending, and after re-sintering [26], in which the silver track cracked as a result of the thermal expansion problem. All these problems and issues account for the current situation that most of the stretchable conductors are not directly fabricated on elastomers. This should be addressed in order to realize the direct printing of stretchable conductors.

4.4. Transfer Printing

The challenges involved in direct printing of stretchable conductors are discussed above. Transfer printing is an alternative approach, which effectively circumvents such difficulties. In a general sense, transfer means switching the substrate of the conductive patterns. As mentioned in Section 4.1, coating the liquid elastomer precursor onto the donor substrate, curing, and peeling off is a typical transfer process [82,104,165].

Transfer is also achievable via intermediate media. Poly(methyl methacrylate) (PMMA) is one of the common transfer media [46,132]. Conductive materials can be embedded into PMMA like in PDMS and transferred to the target substrate, followed by dissolving PMMA in hot acetone. Some other materials, such as functional tapes, were also reported as the transfer media. Water-soluble tape is an adhesive tape which is dissolvable in water. Conductive materials on donor substrate could be archived by using such tape and releasing them on the target substrate. The tape would totally disappear when dipped in water, leaving only conductive materials on the target substrate [27,41,56]. Thermal release tape is an adhesive tape which loses its adhesion upon heating. Thermal release tape was reported as a temporary holder to transfer NW-based devices [178]. Besides the transfer media, a sacrificial layer is an essential assistant in transfer printing techniques. A sacrificial layer is typically fabricated as a structural support between the donor substrate and the conductive materials. This layer can be removed by solvent or etchant, thus lifting off the conductive materials and making them easily transferable. The commonly used sacrificial materials include copper [101], gold [177,179], PMMA [160], SiO_2 [46,157], and hard-baked photoresist [164,180].

Transfer by PDMS stamp is extensively used in soft lithography, where the conductive materials can be selectively transferred from donor substrate to the target substrate using a mold-patterned PDMS stamp. Transfer printing by such PDMS stamps is either additive or subtractive [32]. The key to the PDMS stamp-based transfer process is controlling the adhesion. The adhesion of the PDMS stamp to the donor or the target substrate can be controlled kinetically, owing to the rate-independent viscoelastic property of PDMS [33]. The adhesion can also be controlled by manipulating the surface chemistry and the interfacial properties of the PDMS stamp. [34].

As an example of transfer printing, Figure 6e illustrates the fabrication of stretchable vertical-aligned carbon nanotube (VACNT) film on PDMS substrate, in which two transfer processes were involved. A subtractive transfer printing technique was used to remove the undesired part of VACNT film. The remaining part was transferred by an intermediate PDMS substrate to a pre-stretched target PDMS substrate.

5. Applicable Devices

Stretchable electronics have been developed into a unique and emerging field of electronics. They contravene the common idea that people would attach to electronics—stiff circuit board and cold metals. They make electronics closer to human. Because of the soft, attachable nature, stretchable electronics show great potential in developing strong intimacy with humans, like conforming to the skin (epidermal electronics) [18,35] and implanting into the body [36,37]. They have an enormous potential in future electronics for biomedical use, such as lightweight physiological monitors and in personalized healthcare [60,61]. Stretchable conductors are the most essential building block of stretchable electronic devices. They could simply serve as interconnects, or as functional electrodes. In this section, the applications of the stretchable conductors are discussed by demonstrating the functional devices based on them.

5.1. Sensors

Stretchable sensors are probably the most widely studied type of devices in the recent development of stretchable electronics. Stretchable sensors with various functions have been fabricated, including strain sensors [181], pressure sensors [20,59,107,182], temperature sensors [18,183], gas sensors [132,184], and UV sensors [53]. These sensors detect the changes in the outside environment and reflect these changes in the electrical properties, such as resistance and capacitance. Sensors are also fabricated based on the performance of devices such as transistors [152] and antennas [21].

Stretchable strain sensors have largely extended the sensing range beyond those of traditional non-stretchable strain sensors from below 1% [136] to several hundred percent [181]. The strain sensors are mostly resistive or capacitive. Stretchable conductors with reversible and repeatable resistance change under strain are suitable for fabricating resistive strain sensors. Carbon-based materials are widely used in this application due to their intrinsic piezoresistivity [46,47]. Network structures fabricated from AgNWs [111] or CNTs [41] are also piezoresistive, with mechanisms of local disconnections and tunneling effects [181]. In capacitive strain sensors, an elastomeric dielectric layer is sandwiched between two stretchable electrodes. At certain strain, the decreased width and thickness of the dielectric layer results in changes of capacitance. Both AgNWs and CNTs were reported as the electrode materials [20,59,80]. Figure 7a shows a resistive stretchable strain sensor based on AgNW/PDMS nanocomposites. More information on the stretchable strain sensors can be found in a recent review [181].

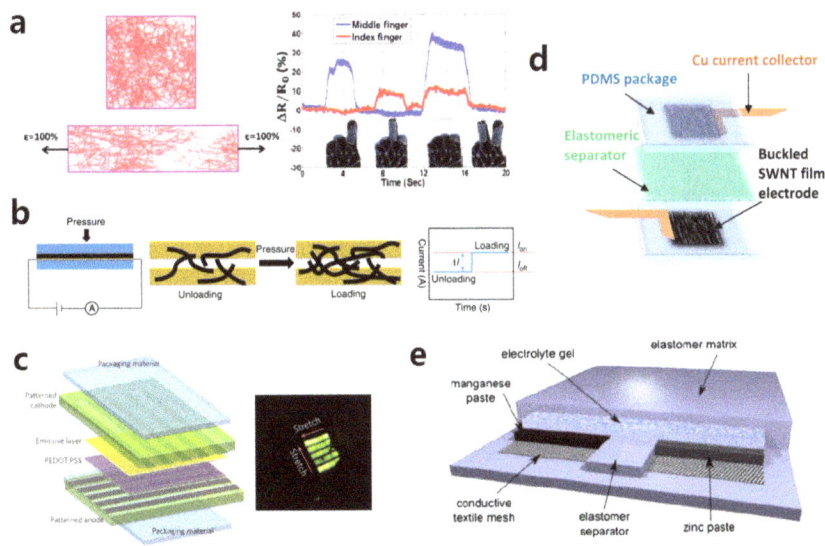

Figure 7. (**a**) Schematic of AgNW network at relaxed and stretched states and the illustration of the motion detect function of an AgNW/PDMS strain sensor. Reprinted with permission from [111]. Copyright 2014 American Chemical Society; (**b**) Schematic of the sensing mechanism of an AuNW-based pressure sensor. Reprinted by permission from Macmillan Publishers Ltd.: *Nat. Commun.* [107], Copyright 2014; (**c**) Schematic and photograph of an encapsulated fully stretchable elastomeric polymer light-emitting devices (EPLED) display with 25 pixels. Reprinted by permission from Macmillan Publishers Ltd.: *Nat. Photonics* [185], Copyright 2013; (**d**) Schematic of the components of a dynamic SWCNT (Single-wall carbon nanotubes)-based stretchable supercapacitor. Reprinted with permission from [57]. Copyright 2012 American Chemical Society; (**e**) Schematic of a rechargeable stretchable alkaline manganese cell. Reproduced from [83] with permission of The Royal Society of Chemistry.

Stretchable pressure sensors detect the vertical deformation caused by pressure and send out differential electrical signals. Capacitive pressure sensors have a similar mechanism to capacitive strain sensors. When the capacitor is under pressure, the distance between the two electrodes decreases, causing changes in capacitance. Capacitive sensing arrays based on AgNWs and various elastomers have been demonstrated with the functionality as strain sensor, pressure sensor, and touch sensor [20,59]. A stretchable resistive pressure sensor was fabricated by sandwiching AuNW coated tissue sheet with an interdigitated electrode between two PDMS sheets, as shown in Figure 7b. The applied pressure was determined by measuring the changes in resistance caused by the different loading of AuNWs on the electrode [107].

Noble metals, such as gold and platinum, are resistance-sensitive under varied temperature. Both traditional temperature sensors and stretchable temperature sensors take advantage of this property. However, these noble metals are not intrinsically stretchable. To incorporate them into stretchable temperature sensors, mechanics to enhance their stretchability were employed. Platinum meander electrodes were employed as temperature sensors in an integrated epidermal electronic system [18]. The buckling strategy was used to fabricate stretchable temperature sensors by transferring gold strips onto pre-stretched PDMS [183].

In most of these stretchable sensors, rough structures, such as NW networks and buckled films/strips, are well encapsulated or packaged with elastomers [20,59,183]. However, the influence of such rough structures on the device performance has not yet been thoroughly discussed, and this is expected to be addressed in the near future.

5.2. Light-Emitting Circuits

Light-emitting circuits on elastomeric substrates are designed to meet the demands of flexible, wearable, and foldable displays and light sources. In these circuits, stretchable conductors serve as interconnects between the power supplies and the light-emitting devices. EGaIn liquid metals embedded into soft elastomeric substrate were used as interconnects to light-emitting diode (LED) arrays on stiff elastomeric islands [53]. A stretchable heart-shaped light-emitting circuit was fabricated by installing LED lights onto hand-written stretchable tracks from an adhesive silver-based conductive ink [147]. In addition, serpentine graphene interconnect bridges were also used as interconnects for microscale inorganic LED array [129].

Besides interconnects, stretchable conductors were used as electrodes in thin film polymer light-emitting devices as well. Liang et al. demonstrated an elastomeric polymer light-emitting devices (EPLED) fabricated by sandwiching an emissive polymer layer in AgNW-PU composite electrodes. Such a structure was further developed into a stretchable EPLED display array of 25 pixels. The schematic and photograph of the display array are shown in Figure 7c [185].

5.3. Transistors

Stretchable conductors are utilized in stretchable transistors as interconnects and electrodes for gate, source, and drain. An SWCNT-based elastic conductor was used to form interconnections between contact pads of an organic transistor-based active matrix [121]. Polymer-sorted and unsorted CNTs were used as the semiconductor and conductor, respectively, in a thermoplastic polyurethane (TPU)-based stretchable transistor [186]. Graphene/AgNW hybrid conductors were used as the drain/source electrodes of an oxide semiconductor transistor array [187]. A stretchable and transparent transistor array was fabricated with patterned graphene thin film as the semiconducting channel and source/drain electrodes. The patterned graphene was transferred to the rubber substrate, followed by aerosol printing of ion gel and conductive polymer as the gate dielectric and gate electrode, respectively. Such graphene transistors showed hole and electron mobilities of 1188 ± 136 and 422 ± 52 cm^2/(V·s), respectively, with stable operation at strain up to 5%, even after more than 1000 cycles [27]. Highly stretchable polymer transistors made entirely of stretchable components were fabricated on SBS fiber mat with stacked Au nanosheets as electrodes and poly(3-hexylthiophene) (P3HT) nanofibers as channel material. The transistors provided a hole mobility of 18 cm^2/(V·s) at a strain of 70% [188]. Recently, an intrinsically stretchable organic thin film transistor was reported with a healable polymeric semiconductor and CNT/PEDOT:PSS electrodes. Such a transistor was able to sustain a field-effect mobility of 1.12 cm^2/(V·s) at a large strain of 100% [189].

5.4. Energy Devices

Almost all of the stretchable electronic devices with amazing functionality were connected to and driven by external power sources. Low cost and compatible internal power supplies for independent stretchable electronic systems are in great demand, as they can facilitate the real-life usages of these stretchable devices. Energy devices are basically divided into energy storage devices and energy conversion devices. Although many research works focused on stretchable energy devices such as supercapacitors [57,58,125,132], batteries [83,84,134,166], solar cells [176], and other environmental energy harvests, challenges still remain to fabricate stretchable energy devices with high performance and reliability. Here, in this part of the review, these devices are briefly introduced with an emphasis on the usage of stretchable conductors. Two recent review papers about stretchable energy storage and conversion devices can be referred to for further information [38,190].

Supercapacitors are commonly used as energy storage devices. Carbon-based nanomaterials are ideal for this application due to their good stability, conductivity, and high surface area. Buckled SWCNT films on PDMS substrate as electrode, together with organic electrolyte and a separator layer were used to fabricate stretchable supercapacitors [57,58]. The schematic of the components of an SWCNT-based supercapacitor is shown in Figure 7d. In addition to this, highly-aligned CNT sheets were fabricated from vertically-grown CNT forest and further used in an all-solid supercapacitor [191]. Micro-supercapacitor array based on P3Ain-wrapped MWCNTs was employed to drive stretchable graphene gas sensors [132]. Graphene microribbons were also utilized as the electrode materials in stretchable supercapacitors [125].

Unlike supercapacitors, batteries store energy in forms of chemical energy. Stretchable batteries of various structures and shapes have been fabricated with several materials till now. Buckled polypyrrole macrofilm cathodes on elastomeric substrate were reported for battery applications [134]. Porous CNT/PDMS nanocomposites fabricated by phase separation were used as the anode in a flexible lithium-ion battery [192]. Intrinsically stretchable and rechargeable cells were fabricated by embedding chemically reactive pastes (zinc anode and manganese dioxide cathode) and electrolyte gels into an elastomer matrix, as shown in Figure 7e [83,84]. Although the abovementioned batteries could easily provide energy by electro-chemical process, they still needed to get recharged by physical contact with external power sources. Xu et al. demonstrated stretchable battery arrays with an integrated wireless recharging system. The wireless recharging system, consisting of wireless coil, Schottky diode, and parallel capacitor, was designed to recharge these batteries without direct physical contact [166].

Energy harvesters exploit environmental energy and generate power for the stretchable devices. Solar cells are one of the well-developed types of energy harvesters. To fabricate stretchable solar cells, various stretching mechanics were exploited to incorporate the stiff photovoltaic materials. Stretchable organic solar cells were fabricated by depositing a PEDOT:PSS electrode, organic photovoltaic materials, and EGaIn as the top contact, sequentially on a pre-stretched PDMS substrate. The 3D wavy buckles appeared in the film upon releasing and imparted the solar cells with reversible stretchability [176,193]. Stretchable dye-sensitized solar cells were fabricated using elastic conductive fiber by spinning MWCNT sheets on rubber fiber. Modified titanium wire was wound onto the fiber as the working electrode, followed by coating with photoactive materials. The devices had open-circuit voltages of 0.71 V and energy conversion efficiency of 7.13% [194]. Besides the solar cell, other energy harvesters that have potential but have not yet been well-studied in stretchable electronics include piezoelectric motion energy harvesters [195,196], RF energy harvesters [60,197,198], the thermal energy harvester [199], and triboelectric energy harvester [200].

6. Summary and Outlook

In this review, we went through the materials, mechanics, and patterning techniques of stretchable conductors and the applicable devices. Stretchable conductors are the basic building block of stretchable electronic devices. Their compliant, deformable virtues have changed the common idea that people would have on rigid silicon-based electronics and opened up a new direction of the next generation of electronics. They have enormous potential in biomedical engineering, which has been demonstrated in many research works [4,18,60,61,94,167,181,201]. Besides that, they have application in broad fields such as wearable communication devices [202], prosthetic electronic skins [203], soft robotics [204], and interactive human–machine interfaces [205]. We foresee a bright future for stretchable electronics as they bring electronics closer to humans.

Despite the huge advancements in stretchable electronics so far, there still exist some issues and challenges that should be solved in order to realize the use of stretchable electronics in everyday life. First, the long-term stable performance of these stretchable electronics, which would be influenced by the stability of conductors and the aging of substrates [206–208], has not yet been demonstrated comprehensively. Second, as many stretchable electronics are aimed at biomedical applications, their biocompatibility should be examined extensively. Although the biocompatibility of some materials has been demonstrated [209], however, thermal management of the stretchable devices, where the heat generated may cause tissue lesioning, are relatively less studied [210,211]. Third, the large mismatches in Young's modulus, elongation at break, and thermal expansion of various materials used in stretchable electronics have posed great challenges in integration with the systems. Several stretching mechanics have been proposed which effectively remedy these integration challenges. However, problems such as cracking and delamination are still reported when the system reaches a certain strain. More fundamental understanding of the causes of failure and interfacial properties between heterogeneous materials is needed in order to achieve better performance of novel stretchable devices. Fourth, some of the stretching mechanisms and structures are hard to implement in large-scale manufacturing, such as pre-strain-induced buckling and out-of-plane design. Efforts are still required to figure out cost-effective, scalable fabrication routes with high reliability, repeatability, and precision for the purpose of commercialization. For stretchable electronics based on nanomaterials, repeatability of the device's performance from batch to batch largely relies on the uniformity of the nanomaterials, which is also challenging in itself. Lastly, despite stretchable electronics with manifold functional devices having been demonstrated, integrated internal power supplies are still in great demand for self-powering stretchable electronic systems.

Based on the current trend, the possible future directions of stretchable electronics are foreseeable. First, the diversity of materials used would continuously expand. The rate of discovery and development of new materials are beyond our expectations. Take graphene as an example, it only took about five years from its first discovery [212] to its first utilization in stretchable conductors [118]. For sure, more and more organic, inorganic, conducting, semiconducting, and dielectric materials would provide opportunities in stretchable electronics. As for the substrate and matrix, besides the widely used PDMS, there are emerging elastomeric materials such as rubber fiber mats. Second, the design of discrete devices and their integration in stretchable systems would grow quickly. Stretchable electronics with sensing functions have gained considerable progresses. However, devices with other functions such as memory [213] and communication [202,214] are still in infancy. There is also an urgent demand for energy devices, and many research groups are working to address this issue. Third, with the increase of complexity and resolution of devices, higher requirements for patterning techniques are expected. Direct printing, as an additive manufacturing method, would satisfy such requirement and offer low cost and high speed in both prototyping and manufacturing. It might be a solution for cost-effective and scalable fabrication of stretchable electronics.

Acknowledgments: This research was supported financially by Interdisciplinary Intercampus Funding Program (IDIC) of University of Missouri System, University of Missouri Research Board (UMRB), Intelligent System Center (ISC) and Material Research Center (MRC) at Missouri University of Science and Technology.

Author Contributions: Xiaowei Yu, Bikram K. Mahajan and Wan Shou conducted the review. Xiaowei Yu wrote the manuscript. Bikram. K. Mahajan and Heng Pan revised and proofread the manuscript. Heng Pan guided on the organization and content.

Conflicts of Interest: The authors declare no conflict of interest.

References

1. Bowden, N.; Brittain, S.; Evans, A.G.; Hutchinson, J.W.; Whitesides, G.M. Spontaneous formation of ordered structures in thin films of metals supported on an elastomeric polymer. *Nature* **1998**, *393*, 146–149.
2. Huang, D.; Liao, F.; Molesa, S.; Redinger, D.; Subramanian, V. Plastic-compatible low resistance printable gold nanoparticle conductors for flexible electronics. *J. Electrochem. Soc.* **2003**, *150*, G412–G417. [CrossRef]
3. Nathan, A.; Ahnood, A.; Cole, M.T.; Lee, S.; Suzuki, Y.; Hiralal, P.; Bonaccorso, F.; Hasan, T.; Garcia-Gancedo, L.; Dyadyusha, A.; et al. Flexible electronics: The next ubiquitous platform. *Proc. IEEE* **2012**, *100*, 1486–1517. [CrossRef]
4. Rogers, J.A.; Someya, T.; Huang, Y. Materials and mechanics for stretchable electronics. *Science* **2010**, *327*, 1603–1607. [CrossRef] [PubMed]
5. Wagner, S.; Bauer, S. Materials for stretchable electronics. *MRS Bull.* **2012**, *37*, 207–213. [CrossRef]
6. Park, M.; Park, J.; Jeong, U. Design of conductive composite elastomers for stretchable electronics. *Nano Today* **2014**, *9*, 244–260. [CrossRef]
7. Cheng, T.; Zhang, Y.; Lai, W.-Y.; Huang, W. Stretchable thin-film electrodes for flexible electronics with high deformability and stretchability. *Adv. Mater.* **2015**, *27*, 3349–3376. [CrossRef] [PubMed]
8. Yao, S.; Zhu, Y. Nanomaterial-enabled stretchable conductors: Strategies, materials and devices. *Adv. Mater.* **2015**, *27*, 1480–1511. [CrossRef] [PubMed]
9. Wang, Y.; Li, Z.; Xiao, J. Stretchable thin film materials: Fabrication, application and mechanics. *J. Electron. Packag.* **2016**, *138*, 020801. [CrossRef]
10. McCoul, D.; Hu, W.; Gao, M.; Mehta, V.; Pei, Q. Recent advances in stretchable and transparent electronic materials. *Adv. Electron. Mater.* **2016**, *2*, 1500407. [CrossRef]
11. Harris, K.D.; Elias, A.L.; Chung, H.-J. Flexible electronics under strain: A review of mechanical characterization and durability enhancement strategies. *J. Mater. Sci.* **2016**, *51*, 2771–2805. [CrossRef]
12. Lacour, S.P.; Wagner, S.; Huang, Z.; Suo, Z. Stretchable gold conductors on elastomeric substrates. *Appl. Phys. Lett.* **2003**, *82*, 2404–2406. [CrossRef]
13. Gray, D.S.; Tien, J.; Chen, C.S. High-conductivity elastomeric electronics. *Adv. Mater.* **2004**, *16*, 393–397. [CrossRef]
14. Rabinowitz, J.; Fritz, G.; Kumar, P.; Lewis, P.; Miller, M.; Dineen, A.; Gray, C. Electrical characterization of traditional and aerosol jet printed conductors under tensile strain. *MRS Adv.* **2016**, *1*, 15–20. [CrossRef]
15. Yang, C.; Gu, H.; Lin, W.; Yuen, M.M.; Wong, C.P.; Xiong, M.; Gao, B. Silver nanowires: From scalable synthesis to recyclable foldable electronics. *Adv. Mater.* **2011**, *23*, 3052–3056. [CrossRef] [PubMed]
16. Park, S.; Vosguerichian, M.; Bao, Z. A review of fabrication and applications of carbon nanotube film-based flexible electronics. *Nanoscale* **2013**, *5*, 1727–1752. [CrossRef] [PubMed]
17. Pease, R.F.; Chou, S.Y. Lithography and other patterning techniques for future electronics. *Proc. IEEE* **2008**, *96*, 248–270. [CrossRef]
18. Kim, D.-H.; Lu, N.; Ma, R.; Kim, Y.-S.; Kim, R.-H.; Wang, S.; Wu, J.; Won, S.M.; Tao, H.; Islam, A.; et al. Epidermal electronics. *Science* **2011**, *333*, 838–843. [CrossRef] [PubMed]
19. Larmagnac, A.; Eggenberger, S.; Janossy, H.; Vörös, J. Stretchable electronics based on Ag-PDMS composites. *Sci. Rep.* **2014**, *4*, 7254. [CrossRef] [PubMed]
20. Yao, S.; Zhu, Y. Wearable multifunctional sensors using printed stretchable conductors made of silver nanowires. *Nanoscale* **2014**, *6*, 2345–2352. [CrossRef] [PubMed]
21. Song, L.; Myers, A.C.; Adams, J.J.; Zhu, Y. Stretchable and reversibly deformable radio frequency antennas based on silver nanowires. *ACS Appl. Mater. Interfaces* **2014**, *6*, 4248–4253. [CrossRef] [PubMed]
22. Lee, M.-T.; Lee, D.; Sherry, A.; Grigoropoulos, C.P. Rapid selective metal patterning on polydimethylsiloxane (PDMS) fabricated by capillarity-assisted laser direct write. *J. Micromech. Microeng.* **2011**, *21*, 095018. [CrossRef]

23. Sun, J.; Jiang, J.; Bao, B.; Wang, S.; He, M.; Zhang, X.; Song, Y. Fabrication of bendable circuits on a polydimethylsiloxane (PDMS) surface by inkjet printing semi-wrapped structures. *Materials* **2016**, *9*, 253. [CrossRef]

24. Chung, S.; Lee, J.; Song, H.; Kim, S.; Jeong, J.; Hong, Y. Inkjet-printed stretchable silver electrode on wave structured elastomeric substrate. *Appl. Phys. Lett.* **2011**, *98*, 153110. [CrossRef]

25. Wu, J.; Wang, R.; Yu, H.; Li, G.; Xu, K.; Tien, N.C.; Roberts, R.C.; Li, D. Inkjet-printed microelectrodes on PDMS as biosensors for functionalized microfluidic systems. *Lab Chip* **2015**, *15*, 690–695. [CrossRef] [PubMed]

26. Kim, Y.; Ren, X.; Kim, J.W.; Noh, H. Direct inkjet printing of micro-scale silver electrodes on polydimethylsiloxane (PDMS) microchip. *J. Micromech. Microeng.* **2014**, *24*, 115010. [CrossRef]

27. Lee, S.-K.; Kim, B.J.; Jang, H.; Yoon, S.C.; Lee, C.; Hong, B.H.; Rogers, J.A.; Cho, J.H.; Ahn, J.-H. Stretchable graphene transistors with printed dielectrics and gate electrodes. *Nano Lett.* **2011**, *11*, 4642–4646. [CrossRef] [PubMed]

28. Wang, K.; Chang, Y.-H.; Zhang, C.; Wang, B. Conductive-on-demand: Tailorable polyimide/carbon nanotube nanocomposite thin film by dual-material aerosol jet printing. *Carbon* **2016**, *98*, 397–403. [CrossRef]

29. Aga, R.; Jordan, C.; Aga, R.S.; Bartsch, C.M.; Heckman, E.M. Metal electrode work function modification using aerosol jet printing. *IEEE Electron Device Lett.* **2014**, *35*, 1124–1126. [CrossRef]

30. Zhao, D.; Liu, T.; Park, J.G.; Zhang, M.; Chen, J.-M.; Wang, B. Conductivity enhancement of aerosol-jet printed electronics by using silver nanoparticles ink with carbon nanotubes. *Microelectron. Eng.* **2012**, *96*, 71–75. [CrossRef]

31. Robinson, A.P.; Minev, I.; Graz, I.M.; Lacour, S.P. Microstructured silicone substrate for printable and stretchable metallic films. *Langmuir* **2011**, *27*, 4279–4284. [CrossRef] [PubMed]

32. Carlson, A.; Bowen, A.M.; Huang, Y.; Nuzzo, R.G.; Rogers, J.A. Transfer printing techniques for materials assembly and micro/nanodevice fabrication. *Adv. Mater.* **2012**, *24*, 5284–5318. [CrossRef] [PubMed]

33. Meitl, M.A.; Zhu, Z.-T.; Kumar, V.; Lee, K.J.; Feng, X.; Huang, Y.Y.; Adesida, I.; Nuzzo, R.G.; Rogers, J.A. Transfer printing by kinetic control of adhesion to an elastomeric stamp. *Nat. Mater.* **2006**, *5*, 33–38. [CrossRef]

34. Loo, Y.-L.; Willett, R.L.; Baldwin, K.W.; Rogers, J.A. Interfacial chemistries for nanoscale transfer printing. *J. Am. Chem. Soc.* **2002**, *124*, 7654–7655. [CrossRef] [PubMed]

35. Wang, S.; Li, M.; Wu, J.; Kim, D.-H.; Lu, N.; Su, Y.; Kang, Z.; Huang, Y.; Rogers, J.A. Mechanics of epidermal electronics. *J. Appl. Mech.* **2012**, *79*, 031022. [CrossRef]

36. Kim, S.J.; Cho, K.W.; Cho, H.R.; Wang, L.; Park, S.Y.; Lee, S.E.; Hyeon, T.; Lu, N.; Choi, S.H.; Kim, D.-H. Stretchable and transparent biointerface using cell-sheet-graphene hybrid for electrophysiology and therapy of skeletal muscle. *Adv. Funct. Mater.* **2016**, *26*, 3207–3217. [CrossRef]

37. Huang, X.; Leduc, C.; Ravussin, Y.; Li, S.; Davis, E.N.; Song, B.; Li, D.; Xu, K.; Accili, D.; Wang, Q.; et al. A differential dielectric affinity glucose sensor. *Lab Chip* **2014**, *14*, 294–301. [CrossRef] [PubMed]

38. Yan, C.; Lee, P.S. Stretchable energy storage and conversion devices. *Small* **2014**, *10*, 3443–3460. [CrossRef] [PubMed]

39. Song, J.; Wang, S. Theory for Stretchable Interconnects. In *Stretchable Electronics*; Someya, T., Ed.; Wiley-VCH Verlag GmbH & Co. KGaA: Weinheim, Germany, 2012; pp. 1–29.

40. Lee, P.; Lee, J.; Lee, H.; Yeo, J.; Hong, S.; Nam, K.H.; Lee, D.; Lee, S.S.; Ko, S.H. Highly stretchable and highly conductive metal electrode by very long metal nanowire percolation network. *Adv. Mater.* **2012**, *24*, 3326–3332. [CrossRef] [PubMed]

41. Yamada, T.; Hayamizu, Y.; Yamamoto, Y.; Yomogida, Y.; Izadi-Najafabadi, A.; Futaba, D.N.; Hata, K. A stretchable carbon nanotube strain sensor for human-motion detection. *Nat. Nanotechnol.* **2011**, *6*, 296–301. [CrossRef] [PubMed]

42. Xu, F.; Wang, X.; Zhu, Y.; Zhu, Y. Wavy ribbons of carbon nanotubes for stretchable conductors. *Adv. Funct. Mater.* **2012**, *22*, 1279–1283. [CrossRef]

43. Zhu, Y.; Xu, F. Buckling of aligned carbon nanotubes as stretchable conductors: A new manufacturing strategy. *Adv. Mater.* **2012**, *24*, 1073–1077. [CrossRef] [PubMed]

44. Cai, L.; Li, J.; Luan, P.; Dong, H.; Zhao, D.; Zhang, Q.; Zhang, X.; Tu, M.; Zeng, Q.; Zhou, W.; Xie, S. Highly transparent and conductive stretchable conductors based on hierarchical reticulate single-walled carbon nanotube architecture. *Adv. Funct. Mater.* **2012**, *22*, 5238–5244. [CrossRef]

45. Shin, U.H.; Jeong, D.-W.; Kim, S.-H.; Lee, H.W.; Kim, J.-M. Elastomer-infiltrated vertically aligned carbon nanotube film-based wavy-configured stretchable conductors. *ACS Appl. Mater. Interfaces* **2014**, *6*, 12909–12914. [CrossRef] [PubMed]

46. Wang, Y.; Yang, R.; Shi, Z.; Zhang, L.; Shi, D.; Wang, E.; Zhang, G. Super-elastic graphene ripples for flexible strain sensors. *ACS Nano* **2011**, *5*, 3645–3650. [CrossRef] [PubMed]

47. Wang, Y.; Wang, L.; Yang, T.; Li, X.; Zang, X.; Zhu, M.; Wang, K.; Wu, D.; Zhu, H. Wearable and highly sensitive graphene strain sensors for human motion monitoring. *Adv. Funct. Mater.* **2014**, *24*, 4666–4670. [CrossRef]

48. Huang, Y.Y.; Terentjev, E.M. Tailoring the electrical properties of carbon nanotube-polymer composites. *Adv. Funct. Mater.* **2010**, *20*, 4062–4068. [CrossRef]

49. Hansen, T.S.; West, K.; Hassager, O.; Larsen, N.B. Highly stretchable and conductive polymer material made from poly(3,4-ethylenedioxythiophene) and polyurethane elastomers. *Adv. Funct. Mater.* **2007**, *17*, 3069–3073. [CrossRef]

50. Niu, X.; Peng, S.; Liu, L.; Wen, W.; Sheng, P. Characterizing and patterning of PDMS-based conducting composites. *Adv. Mater.* **2007**, *19*, 2682–2686. [CrossRef]

51. Bandodkar, A.J.; Nuñez-Flores, R.; Jia, W.; Wang, J. All-printed stretchable electrochemical devices. *Adv. Mater.* **2015**, *27*, 3060–3065. [CrossRef] [PubMed]

52. Kubo, M.; Li, X.; Kim, C.; Hashimoto, M.; Wiley, B.J.; Ham, D.; Whitesides, G.M. Stretchable microfluidic radiofrequency antennas. *Adv. Mater.* **2010**, *22*, 2749–2752. [CrossRef] [PubMed]

53. Yoon, J.; Hong, S.Y.; Lim, Y.; Lee, S.-J.; Zi, G.; Ha, J.S. Design and fabrication of novel stretchable device arrays on a deformable polymer substrate with embedded liquid-metal interconnections. *Adv. Mater.* **2014**, *26*, 6580–6586. [CrossRef] [PubMed]

54. Lee, J.; Wu, J.; Shi, M.; Yoon, J.; Park, S.-I.; Li, M.; Liu, Z.; Huang, Y.; Rogers, J.A. Stretchable GaAs photovoltaics with designs that enable high areal coverage. *Adv. Mater.* **2011**, *23*, 986–991. [CrossRef] [PubMed]

55. Vohra, A.; Filiatrault, H.L.; Amyotte, S.D.; Carmichael, R.S.; Suhan, N.D.; Siegers, C.; Ferrari, L.; Davidson, G.J.E.; Carmichael, T.B. Reinventing butyl rubber for stretchable electronics. *Adv. Funct. Mater.* **2016**, *26*, 5222–5229. [CrossRef]

56. Huang, X.; Liu, Y.; Kong, G.W.; Seo, J.H.; Ma, Y.; Jang, K.-I.; Fan, J.A.; Mao, S.; Chen, Q.; Li, D.; et al. Epidermal radio frequency electronics for wireless power transfer. *Microsyst. Nanoeng.* **2016**, *2*, 16052. [CrossRef]

57. Li, X.; Gu, T.; Wei, B. Dynamic and galvanic stability of stretchable supercapacitors. *Nano Lett.* **2012**, *12*, 6366–6371. [CrossRef] [PubMed]

58. Yu, C.; Masarapu, C.; Rong, J.; Wei, B.; Jiang, H. Stretchable supercapacitors based on buckled single-walled carbon nanotube macrofilms. *Adv. Mater.* **2009**, *21*, 4793–4797. [CrossRef] [PubMed]

59. Hu, W.; Niu, X.; Zhao, R.; Pei, Q. Elastomeric transparent capacitive sensors based on an interpenetrating composite of silver nanowires and polyurethane. *Appl. Phys. Lett.* **2013**, *102*, 083303. [CrossRef]

60. Kim, D.-H.; Ghaffari, R.; Lu, N.; Rogers, J.A. Flexible and stretchable electronics for biointegrated devices. *Annu. Rev. Biomed. Eng.* **2012**, *14*, 113–128. [CrossRef] [PubMed]

61. Zhang, Y. Mechanics and designs of stretchable bioelectronics. In *Stretchable Bioelectronics for Medical Devices and Systems*; Rogers, J.A., Ghaffari, R., Kim, D.-H., Eds.; Springer: Zug, Switzerland, 2016; pp. 53–68.

62. Lee, J.N.; Park, C.; Whitesides, G.M. Solvent compatibility of poly(dimethylsiloxane)-based microfluidic devices. *Anal. Chem.* **2003**, *75*, 6544–6554. [CrossRef] [PubMed]

63. Fortunato, G.; Pecora, A.; Maiolo, L. Polysilicon thin-film transistors on polymer substrates. *Mater. Sci. Semicond. Process.* **2012**, *15*, 627–641. [CrossRef]

64. Dangla, R.; Gallaire, F.; Baroud, C.N. Microchannel deformations due to solvent-induced PDMS swelling. *Lab Chip* **2010**, *10*, 2972–2978. [CrossRef] [PubMed]

65. Chan, E.P.; Crosby, A.J. Fabricating microlens arrays by surface wrinkling. *Adv. Mater.* **2006**, *18*, 3238–3242. [CrossRef]

66. Chan, E.P.; Crosby, A.J. Spontaneous formation of stable aligned wrinkling patterns. *Soft Matter* **2006**, *2*, 324–328. [CrossRef]

67. Mata, A.; Fleischman, A.J.; Roy, S. Characterization of polydimethylsiloxane (PDMS) properties for biomedical micro/nanosystems. *Biomed. Microdevices* **2005**, *7*, 281–293. [CrossRef] [PubMed]

68. Bodas, D.; Khan-Malek, C. Hydrophilization and hydrophobic recovery of PDMS by oxygen plasma and chemical treatment-An SEM investigation. *Sens. Actuators B Chem.* **2007**, *123*, 368–373. [CrossRef]

69. Wu, J.; Roberts, R.C.; Tien, N.C.; Li, D. Inkjet Printed Silver Patterning on PDMS to Fabricate Microelectrodes for Microfluidic Sensing. In Proceedings of the 2014 IEEE Sensors, Valencia, Spain, 2–5 November 2014; pp. 1100–1103.

70. Efimenko, K.; Rackaitis, M.; Manias, E.; Vaziri, A.; Mahadevan, L.; Genzer, J. Nested self-similar wrinkling patterns in skins. *Nat. Mater.* **2005**, *4*, 293–297. [CrossRef] [PubMed]

71. Lipomi, D.J.; Lee, J.A.; Vosgueritchian, M.; Tee, B.C.-K.; Bolander, J.A.; Bao, Z. Electronic properties of transparent conductive films of PEDOT:PSS on stretchable substrates. *Chem. Mater.* **2012**, *24*, 373–382. [CrossRef]

72. Huck, W.T.S.; Bowden, N.; Onck, P.; Pardoen, T.; Hutchinson, J.W.; Whitesides, G.M. Ordering of spontaneously formed buckles on planar surfaces. *Langmuir* **2000**, *16*, 3497–3501. [CrossRef]

73. Xia, Y.; Whitesides, G.M. Soft lithography. *Annu. Rev. Mater. Sci.* **1998**, *28*, 153–184. [CrossRef]

74. Lee, P.; Ham, J.; Lee, J.; Hong, S.; Han, S.; Suh, Y.D.; Lee, S.E.; Yeo, J.; Lee, S.S.; Lee, D.; et al. Highly stretchable or transparent conductor fabrication by a hierarchical multiscale hybrid nanocomposite. *Adv. Funct. Mater.* **2014**, *24*, 5671–5678. [CrossRef]

75. Woo, J.Y.; Kim, K.K.; Lee, J.; Kim, J.T.; Han, C.-S. Highly conductive and stretchable Ag nanowire/carbon nanotube hybrid conductors. *Nanotechnology* **2014**, *25*, 285203. [CrossRef] [PubMed]

76. Shin, M.K.; Oh, J.; Lima, M.; Kozlov, M.E.; Kim, S.J.; Baughman, R.H. Elastomeric conductive composites based on carbon nanotube forests. *Adv. Mater.* **2010**, *22*, 2663–2667. [CrossRef] [PubMed]

77. Engel, J.; Chen, J.; Chen, N.; Pandya, S.; Liu, C. Multi-walled carbon nanotube filled conductive elastomers: Materials and application to micro transducers. In Proceedings of the 19th IEEE International Conference on Micro Electro Mechanical Systems, Istanbul, Turkey, 22–26 January 2006; pp. 246–249.

78. Ma, Z.; Su, B.; Gong, S.; Wang, Y.; Yap, L.W.; Simon, G.P.; Cheng, W. Liquid-wetting-solid strategy to fabricate stretchable sensors for human-motion detection. *ACS Sens.* **2016**, *1*, 303–311. [CrossRef]

79. Park, M.; Im, J.; Shin, M.; Min, Y.; Park, J.; Cho, H.; Park, S.; Shim, M.-B.; Jeon, S.; Chung, D.-Y.; et al. Highly stretchable electric circuits from a composite material of silver nanoparticles and elastomeric fibres. *Nat. Nanotechnol.* **2012**, *7*, 803–809. [CrossRef] [PubMed]

80. Cai, L.; Song, L.; Luan, P.; Zhang, Q.; Zhang, N.; Gao, Q.; Zhao, D.; Zhang, X.; Tu, M.; Yang, F.; et al. Super-stretchable, transparent carbon nanotube-based capacitive strain sensors for human motion detection. *Sci. Rep.* **2013**, *3*, 3048. [CrossRef] [PubMed]

81. Huang, X.; Cheng, H.; Chen, K.; Zhang, Y.; Zhang, Y.; Liu, Y.; Zhu, C.; Ouyang, S.-C.; Kong, G.-W.; Yu, C.; et al. Epidermal impedance sensing sheets for precision hydration assessment and spatial mapping. *IEEE Trans. Biomed. Eng.* **2013**, *60*, 2848–2857. [CrossRef] [PubMed]

82. Hu, W.; Niu, X.; Li, L.; Yun, S.; Yu, Z.; Pei, Q. Intrinsically stretchable transparent electrodes based on silver-nanowire–crosslinked-polyacrylate composites. *Nanotechnology* **2012**, *23*, 344002. [CrossRef] [PubMed]

83. Kettlgruber, G.; Kaltenbrunner, M.; Siket, C.M.; Moser, R.; Graz, I.M.; Schwödiauer, R.; Bauer, S. Intrinsically stretchable and rechargeable batteries for self-powered stretchable electronics. *J. Mater. Chem. A* **2013**, *1*, 5505–5508. [CrossRef]

84. Kaltenbrunner, M.; Kettlgruber, G.; Siket, C.; Schwödiauer, R.; Bauer, S. Arrays of ultracompliant electrochemical dry gel cells for stretchable electronics. *Adv. Mater.* **2010**, *22*, 2065–2067. [CrossRef] [PubMed]

85. Wu, J.; Zang, J.; Rathmell, A.R.; Zhao, X.; Wiley, B.J. Reversible sliding in networks of nanowires. *Nano Lett.* **2013**, *13*, 2381–2386. [CrossRef] [PubMed]

86. Chun, K.-Y.; Oh, Y.; Rho, J.; Ahn, J.-H.; Kim, Y.-J.; Choi, H.R.; Baik, S. Highly conductive, printable and stretchable composite films of carbon nanotubes and silver. *Nat. Nanotechnol.* **2010**, *5*, 853–857. [CrossRef] [PubMed]

87. Ecoflex® Series—Smooth-On. Available online: https://www.smooth-on.com/tb/files/ECOFLEX_SERIES_TB.pdf (accessed on 10 October 2016).

88. Dragon Skin® Series—Smooth-On. Available online: https://www.smooth-on.com/tb/files/DRAGON_SKIN_SERIES_TB.pdf (accessed on 10 October 2016).

89. Solaris®—Smooth-On. Available online: https://www.smooth-on.com/tb/files/Solaris_TB.pdf (accessed on 10 October 2016).

90. VHBTM Tape Specialty Tapes—3M.com. Available online: http://multimedia.3m.com/mws/media/9866950/3m-vhb-tape-specialty-tapes.pdf (accessed on 10 October 2016).

91. Nitrile Butadiene Rubber (NBR)—Romac. Available online: http://www.romac.com/Submittals/RUBBER/NBR-SUB.pdf (accessed on 10 October 2016).

92. Lacour, S.P.; Jones, J.; Suo, Z.; Wagner, S. Design and performance of thin metal film interconnects for skin-like electronic circuits. *IEEE Electron Device Lett.* **2004**, *25*, 179–181. [CrossRef]

93. Lacour, S.P.; Jones, J.; Wagner, S.; Li, T.; Suo, Z. Stretchable interconnects for elastic electronic surfaces. *Proc. IEEE* **2005**, *93*, 1459–1467. [CrossRef]

94. Yeo, W.-H.; Kim, Y.-S.; Lee, J.; Ameen, A.; Shi, L.; Li, M.; Wang, S.; Ma, R.; Jin, S.H.; Kang, Z.; et al. Multifunctional epidermal electronics printed directly onto the skin. *Adv. Mater.* **2013**, *25*, 2773–2778. [CrossRef] [PubMed]

95. Kim, D.-H.; Ahn, J.-H.; Choi, W.M.; Kim, H.-S.; Kim, T.-H.; Song, J.; Huang, Y.Y.; Liu, Z.; Lu, C.; Rogers, J.A. Stretchable and foldable silicon integrated circuits. *Science* **2008**, *320*, 507–511. [CrossRef] [PubMed]

96. He, J.; Nuzzo, R.G.; Rogers, J.A. Inorganic materials and assembly techniques for flexible and stretchable electronics. *Proc. IEEE* **2015**, *103*, 619–632. [CrossRef]

97. Leenen, M.A.M.; Arning, V.; Thiem, H.; Steiger, J.; Anselmann, R. Printable electronics: Flexibility for the future. *Phys. Status Solidi A* **2009**, *206*, 588–597. [CrossRef]

98. Perelaer, J.; Smith, P.J.; Mager, D.; Soltman, D.; Volkman, S.K.; Subramanian, V.; Korvink, J.G.; Schubert, U.S. Printed electronics: The challenges involved in printing devices, interconnects, and contacts based on inorganic materials. *J. Mater. Chem.* **2010**, *20*, 8446–8453. [CrossRef]

99. Lee, K.J.; Jun, B.H.; Kim, T.H.; Joung, J. Direct synthesis and inkjetting of silver nanocrystals toward printed electronics. *Nanotechnology* **2006**, *17*, 2424–2428. [CrossRef]

100. Van der Sluis, O.; Hsu, Y.Y.; Timmermans, P.H.M.; Gonzalez, M.; Hoefnagels, J.P.M. Stretching-induced interconnect delamination in stretchable electronic circuits. *J. Phys. D Appl. Phys.* **2011**, *44*, 34008. [CrossRef]

101. Gonzalez, M.; Axisa, F.; Bulcke, M.V.; Brosteaux, D.; Vandevelde, B.; Vanfleteren, J. Design of metal interconnects for stretchable electronic circuits. *Microelectron. Reliab.* **2008**, *48*, 825–832. [CrossRef]

102. Fan, J.A.; Yeo, W.-H.; Su, Y.; Hattori, Y.; Lee, W.; Jung, S.-Y.; Zhang, Y.; Liu, Z.; Cheng, H.; Falgout, L.; et al. Fractal design concepts for stretchable electronics. *Nat. Commun.* **2014**, *5*, 3266. [CrossRef] [PubMed]

103. Hsu, Y.-Y.; Gonzalez, M.; Bossuyt, F.; Vanfleteren, J.; de Wolf, I. Polyimide-enhanced stretchable interconnects: Design, fabrication, and Characterization. *IEEE Trans. Electron Devices* **2011**, *58*, 2680–2688. [CrossRef]

104. Xu, F.; Zhu, Y. Highly conductive and stretchable silver nanowire conductors. *Adv. Mater.* **2012**, *24*, 5117–5122. [CrossRef] [PubMed]

105. Han, S.; Hong, S.; Ham, J.; Yeo, J.; Lee, J.; Kang, B.; Lee, P.; Kwon, J.; Lee, S.S.; Yang, M.-Y.; et al. Fast plasmonic laser nanowelding for a Cu-nanowire percolation network for flexible transparent conductors and stretchable electronics. *Adv. Mater.* **2014**, *26*, 5808–5814. [CrossRef] [PubMed]

106. Chen, Y.; Ouyang, Z.; Gu, M.; Cheng, W. Mechanically strong, optically transparent, giant metal superlattice nanomembranes from ultrathin gold nanowires. *Adv. Mater.* **2013**, *25*, 80–85. [CrossRef] [PubMed]

107. Gong, S.; Schwalb, W.; Wang, Y.; Chen, Y.; Tang, Y.; Si, J.; Shirinzadeh, B.; Cheng, W. A wearable and highly sensitive pressure sensor with ultrathin gold nanowires. *Nat. Commun.* **2014**, *5*, 3132. [CrossRef] [PubMed]

108. Langley, D.; Giusti, G.; Mayousse, C.; Celle, C.; Bellet, D.; Simonato, J.-P. Flexible transparent conductive materials based on silver nanowire networks: A review. *Nanotechnology* **2013**, *24*, 452001. [CrossRef] [PubMed]

109. Hu, L.; Kim, H.S.; Lee, J.-Y.; Peumans, P.; Cui, Y. Scalable coating and properties of transparent, flexible, silver nanowire electrodes. *ACS Nano* **2010**, *4*, 2955–2963. [CrossRef] [PubMed]

110. De, S.; Higgins, T.M.; Lyons, P.E.; Doherty, E.M.; Nirmalraj, P.N.; Blau, W.J.; Boland, J.J.; Coleman, J.N. Silver nanowire networks as flexible, transparent, conducting films: Extremely high DC to optical conductivity ratios. *ACS Nano* **2009**, *3*, 1767–1774. [CrossRef] [PubMed]

111. Amjadi, M.; Pichitpajongkit, A.; Lee, S.; Ryu, S.; Park, I. Highly stretchable and sensitive strain sensor based on silver nanowire-elastomer nanocomposite. *ACS Nano* **2014**, *8*, 5154–5163. [CrossRef] [PubMed]

112. Akter, T.; Kim, W.S. Reversibly stretchable transparent conductive coatings of spray-deposited silver nanowires. *ACS Appl. Mater. Interfaces* **2012**, *4*, 1855–1859. [CrossRef] [PubMed]

113. Huang, G.-W.; Xiao, H.-M.; Fu, S.-Y. Wearable electronics of silver-nanowire/poly(dimethylsiloxane) nanocomposite for smart clothing. *Sci. Rep.* **2015**, *5*, 13971. [CrossRef] [PubMed]

114. Madaria, A.R.; Kumar, A.; Ishikawa, F.N.; Zhou, C. Uniform, highly conductive, and patterned transparent films of a percolating silver nanowire network on rigid and flexible substrates using a dry transfer technique. *Nano Res.* **2010**, *3*, 564–573. [CrossRef]

115. Liang, J.; Li, L.; Tong, K.; Ren, Z.; Hu, W.; Niu, X.; Chen, Y.; Pei, Q. Silver nanowire percolation network soldered with graphene oxide at room temperature and its application for fully stretchable polymer light-emitting diodes. *ACS Nano* **2014**, *8*, 1590–1600. [CrossRef] [PubMed]

116. Chen, S.-P.; Liao, Y.-C. Highly stretchable and conductive silver nanowire thin films formed by soldering nanomesh junctions. *Phys. Chem. Chem. Phys.* **2014**, *16*, 19856–19860. [CrossRef] [PubMed]

117. Tamai, T. Electrical properties of conductive elastomer as electrical contact material. *IEEE Trans. Compon. Hybrids Manuf. Technol.* **1982**, *5*, 56–61. [CrossRef]

118. Kim, K.S.; Zhao, Y.; Jang, H.; Lee, S.Y.; Kim, J.M.; Kim, K.S.; Ahn, J.-H.; Kim, P.; Choi, J.-Y.; Hong, B.H. Large-scale pattern growth of graphene films for stretchable transparent electrodes. *Nature* **2009**, *457*, 706–710. [CrossRef] [PubMed]

119. Rwei, S.-P.; Ku, F.-H.; Cheng, K.-C. Dispersion of carbon black in a continuous phase: Electrical, rheological, and morphological studies. *Colloid Polym. Sci.* **2002**, *280*, 1110–1115.

120. Kujawski, M.; Pearse, J.D.; Smela, E. Elastomers filled with exfoliated graphite as compliant electrodes. *Carbon* **2010**, *48*, 2409–2417. [CrossRef]

121. Sekitani, T.; Noguchi, Y.; Hata, K.; Fukushima, T.; Aida, T.; Someya, T. A rubberlike stretchable active matrix using elastic conductors. *Science* **2008**, *321*, 1468–1472. [CrossRef] [PubMed]

122. Kim, K.H.; Vural, M.; Islam, M.F. Single-walled carbon nanotube aerogel-based elastic conductors. *Adv. Mater.* **2011**, *23*, 2865–2869. [CrossRef] [PubMed]

123. Zhou, Y.; Bayram, Y.; Du, F.; Dai, L.; Volakis, J.L. Polymer-carbon nanotube sheets for conformal load bearing antennas. *IEEE Trans. Antennas Propag.* **2010**, *58*, 2169–2175. [CrossRef]

124. Zhang, Y.; Sheehan, C.J.; Zhai, J.; Zou, G.; Luo, H.; Xiong, J.; Zhu, Y.T.; Jia, Q.X. Polymer-embedded carbon nanotube ribbons for stretchable conductors. *Adv. Mater.* **2010**, *22*, 3027–3031. [CrossRef] [PubMed]

125. Liu, K.; Sun, Y.; Liu, P.; Lin, X.; Fan, S.; Jiang, K. Cross-stacked superaligned carbon nanotube films for transparent and stretchable conductors. *Adv. Funct. Mater.* **2011**, *21*, 2721–2728. [CrossRef]

126. Shang, Y.; He, X.; Li, Y.; Zhang, L.; Li, Z.; Ji, C.; Shi, E.; Li, P.; Zhu, K.; Peng, Q.; et al. Super-stretchable spring-like carbon nanotube ropes. *Adv. Mater.* **2012**, *24*, 2896–2900. [CrossRef] [PubMed]

127. Li, X.; Cai, W.; An, J.; Kim, S.; Nah, J.; Yang, D.; Piner, R.; Velamakanni, A.; Jung, I.; Tutuc, E.; et al. Large-area synthesis of high quality and uniform graphene films on copper foils. *Science* **2009**, *324*, 1312–1314. [CrossRef] [PubMed]

128. Qi, D.; Liu, Z.; Liu, Y.; Leow, W.R.; Zhu, B.; Yang, H.; Yu, J.; Wang, W.; Wang, H.; Yin, S.; et al. Suspended wavy graphene microribbons for highly stretchable microsupercapacitors. *Adv. Mater.* **2015**, *27*, 5559–5566. [CrossRef] [PubMed]

129. Kim, R.-H.; Bae, M.-H.; Kim, D.G.; Cheng, H.; Kim, B.H.; Kim, D.-H.; Li, M.; Wu, J.; Du, F.; Kim, H.-S.; et al. Stretchable, transparent graphene interconnects for arrays of microscale inorganic light emitting diodes on rubber substrates. *Nano Lett.* **2011**, *11*, 3881–3886. [CrossRef] [PubMed]

130. Sepúlveda, A.T.; Guzman de Villoria, R.; Viana, J.C.; Pontes, A.J.; Wardle, B.L.; Rocha, L.A. Full elastic constitutive relation of non-isotropic aligned-CNT/PDMS flexible nanocomposites. *Nanoscale* **2013**, *5*, 4847–4854. [CrossRef] [PubMed]

131. Zhang, Y.; Ronning, F.; Gofryk, K.; Mara, N.A.; Haberkorn, N.; Zou, G.; Wang, H.; Lee, J.H.; Bauer, E.; McCleskey, T.M.; et al. Aligned carbon nanotubes sandwiched in epitaxial NbC film for enhanced superconductivity. *Nanoscale* **2012**, *4*, 2268–2271. [CrossRef] [PubMed]

132. Yun, J.; Lim, Y.; Jang, G.N.; Kim, D.; Lee, S.-J.; Park, H.; Hong, S.Y.; Lee, G.; Zi, G.; Ha, J.S. Stretchable patterned graphene gas sensor driven by integrated micro-supercapacitor array. *Nano Energy* **2016**, *19*, 401–414. [CrossRef]

133. Watanabe, M.; Shirai, H.; Hirai, T. Wrinkled polypyrrole electrode for electroactive polymer actuators. *J. Appl. Phys.* **2002**, *92*, 4631–4637. [CrossRef]

134. Wang, C.; Zheng, W.; Yue, Z.; Too, C.O.; Wallace, G.G. Buckled, stretchable polypyrrole electrodes for battery applications. *Adv. Mater.* **2011**, *23*, 3580–3584. [CrossRef] [PubMed]

135. Zeng, W.; Shu, L.; Li, Q.; Chen, S.; Wang, F.; Tao, X.-M. Fiber-based wearable electronics: A review of materials, fabrication, devices, and applications. *Adv. Mater.* **2014**, *26*, 5310–5336. [CrossRef] [PubMed]

136. Noh, J.-S. Conductive elastomers for stretchable electronics, sensors and energy harvesters. *Polymers* **2016**, *8*, 123. [CrossRef]

137. Savagatrup, S.; Chan, E.; Renteria-Garcia, S.M.; Printz, A.D.; Zaretski, A.V.; O'Connor, T.F.; Rodriquez, D.; Valle, E.; Lipomi, D.J. Plasticization of PEDOT:PSS by common additives for mechanically robust organic solar cells and wearable sensors. *Adv. Funct. Mater.* **2015**, *25*, 427–436. [CrossRef]

138. Lee, Y.-Y.; Kang, H.-Y.; Gwon, S.H.; Choi, G.M.; Lim, S.-M.; Sun, J.-Y.; Joo, Y.-C. A Strain-Insensitive Stretchable Electronic Conductor: PEDOT:PSS/Acrylamide Organogels. *Adv. Mater.* **2016**, *28*, 1636–1643. [CrossRef] [PubMed]

139. Ding, H.; Zhong, M.; Wu, H.; Park, S.; Mohin, J.W.; Klosterman, L.; Yang, Z.; Yang, H.; Matyjaszewski, K.; Bettinger, C.J. Elastomeric Conducting Polyaniline Formed Through Topological Control of Molecular Templates. *ACS Nano* **2016**, *10*, 5991–5998. [CrossRef] [PubMed]

140. Palleau, E.; Reece, S.; Desai, S.C.; Smith, M.E.; Dickey, M.D. Self-healing stretchable wires for reconfigurable circuit wiring and 3D microfluidics. *Adv. Mater.* **2013**, *25*, 1589–1592. [CrossRef] [PubMed]

141. Dickey, M.D.; Chiechi, R.C.; Larsen, R.J.; Weiss, E.A.; Weitz, D.A.; Whitesides, G.M. Eutectic gallium-indium (EGaIn): A liquid metal alloy for the formation of stable structures in microchannels at room temperature. *Adv. Funct. Mater.* **2008**, *18*, 1097–1104. [CrossRef]

142. Tabatabai, A.; Fassler, A.; Usiak, C.; Majidi, C. Liquid-phase gallium-indium alloy electronics with microcontact printing. *Langmuir* **2013**, *29*, 6194–6200. [CrossRef] [PubMed]

143. Park, C.W.; Moon, Y.G.; Seong, H.; Jung, S.W.; Oh, J.-Y.; Na, B.S.; Park, N.-M.; Lee, S.S.; Im, S.G.; Koo, J.B. Photolithography-based patterning of liquid metal interconnects for monolithically integrated stretchable circuits. *ACS Appl. Mater. Interfaces* **2016**, *8*, 15459–15465. [CrossRef] [PubMed]

144. Kim, H.J.; Son, C.; Ziaie, B. A multiaxial stretchable interconnect using liquid-alloy-filled elastomeric microchannels. *Appl. Phys. Lett.* **2008**, *92*, 011904. [CrossRef]

145. Armand, M.; Endres, F.; MacFarlane, D.R.; Ohno, H.; Scrosati, B. Ionic-liquid materials for the electrochemical challenges of the future. *Nat. Mater.* **2009**, *8*, 621–629. [CrossRef] [PubMed]

146. Cong, H.; Pan, T. Photopatternable conductive PDMS materials for microfabrication. *Adv. Funct. Mater.* **2008**, *18*, 1912–1921. [CrossRef]

147. Hu, M.; Cai, X.; Guo, Q.; Bian, B.; Zhang, T.; Yang, J. Direct pen writing of adhesive particle-free ultrahigh silver salt-loaded composite ink for stretchable circuits. *ACS Nano* **2016**, *10*, 396–404. [CrossRef] [PubMed]

148. Duffy, D.C.; McDonald, J.C.; Schueller, O.J.A.; Whitesides, G.M. Rapid prototyping of microfluidic systems in poly(dimethylsiloxane). *Anal. Chem.* **1998**, *70*, 4974–4984. [CrossRef] [PubMed]

149. Siegel, A.C.; Bruzewicz, D.A.; Weibel, D.B.; Whitesides, G.M. Microsolidics: Fabrication of three-dimensional metallic microstructures in poly(dimethylsiloxane). *Adv. Mater.* **2007**, *19*, 727–733. [CrossRef]

150. Kim, H.-J.; Maleki, T.; Wei, P.; Ziaie, B. A biaxial stretchable interconnect with liquid-alloy-covered joints on elastomeric substrate. *J. Microelectromech. Syst.* **2009**, *18*, 138–146. [CrossRef]

151. Vohra, A.; Imin, P.; Imit, M.; Carmichael, R.S.; Meena, J.S.; Adronov, A.; Carmichael, T.B. Transparent, stretchable, and conductive SWNT films using supramolecular functionalization and layer-by-layer self-assembly. *RSC Adv.* **2016**, *6*, 29254–29263. [CrossRef]

152. Someya, T.; Kato, Y.; Sekitani, T.; Iba, S.; Noguchi, Y.; Murase, Y.; Kawaguchi, H.; Sakurai, T. Conformable, flexible, large-area networks of pressure and thermal sensors with organic transistor active matrixes. *Proc. Natl. Acad. Sci. USA* **2005**, *102*, 12321–12325. [CrossRef] [PubMed]

153. Takahashi, T.; Takei, K.; Gillies, A.G.; Fearing, R.S.; Javey, A. Carbon nanotube active-matrix backplanes for conformal electronics and sensors. *Nano Lett.* **2011**, *11*, 5408–5413. [CrossRef] [PubMed]

154. Zhang, Y.; Wang, S.; Li, X.; Fan, J.A.; Xu, S.; Song, Y.M.; Choi, K.-J.; Yeo, W.-H.; Lee, W.; Nazaar, S.N.; et al. Experimental and theoretical studies of serpentine microstructures bonded to prestrained elastomers for stretchable electronics. *Adv. Funct. Mater.* **2014**, *24*, 2028–2037. [CrossRef]

155. Hsu, Y.-Y.; Gonzalez, M.; Bossuyt, F.; Axisa, F.; Vanfleteren, J.; de Wolf, I. The effects of encapsulation on deformation behavior and failure mechanisms of stretchable interconnects. *Thin Solid Films* **2011**, *519*, 2225–2234. [CrossRef]

156. Zhang, Y.; Xu, S.; Fu, H.; Lee, J.; Su, J.; Hwang, K.-C.; Rogers, J.A.; Huang, Y. Buckling in serpentine microstructures and applications in elastomer-supported ultra-stretchable electronics with high areal coverage. *Soft Matter* **2013**, *9*, 8062–8070. [CrossRef] [PubMed]

157. Khang, D.-Y.; Jiang, H.; Huang, Y.; Rogers, J.A. A stretchable form of single-crystal silicon for high-performance electronics on rubber substrates. *Science* **2006**, *311*, 208–212. [CrossRef] [PubMed]

158. Song, J.; Jiang, H.; Choi, W.M.; Khang, D.Y.; Huang, Y.; Rogers, J.A. An analytical study of two-dimensional buckling of thin films on compliant substrates. *J. Appl. Phys.* **2008**, *103*, 014303. [CrossRef]

159. Khang, D.-Y.; Rogers, J.A.; Lee, H.H. Mechanical buckling: Mechanics, metrology, and stretchable electronics. *Adv. Funct. Mater.* **2008**, *19*, 1526–1536. [CrossRef]

160. Kim, D.-H.; Song, J.; Choi, W.M.; Kim, H.-S.; Kim, R.-H.; Liu, Z.; Huang, Y.Y.; Hwang, K.-C.; Zhang, Y.; Rogers, J.A. Materials and noncoplanar mesh designs for integrated circuits with linear elastic responses to extreme mechanical deformations. *Proc. Natl. Acad. Sci. USA* **2008**, *105*, 18675–18680. [CrossRef] [PubMed]

161. Xiang, Y.; Li, T.; Suo, Z.; Vlassak, J.J. High ductility of a metal film adherent on a polymer substrate. *Appl. Phys. Lett.* **2005**, *87*, 161910. [CrossRef]

162. Hung, P.J.; Jeong, K.; Liu, G.L.; Lee, L.P. Microfabricated suspensions for electrical connections on the tunable elastomer membrane. *Appl. Phys. Lett.* **2004**, *85*, 6051–6053. [CrossRef]

163. Hyun, D.C.; Park, M.; Park, C.; Kim, B.; Xia, Y.; Hur, J.H.; Kim, J.M.; Park, J.J.; Jeong, U. Ordered zigzag stripes of polymer gel/metal nanoparticle composites for highly stretchable conductive electrodes. *Adv. Mater.* **2011**, *23*, 2946–2950. [CrossRef] [PubMed]

164. Zhu, Y.; Moran-Mirabal, J. Highly bendable and stretchable electrodes based on micro/nanostructured gold films for flexible sensors and electronics. *Adv. Electron. Mater.* **2016**, *2*, 1500345. [CrossRef]

165. Lee, K.J.; Fosser, K.A.; Nuzzo, R.G. Fabrication of stable metallic patterns embedded in poly (dimethylsiloxane) and model applications in non-planar electronic and lab-on-a-chip device patterning. *Adv. Funct. Mater.* **2005**, *15*, 557–566. [CrossRef]

166. Xu, S.; Zhang, Y.; Cho, J.; Lee, J.; Huang, X.; Jia, L.; Fan, J.A.; Su, Y.; Su, J.; Zhang, H.; et al. Stretchable batteries with self-similar serpentine interconnects and integrated wireless recharging systems. *Nat. Commun.* **2013**, *4*, 1543. [CrossRef] [PubMed]

167. Huang, X.; Liu, Y.; Cheng, H.; Shin, W.-J.; Fan, J.A.; Liu, Z.; Lu, C.-J.; Kong, G.-W.; Chen, K.; Patnaik, D.; et al. Materials and designs for wireless epidermal sensors of hydration and strain. *Adv. Funct. Mater.* **2014**, *24*, 3846–3854. [CrossRef]

168. Kramer, R.K.; Majidi, C.; Wood, R.J. Masked deposition of gallium-indium alloys for liquid-embedded elastomer conductors. *Adv. Funct. Mater.* **2013**, *23*, 5292–5296. [CrossRef]

169. Skylar-Scott, M.A.; Gunasekaran, S.; Lewis, J.A. Laser-assisted direct ink writing of planar and 3D metal architectures. *Proc. Natl. Acad. Sci. USA* **2016**, *113*, 6137–6142. [CrossRef] [PubMed]

170. Seifert, T.; Sowade, E.; Roscher, F.; Wiemer, M.; Gessner, T.; Baumann, R.R. Additive manufacturing technologies compared: Morphology of deposits of silver ink using inkjet and aerosol jet printing. *Ind. Eng. Chem. Res.* **2015**, *54*, 769–779. [CrossRef]

171. Saunders, R.E.; Derby, B. Inkjet printing biomaterials for tissue engineering: Bioprinting. *Int. Mater. Rev.* **2014**, *59*, 430–448. [CrossRef]

172. Vatani, M.; Lu, Y.; Lee, K.-S.; Kim, H.-C.; Choi, J.-W. Direct-write stretchable sensors using single-walled carbon nanotube/polymer matrix. *J. Electron. Packag.* **2013**, *135*, 11009. [CrossRef]

173. Srichan, C.; Saikrajang, T.; Lomas, T.; Jomphoak, A.; Maturos, T.; Phokaratkul, D.; Kerdcharoen, T.; Tuantranont, A. Inkjet printing PEDOT:PSS using desktop inkjet printer. In Proceedings of the 6th International Conference on Electrical Engineering/Electronics, Computer, Telecommunications and Information Technology, Pattaya, Chonburi, Thailand, 6–9 May 2009; Volume 1, pp. 465–468.

174. Jones, C.S.; Lu, X.; Renn, M.; Stroder, M.; Shih, W.-S. Aerosol-jet-printed, high-speed, flexible thin-film transistor made using single-walled carbon nanotube solution. *Microelectron. Eng.* **2010**, *87*, 434–437. [CrossRef]

175. Bfahy, S.; Yunus, S.; Pardoen, T.; Bertrand, P.; Troosters, M. Stretchable helical gold conductor on silicone rubber microwire. *Appl. Phys. Lett.* **2007**, *91*, 141911. [CrossRef]

176. Lipomi, D.J.; Chong, H.; Vosgueritchian, M.; Mei, J.; Bao, Z. Toward mechanically robust and intrinsically stretchable organic solar cells: Evolution of photovoltaic properties with tensile strain. *Sol. Energy Mater. Sol. Cells* **2012**, *107*, 355–365. [CrossRef]

177. Lim, K.S.; Chang, W.-J.; Koo, Y.-M.; Bashir, R. Reliable fabrication method of transferable micron scale metal pattern for poly(dimethylsiloxane) metallization. *Lab Chip* **2006**, *6*, 578–580. [CrossRef] [PubMed]

178. Lee, C.H.; Kim, D.R.; Zheng, X. Fabrication of nanowire electronics on nonconventional substrates by water-assisted transfer printing method. *Nano Lett.* **2011**, *11*, 3435–3439. [CrossRef] [PubMed]

179. Song, L.; Ci, L.; Gao, W.; Ajayan, P.M. Transfer printing of graphene using gold film. *ACS Nano* **2009**, *3*, 1353–1356. [CrossRef] [PubMed]

180. Park, J.; Wang, S.; Li, M.; Ahn, C.; Hyun, J.K.; Kim, D.S.; Kim, D.K.; Rogers, J.A.; Huang, Y.; Jeon, S. Three-dimensional nanonetworks for giant stretchability in dielectrics and conductors. *Nat. Commun.* **2012**, *3*, 916. [CrossRef] [PubMed]

181. Amjadi, M.; Kyung, K.-U.; Park, I.; Sitti, M. Stretchable, skin-mountable, and wearable strain sensors and their potential applications: A review. *Adv. Funct. Mater.* **2016**, *26*, 1678–1698. [CrossRef]

182. Mannsfeld, S.C.B.; Tee, B.C.-K.; Stoltenberg, R.M.; Chen, C.V.H.-H.; Barman, S.; Muir, B.V.O.; Sokolov, A.N.; Reese, C.; Bao, Z. Highly sensitive flexible pressure sensors with microstructured rubber dielectric layers. *Nat. Mater.* **2010**, *9*, 859–864. [CrossRef] [PubMed]

183. Yu, C.; Wang, Z.; Yu, H.; Jiang, H. A stretchable temperature sensor based on elastically buckled thin film devices on elastomeric substrates. *Appl. Phys. Lett.* **2009**, *95*, 141912. [CrossRef]

184. Chung, M.G.; Kim, D.H.; Lee, H.M.; Kim, T.; Choi, J.H.; Seo, D.K.; Yoo, J.-B.; Hong, S.-H.; Kang, T.J.; Kim, Y.H. Highly sensitive NO_2 gas sensor based on ozone treated graphene. *Sens. Actuators B Chem.* **2012**, *166–167*, 172–176. [CrossRef]

185. Liang, J.; Li, L.; Niu, X.; Yu, Z.; Pei, Q. Elastomeric polymer light-emitting devices and displays. *Nat. Photonics* **2013**, *7*, 817–824. [CrossRef]

186. Chortos, A.; Koleilat, G.I.; Pfattner, R.; Kong, D.; Lin, P.; Nur, R.; Lei, T.; Wang, H.; Liu, N.; Lai, Y.-C.; et al. Mechanically Durable and Highly Stretchable Transistors Employing Carbon Nanotube Semiconductor and Electrodes. *Adv. Mater.* **2016**, *28*, 4441–4448. [CrossRef] [PubMed]

187. Lee, M.-S.; Lee, K.; Kim, S.-Y.; Lee, H.; Park, J.; Choi, K.-H.; Kim, H.-K.; Kim, D.-G.; Lee, D.-Y.; Nam, S.; et al. High-performance, transparent, and stretchable electrodes using graphene-metal nanowire hybrid structures. *Nano Lett.* **2013**, *13*, 2814–2821. [CrossRef] [PubMed]

188. Shin, M.; Song, J.H.; Lim, G.-H.; Lim, B.; Park, J.-J.; Jeong, U. Highly stretchable polymer transistors consisting entirely of stretchable device components. *Adv. Mater.* **2014**, *26*, 3706–3711. [CrossRef] [PubMed]

189. Oh, J.Y.; Rondeau-Gagné, S.; Chiu, Y.-C.; Chortos, A.; Lissel, F.; Wang, G.-J.N.; Schroeder, B.C.; Kurosawa, T.; Lopez, J.; Katsumata, T.; et al. Intrinsically stretchable and healable semiconducting polymer for organic transistors. *Nature* **2016**, *539*, 411–415. [CrossRef] [PubMed]

190. Xie, K.; Wei, B. Materials and structures for stretchable energy storage and conversion devices. *Adv. Mater.* **2014**, *26*, 3592–3617. [CrossRef] [PubMed]

191. Chen, T.; Peng, H.; Durstock, M.; Dai, L. High-performance transparent and stretchable all-solid supercapacitors based on highly aligned carbon nanotube sheets. *Sci. Rep.* **2014**, *4*, 3612. [CrossRef] [PubMed]

192. Lee, H.; Yoo, J.-K.; Park, J.-H.; Kim, J.H.; Kang, K.; Jung, Y.S. A stretchable polymer-carbon nanotube composite electrode for flexible lithium-ion batteries: Porosity engineering by controlled phase separation. *Adv. Energy Mater.* **2012**, *2*, 976–982. [CrossRef]

193. Lipomi, D.J.; Tee, B.C.-K.; Vosgueritchian, M.; Bao, Z. Stretchable organic solar cells. *Adv. Mater.* **2011**, *23*, 1771–1775. [CrossRef] [PubMed]

194. Yang, Z.; Deng, J.; Sun, X.; Li, H.; Peng, H. Stretchable, wearable dye-sensitized solar cells. *Adv. Mater.* **2014**, *26*, 2643–2647. [CrossRef] [PubMed]

195. Dagdeviren, C.; Yang, B.D.; Su, Y.; Tran, P.L.; Joe, P.; Anderson, E.; Xia, J.; Doraiswamy, V.; Dehdashti, B.; Feng, X.; et al. Conformal piezoelectric energy harvesting and storage from motions of the heart, lung, and diaphragm. *Proc. Natl. Acad. Sci. USA* **2014**, *111*, 1927–1932. [CrossRef] [PubMed]

196. Qi, Y.; Kim, J.; Nguyen, T.D.; Lisko, B.; Purohit, P.K.; McAlpine, M.C. Enhanced piezoelectricity and stretchability in energy harvesting devices fabricated from buckled PZT ribbons. *Nano Lett.* **2011**, *11*, 1331–1336. [CrossRef] [PubMed]

197. Bito, J.; Hester, J.G.; Tentzeris, M.M. Ambient RF energy harvesting from a two-way talk radio for flexible wearable wireless sensor devices utilizing inkjet printing technologies. *IEEE Trans. Microw. Theory Tech.* **2015**, *63*, 4533–4543. [CrossRef]

198. Kim, S.; Bito, I.; Jeong, S.; Georgiadis, A.; Tentzeris, M.M. A flexible hybrid printed RF energy harvester utilizing catalyst-based copper printing technologies for far-field RF energy harvesting applications. In Proceedings of the 2015 IEEE MTT-S International Microwave Symposium, Phoenix, AZ, USA, 17–22 May 2015; pp. 1–4.

199. Lee, J.-H.; Ryu, H.; Kim, T.-Y.; Kwak, S.-S.; Yoon, H.-J.; Kim, T.-H.; Seung, W.; Kim, S.-W. Thermally induced strain-coupled highly stretchable and sensitive pyroelectric nanogenerators. *Adv. Energy Mater.* **2015**, *5*, 1500704. [CrossRef]

200. Yi, F.; Lin, L.; Niu, S.; Yang, P.K.; Wang, Z.; Chen, J.; Zhou, Y.; Zi, Y.; Wang, J.; Liao, Q.; et al. Stretchable-rubber-based triboelectric nanogenerator and its application as self-powered body motion sensors. *Adv. Funct. Mater.* **2015**, *25*, 3688–3696. [CrossRef]

201. Honda, W.; Harada, S.; Arie, T.; Akita, S.; Takei, K. Wearable, human-interactive, health-monitoring, wireless devices fabricated by macroscale printing techniques. *Adv. Funct. Mater.* **2014**, *24*, 3299–3304. [CrossRef]

202. Hussain, A.M.; Ghaffar, F.A.; Park, S.I.; Rogers, J.A.; Shamim, A.; Hussain, M.M. Metal/polymer based stretchable antenna for constant frequency far-field communication in wearable electronics. *Adv. Funct. Mater.* **2015**, *25*, 6565–6575. [CrossRef]

203. Chortos, A.; Liu, J.; Bao, Z. Pursuing prosthetic electronic skin. *Nat. Mater.* **2016**, *15*, 937–950. [CrossRef] [PubMed]

204. Yeo, J.C.; Yap, H.K.; Xi, W.; Wang, Z.; Yeow, C.-H.; Lim, C.T. Flexible and stretchable strain sensing actuator for wearable soft robotic applications. *Adv. Mater. Technol.* **2016**, *1*, 1600018. [CrossRef]

205. Lim, S.; Son, D.; Kim, J.; Lee, Y.B.; Song, J.-K.; Choi, S.; Lee, D.J.; Kim, J.H.; Lee, M.; Hyeon, T.; Kim, D.-H. Transparent and stretchable interactive human machine interface based on patterned graphene heterostructures. *Adv. Funct. Mater.* **2015**, *25*, 375–383. [CrossRef]

206. Hillborg, H.; Gedde, U.W. Hydrophobicity changes in silicone rubbers. *IEEE Trans. Dielectr. Electr. Insul.* **1999**, *6*, 703–717. [CrossRef]

207. Grassie, N.; Macfarlane, I.G. The thermal degradation of polysiloxanes-I. Poly(dimethylsiloxane). *Eur. Polym. J.* **1978**, *14*, 875–884. [CrossRef]

208. Morra, M.; Occhiello, E.; Marola, R.; Garbassi, F.; Humphrey, P.; Johnson, D. On the aging of oxygen plasma-treated PDMS surfaces. *J. Colloid Interface Sci.* **1990**, *137*, 11–24. [CrossRef]

209. Park, G.; Chung, H.-J.; Kim, K.; Lim, S.A.; Kim, J.; Kim, Y.-S.; Liu, Y.; Yeo, W.-H.; Kim, R.-H.; Kim, S.S.; et al. Immunologic and tissue biocompatibility of flexible/stretchable electronics and optoelectronics. *Adv. Healthc. Mater.* **2014**, *3*, 515–525. [CrossRef] [PubMed]

210. Song, J.; Feng, X.; Huang, Y. Mechanics and thermal management of stretchable inorganic electronics. *Natl. Sci. Rev.* **2016**, *3*, 128–143. [CrossRef] [PubMed]

211. Li, Y.; Gao, Y.; Song, J. Recent advances on thermal analysis of stretchable electronics. *Theor. Appl. Mech. Lett.* **2016**, *6*, 32–37. [CrossRef]

212. Novoselov, K.S.S.; Geim, A.K.K.; Morozov, S.V.; Jiang, D.; Zhang, Y.; Dubonos, S.V.; Grigorieva, I.V.; Firsov, A.A. Electric field effect in atomically thin carbon films. *Science* **2004**, *306*, 666–669. [CrossRef] [PubMed]

213. Lai, Y.-C.; Huang, Y.-C.; Lin, T.-Y.; Wang, Y.-X.; Chang, C.-Y.; Li, Y.; Lin, T.-Y.; Ye, B.-W.; Hsieh, Y.-P.; Su, W.-F.; et al. Stretchable organic memory: Toward learnable and digitized stretchable electronic applications. *NPG Asia Mater.* **2014**, *6*, e87. [CrossRef]

214. Moon, J.-H.; Baek, D.-H.; Choi, Y.Y.; Lee, K.H.; Kim, H.C.; Lee, S.-H. Wearable polyimide-PDMS electrodes for intrabody communication. *J. Micromech. Microeng.* **2010**, *20*, 25032. [CrossRef]

micromachines

MDPI

Review

Recent Advancements in Liquid Metal Flexible Printed Electronics: Properties, Technologies, and Applications

Xuelin Wang [1] and Jing Liu [1,2,*]

[1] Department of Biomedical Engineering, School of Medicine, Tsinghua University, Beijing 100084, China; wang-xl15@mails.tsinghua.edu.cn

[2] Beijing Key Lab of CryoBiomedical Engineering and Key Lab of Cryogenics, Technical Institute of Physics and Chemistry, Chinese Academy of Sciences, Beijing 100190, China

*   Correspondence: jliubme@tsinghua.edu.cn; Tel.: +86-10-6279-4896; Fax: +86-10-8254-3767

Academic Editors: Seung Hwan Ko, Daeho Lee and Zhigang Wu
Received: 26 September 2016; Accepted: 27 October 2016; Published: 30 November 2016

Abstract: This article presents an overview on typical properties, technologies, and applications of liquid metal based flexible printed electronics. The core manufacturing material—room-temperature liquid metal, currently mainly represented by gallium and its alloys with the properties of excellent resistivity, enormous bendability, low adhesion, and large surface tension, was focused on in particular. In addition, a series of recently developed printing technologies spanning from personal electronic circuit printing (direct painting or writing, mechanical system printing, mask layer based printing, high-resolution nanoimprinting, etc.) to 3D room temperature liquid metal printing is comprehensively reviewed. Applications of these planar or three-dimensional printing technologies and the related liquid metal alloy inks in making flexible electronics, such as electronical components, health care sensors, and other functional devices were discussed. The significantly different adhesions of liquid metal inks on various substrates under different oxidation degrees, weakness of circuits, difficulty of fabricating high-accuracy devices, and low rate of good product—all of which are challenges faced by current liquid metal flexible printed electronics—are discussed. Prospects for liquid metal flexible printed electronics to develop ending user electronics and more extensive applications in the future are given.

Keywords: liquid metal; flexible printed electronics; 3D printing; functional device; additive manufacture; consumer electronics

1. Introduction

Printed electronics is the technology to fabricate electronic devices based on the principle of printing [1]. Unlike traditional printing technology, the inks used in the printing machine are electronic materials with the properties of being conductive, dielectric, semi-conductive, or magnetic. Printed electronics is closely related to diverse fields such as organic electronics, plastic electronics, flexible electronics, and paper electronics [2], which indicates that one can not only print circuits on silicon and glass, but also on plastic, paper, and more flexible substrates. One of the methods to produce flexible electronics is to directly print conductive materials onto flexible substrates.

Classical flexible electronics refers to the technology by which organic/inorganic electronic devices are deposited on a flexible substrate to form a circuit [3]. Although the rigid circuit board can protect the electronic components from being damaged, it restricts the ductility and flexibility of the electronics. With intriguing properties such as softness, ductility, and low-cost fabrication, flexible electronics has broad application prospects in the information, energy, medicine, and defense technology fields through providing smart sensors, actuators, flexible displays, organic light-emitting diodes (OLEDs), and so on. The most obvious characteristic of flexible electronics lies in their flexibility compared with traditional rigid microelectronics, which makes them stretchable, conformal, portable, wearable, and easy to print quickly [4–8]. Because of their unique merits in terms of electrical, printable, biomedical, and sensing properties, flexible electronics can find diverse applications in electronic components [9], printing technology, implantable devices [10], and health monitors [11], with specific uses such as antennas [12], eyeball cameras [13] and pressure sensors [14], etc. Nowadays, two common strategies have been adopted to improve the softness of the electronics [15]. One is embedding conductive materials, which are stiff and rigid, onto a stretchable substrate, such as poly(dimethylsiloxane) (PDMS). Rogers and coworkers introduced a complex wavy structure to keep the circuit stretchable, which can absorb the major tension when stress acts on the soft substrate [16,17]. The other is using inherently stretchable conductors to form the circuit [18]. Russo and coworkers demonstrated a way to connect the circuit using conductive silver ink, which can directly write conductive text to interconnect light-emitting diode (LED) arrays and three-dimensional (3D) antennas on paper [12].

Clearly, flexible printed electronics (FPE) combines features of flexible electronics and printed electronics. In this way, one can quickly manufacture functional flexible electronic devices. Along with the maturation of printing technology, flexible printing has become a hot topic in research. Several typical printing technologies have been emerging, such as tapping mode printing [18], aerosol jet printing, roll-to-roll technology [19], inkjet printing [20], and micropen and brush printing [21,22]. Among them, micropen is perhaps the simplest method: pump the electrical ink into a cartridge to directly write out conductive texts. This makes it possible to draw a circuit diagram on an A4 sheet of paper. A micropen evidently reduces the complexity of circuit production and improves the efficiency of electronic circuit manufacturing. FPE have contributed to significant achievements in different fields, such as flexible display devices [23], thin-film solar cells [24,25], large area sensors and drivers [26,27], electronic skin [28], wearable electronics and biological prosthetic devices [29], self-charging system [30], self-powered wireless monitoring [31], etc. If a circuit benefits from the properties of softness, conformability, stretchability and portability, one can call it a flectional electronic device. Meanwhile, many laboratories have developed electronic devices with flexibility, such as warped display [32], folding battery [9], soft field effect transistor [33], stretchable wire [34], and so on. FPE can also be used in health monitoring, medical examination, vital signs detection, and other daily life needs. Flexible biomedical equipment has been widely utilized in the field of implantable devices [10], nerve connection [35,36], health monitoring [11], and wearable medical devices. FPE has also been used for biological sensing, the most important application of which is electronic skin (E-skin), such as pressure mapping [37], self-healing sensor [38], prosthesis [39], pressure sensor [14], etc. There are many research organizations working on different materials applied on flexible electronic skin. In Bao's laboratory, they invented a skin-inspired artificial mechanoreceptor system and the receptor can transform the pressure stimulation signal into electrophysiological signal that human neurons can perceive [40]. In Rogers's laboratory, they initiated a wavy structure to realize stretchable epidermal electronics that can maintain the original shape and function after compressing or stretching [16,41,42], In Chiolerio's laboratory, the researchers invented a printable spin-coated silver nanocomposite ink to manufacture a resistive switching devices for neuromorphic applications and a nanocomposite flexible liquid state device in a synthetic colloidal suspension [43,44].

Generally, traditional nanoparticle-based conductive ink does not have intrinsic conductivity, and needs special post-processing (e.g., sintering, annealing) to remove solvent from the conductive ink to achieve conductive capacity, such as silver nanoparticle, PEDOT:PSS, polyaniline-based ink, nickel and copper conductive ink [45–49]. The newly emerging liquid metal ink has intrinsically high conductivity, which enables it to be a kind of ideal conductive ink (Table 1). (In this paper, EGaIn refers to Ga–In, which contains the metal elements gallium and indium, commonly in a mixture of 75.5% gallium and 24.5% indium by weight; Galinstan refers to Ga–In–Sn, which contains the metal elements gallium, indium, and tin, commonly in a mixture of 62.5% gallium, 21.5% indium and 16% tin by weight.)

Liquid metal, just as its name suggests, is a kind of metal that remains in liquid phase from room temperature up to 2000 °C. (However, 2000 °C is a generic number; it depends upon the composition.) It is superior to many other liquid materials in terms of thermal conductivity and resistivity at low temperatures or around room temperature [50]. One traditional liquid metal—mercury—is not accepted on account of its high toxicity to the human body. However, because of its non-toxicity and benign biocompatibility, a liquid metal eutectic alloy of gallium indium has made important progresses in the biomedical arena such as being used as the material of bone cement [51], as a vascular contrast agent [52], and in drug delivery nanomedicine [53]. All of these indicate that liquid metal can be widely used in the fields of electronics, materials, and biology, which significantly broadens its impact. Figure 1 shows the typical applications of personalized flexible printed electronics based on liquid metal, such as implantable devices [35,36,54], electrical skin [55–57] and wearable bioelectronics [58–60] etc. Along with the development of material science and technology, liquid metal flexible printed electronics are quickly shaping the field of flexible electronic circuits and allied machines, serving as a basic way of quickly making functional devices. In this review, we are dedicated to integrating the liquid metal alloy inks with flexible printed electronics and focus our attention on interpreting the recent advancements in liquid metal flexible printed electronics.

Table 1. A comparison of the electrical conductivities of several typical conductive inks [61].

Ink Type	Ink Composition	Conductivity
Carbon conductive ink	Carbon	1.8×10^3 S/m
	CNT	$(5.03 \pm 0.05) \times 10^3$ S/m
Polymer conductive ink	PEDOT:PSS	8.25×10^3 S/m
Nano-silver ink	Ag-DDA	3.45×10^7 S/m
	Ag-PVP	6.25×10^6 S/m
Liquid metal ink	EGaIn	3.4×10^6 S/m
	$Bi_{35}In_{48.6}Sn_{16}Zn_{0.4}$	7.3×10^6 S/m

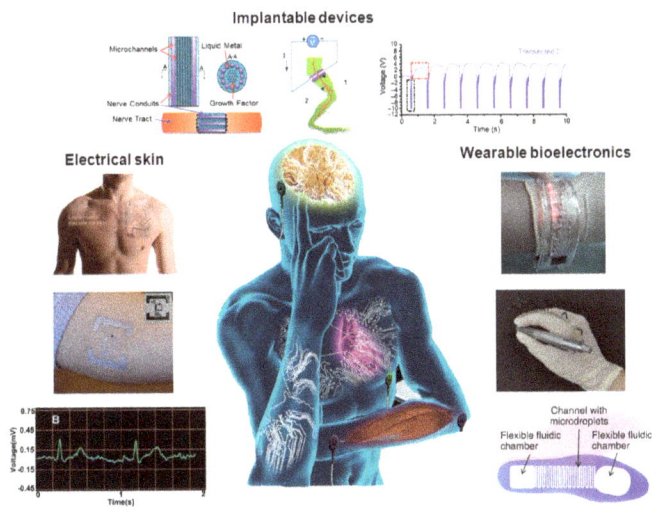

Figure 1. Personalized flexible printed electronics based on liquid metal. "Implantable devices": Liquid metal nerve connection [35] (https://arxiv.org/abs/1404.5931); Injectable 3D bioelectrode [54] (Reproduced with permission from Jin, C.; Liu, J. et al., Injectable 3-D fabrication of medical electronics at the target biological tissues; published by Scientific Reports, 2013.); Electrophysiological measurement of liquid metal reconnected nerve [36]; "Electrical skin": Liquid metal circuits on human body [55] (Reproduced with permission from Guo, C.; Liu, J. et al., Rapidly patterning conductive components

on skin substrates as physiological testing devices via liquid metal spraying and pre-designed mask; published by Journal of Mterials Chemistry B, 2014.); Liquid metal tag on human skin [56] (Reproduced with permission from Jeong, S.H.; Wu, Z. et al., Liquid alloy printing of microfluidic stretchable electronics; published by Lab on a Chip, 2012.); ECG test by liquid metal electrode [57] (Reproduced with permission from Yu, Y.; Zhang, J.; Liu, J., Biomedical implementation of liquid metal ink as drawable ECG electrode and skin circuit; published by PLoS ONE, 2013.); "Wearable bioelectronics": Liquid metal wristband [58] (Reproduced with permission from Wang, Q.; Yu, Y.; Yang, J.; Liu, J., Fast fabrication of flexible functional circuits based on liquid metal dual-trans printing; published by Advanced Material, 2015.); A wearable data glove [59] (Reproduced with permission from Matsuzaki, R.; Tabayashi, K., Highly stretchable, global, and distributed local strain sensing line using gainsn electrodes for wearable electronics; published by Advanced Functional Material, 2015.); Footwear-embedded microfluidic energy harvest [60] (Reproduced with permission from Krupenkin, T.; Taylor, J.A., Reverse electrowetting as a new approach to high-power energy harvesting; published by Nature Communication, 2011.).

2. Basic Properties of Liquid Metal

Generally speaking, alloy elements of low melting point include gallium, bismuth, lead, tin, cadmium, and indium. Among them, $GaIn_{24.5}$ is commonly used, which is a mixture of 75.5% gallium and 24.5% indium by weight. For the same composition, a different proportion may lead to quite varied properties in the alloys. Sometimes, even a small change of mass ratio would cause strong variation in the material behaviors. If loaded with certain microelements, the alloy may also display additional properties accordingly. Therefore, one can change the ratio of chemical materials of alloys or add some microelements to adjust the melting point and other properties to fulfill various specific needs. Room-temperature liquid metal material, generally represented by gallium and its alloys, offers a unique for to manufacturing flexible circuits, due to the combined nature of metallicity and fluidity [62,63], which makes it especially suitable for printing on soft substrates. (In this paper, the liquid metal alloy refers to gallium and its alloys.) From [62], one can get information on the main physical properties of frequently used liquid metal alloys, such as combination, melting point, density, resistivity, and thermal conductivity. In this review, we mainly discuss the resistivity, flexibility, adhesion, and wettability of liquid metal in flexible printed electronics.

2.1. Resistivity

Resistivity is the ability of an object to conduct an electric current, and generally the conductivity of metal is better than that of a non-metal. If the liquid metal is applied to print a flexible circuit, we can calculate the resistance value of a liquid metal wire by the formula:

$$R = \frac{\rho \cdot L}{A},$$ (1)

where R is the resistance; ρ is the resistivity and for the same material ρ is a fixed value; L is the length of measured liquid metal conductive wire; and A is the cross-sectional area. As can be seen from Equation (1), the resistance of the material is directly proportional to the length of the material and the resistivity, and is inversely proportional to its cross-sectional area.

The relationship between the volume and length is:

$$A = \frac{V}{L}.$$ (2)

Therefore, one can get the factors associated with resistance change:

$$\frac{R}{R_0} = \frac{\frac{\rho \cdot L}{A}}{\frac{\rho_0 \cdot L_0}{A_0}} = \frac{\rho V_0}{\rho_0 V} \left(\frac{L}{L_0}\right)^2.$$ (3)

The volume of the liquid metal conductor is constant when drawing $V = V_0$; the resistivity is also considered constant, $\rho = \rho_0$.

Consequently, one can simplify Equation (3) to:

$$\frac{R}{R_0} = \left(\frac{L}{L_0}\right)^2. \tag{4}$$

The resistance variation of a liquid metal conductor has a linear relationship with the square of the length change of the conductor before and after stretching, confirming the relationship between resistance and tensile length, which provides theoretical support for the development of a liquid metal flexible sensor [55].

However, the resistivity c of the liquid metal has a significant relationship with the thermal conductivity [64,65]:

$$\frac{\Lambda}{\rho T} = \frac{\pi^2 \kappa_B^2}{3e^2} = 2.45 \times 10^{-8} (\text{W} \cdot \Omega/\text{K}^2). \tag{5}$$

Here, Λ is the thermal conductivity; ρ is the resistivity; and T is the thermodynamic temperature. Figure 2B shows the $\rho - T$ and the $\Lambda - T$ curves for $Bi_{35}In_{48.6}Sn_{16}Zn_{0.4}$. With the increase in temperature from $-150\,^{\circ}\text{C}$ to $50\,^{\circ}\text{C}$, Λ shows a linear increase while ρ shows a linear decrease. ρ_B is the Boltzmann constant. e is the electron charge, $e = -1.6 \times 10^{-19}\text{C}$. Equation (5) is the famous Wiedemann–Franz–Lorenz equation, from which one can deduce that the thermal conductivity of the liquid metal is proportional to the product of the resistivity and the thermodynamic temperature. One can get the resistance value of some number of wires connected in parallel by Ohm's Law [66]:

$$R = \left(\sum_{i=1}^{n} \frac{1}{R_i}\right)^{-1}. \tag{6}$$

Here, n is the number of wires connected in parallel. R_i is the individual resistance values, i.e.,

$$R_i = \rho \int_0^L \frac{dx}{A_i(x)}, \tag{7}$$

where ρ is the resistivity and the resistivity of EGaIn is $29.4 \times 10^{-6}\,\Omega \cdot \text{cm}$. Gozen et al. demonstrates that the measured and predicted resistance values show a high degree of agreement with the number of wires changing [66]. However, there are some non-Ohm's Law phenomena that the current does not linearly vary with the voltage (Figure 3A) [67,68]. In Reference [67], the liquid metal marble is coated with WO_3 and the substrate is gold. It forms an electrical interface across the n-type semiconducting coating (WO_3) without the native oxide layer on the surface of the liquid metal, and the flow of current is owing to the combination of electron and hole carriers, which forms a n-type semiconductor layer. Therefore the measured current–voltage (I–V) shows the non-Ohm's Law phenomena. They have similar I–V curves between Reference [67] and Figure 3A, however different in theory. In Figure 3A, the electronic device is composed of one gallium droplet and a layer of gold nanoparticles, and the gallium droplet is the electrode. The nanoparticle and its ligands form a capacitor. This is the process of capacitor charging and discharging. Under the threshold voltage, the circuit is not on and the current is zero. Once up the threshold voltage, the circuit is switched on. Meanwhile, some data demonstrate the relationships between the resistance, the reactance, and the frequency [57]. Figure 2A shows that when the frequency changes from 1 Hz to 10 kHz, Ga resistance remains at 0.225 Ω. However, Ga reactance closes to 0 Ω from 1 Hz to 100 Hz and has analogously exponential growth from 100 Hz to 10 kHz. Liquid metal resistance invariance under different frequencies can be applied to measure some circuit frequency changes.

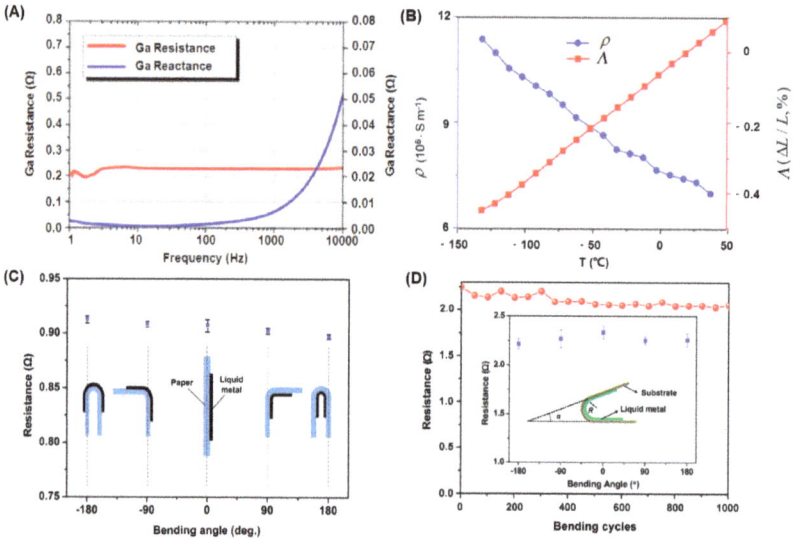

Figure 2. (**A**) The resistance and reactance curves from 1 Hz to 10 kHz of the liquid metal Ga [57] (Reproduced with permission from Yu, Y.; Zhang, J.; Liu, J., Biomedical implementation of liquid metal ink as drawable ECG electrode and skin circuit; published by PLoS ONE, 2013.); (**B**) ρ (the resistivity)-*T* (temperature) and Λ (the thermal conductivity)-*T* (temperature) curves of liquid metal $Bi_{35}In_{48.6}Sn_{16}Zn_{0.4}$ [64] (Reproduced with permission from Wang, L.; Liu, J., Compatible hybrid 3D printing of metal and nonmetal inks for direct manufacture of end functional devices; published by Science China Technological Sciences, 2014.); (**C**) resistance values of liquid metal line at five different bending angles [69] (Reproduced with permission from Zheng, Y.; He, Z.; Gao, Y.; Liu, J., Direct desktop printed-circuits-on-paper flexible electronics; published by Scientific Reports, 2013.); (**D**) resistance values of liquid metal line at different bending angles and different bending cycles [70] (Reproduced with permission from Wang, L.; Liu, J., Pressured liquid metal screen printing for rapid manufacture of high resolution electronic patterns; published by RSC Advances, 2015.).

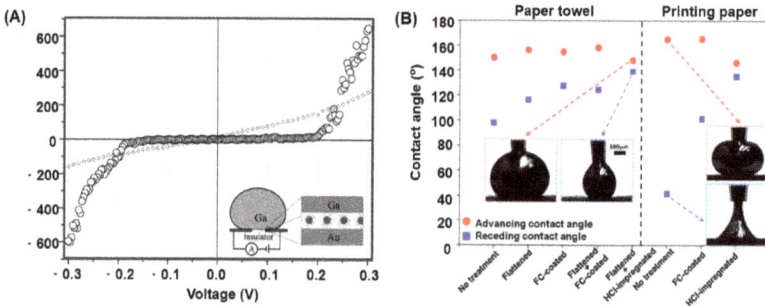

Figure 3. (**A**) Plot of the measured current and voltage of the electronic device composed of one gallium droplet and a layer of gold nanoparticles [68] (Reproduced with permission from Du, K.; Glogowski, E.; Tuominen, M.T.; Emrick, T.; Russell, T.P.; Dinsmore, A.D., Self-assembly of gold nanoparticles on gallium droplets: Controlling charge transport through microscopic devices; published by Langmuir, 2013.); (**B**) the advancing and receding contact angle of the liquid metal droplets on the paper towel and printing paper [71] (Reproduced with permission from Kim, D.; Lee, Y.; Lee, D.-W.; Choi, W.; Yoo, K.; Lee, J.-B., Hydrochloric acid-impregnated paper for gallium-based liquid metal microfluidics; published by Sensors and Actuators B: Chemical, 2015.).

2.2. Superior Flexibility

In order to illustrate the feasibility of liquid metal in manufacturing flexible circuits, Zheng et al. carried out a test to study the influence of bending on the stability of the circuit through measuring the resistance value when bending the printed liquid metal wire at $-180°$, $-90°$, $0°$, $90°$, and $180°$. Figure 2C evidently indicates that the resistance variation is very little and the resistance stability of the liquid metal when bending manifests that liquid metal printed wire can be well applied for making flexible electronics [69]. Furthermore, Wang et al. bent the PVC plastic film printed by liquid metal at $-180°$, $-90°$, $0°$, $90°$, and $180°$ for one cycle, and the resistance value was measured every 50 bending cycles (Figure 2D). The resistance values fluctuated slightly from 1 to 1000 cycles, which demonstrated the stability of the liquid metal resistor line [70]. References [72,73] demonstrate that when straining the liquid metal circuits, the resonance frequency and the resistance value can be changed. All the bending and stretching tests suggest that liquid metal printed wire has reliable flexibility when employed in flexible printed electronics.

2.3. Tunable Adhesion

Adhesion is the attraction among different molecules, such as the adhesion between the particle and the substrate [64–66]. When printing a circuit, the liquid metal is ejected from the spray gun in the form of droplets [65]. When put in air, the surface of liquid metal droplets may generate a layer of oxide whose composition is Ga_2O_3/Ga_2O [74]. As is disclosed on the proportional relationship between the surface tension of liquid metal droplets and oxide content [75], the smaller the metal droplets the higher the oxide content, thus the larger the surface tension and the better the substrate adhesion. Starting from this point, Zhang et al. established a generalized methodology for ubiquitous printed electronics [65] whereby, through atomized spraying of liquid metal droplets, a circuit can be quickly fabricated on almost any desired solid substrate surface, whether smooth or rough [65]. Additionally, the greater the oxide content of the metal droplets, the smaller the contact angle between the droplet and substrate [76].

Based on the Young–Dupre equation [77], one has

$$W_{SL} = \gamma_L \left(1 + \cos\theta\right), \tag{8}$$

where W_{SL} is the adhesion work of the liquid metal drop, γ_L is its surface tension, and θ is the contact angle between the droplet and substrate. The smaller the droplet, the larger the γ_L and the smaller the θ, hence the greater the W_{SL}, representing better adhesion. In other words, one can regulate the adhesion of liquid metal ink on a substrate by controlling its oxide layer [78]. Another possible strategy would be realizing a controlled de-wetting after the metal layer has solidified by a slight temperature increase [79]. However, the liquid metal alloy ink has strong adhesion to a majority of substrates, which makes it a challenge to manufacture micro-precision circuits [80].

2.4. Prominent Wettability

Wettability is the ability of liquid to spread on a solid surface. The contact angle usually expresses the wettability: the smaller the contact angle the better the wettability. The dynamic contact angle is the contact angle of a liquid moving over a surface. The dynamic contact angle (sliding angle and advancing-receding angle) plays an important role in quantifying the wetting property of the oxidized Galinstan droplets compared with the static contact angle [81,82]:

$$\sin\alpha = \gamma_{LG} \frac{Rk}{mg} \left[\cos\theta_{rec} - \cos\theta_{adv}\right]. \tag{9}$$

Here, α is the sliding angle; R is the liquid metal droplet radius; m is the mass of the droplet; θ_{rec} is the receding contact angle; and θ_{adv} is the advancing contact angle. The liquid metal droplet on the

HCl-impregnated flattened paper has the lowest contact angle, while for the non-treated printing paper it has the highest contact angle (Figure 3B) [71]. Gao et al. have studied the excellent wettability of liquid metal $GaIn_{10}$ with different substrate materials including smooth polyvinylchloride (Figure 4A), porous rubber (Figure 4B), rough polyvinylchloride (Figure 4C), tree leaf (Figure 4D), epoxy resin board (Figure 4E), typing paper (Figure 4F), glass (Figure 4G), cotton paper (Figure 4H), plastic (Figure 4I), cotton cloth (Figure 4J), silica gel plate (Figure 4K), glass fiber cloth (Figure 4L), and other substrates with different surface roughness and material properties [63,75]. Kramer et al. investigated the relationship between the microtextured surface topography of the liquid metal marbles and the wetting behaviors [83]. Doudrick et al. demonstrated that the oxide surface of the liquid metal alloy can improve the wettability between the metal material and substrate [80]. When increasing the oxygen content, the resistivity of the liquid metal alloy decreases while the viscosity and the wettability increase. For this reason, we can control the thickness of the oxide layer and then the wettability of the liquid metal wire on the soft substrate to obtain a superior performance of the flexible printed circuit.

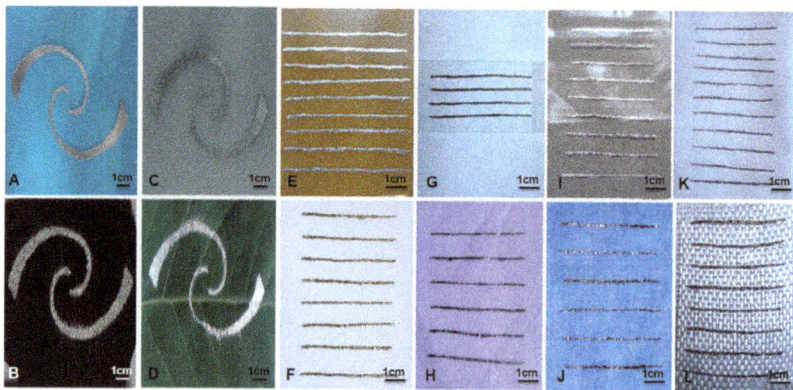

Figure 4. Demonstrated wettability of liquid metal printed on different substrate materials [63,75]; (**A**) Smooth polyvinylchloride; (**B**) porous rubber; (**C**) rough polyvinylchloride; (**D**) tree leaf; (**E**) epoxy resin board; (**F**) typing paper; (**G**) glass; (**H**) cotton paper; (**I**) plastic; (**J**) cotton cloth; (**K**) silica gel plate; (**L**) glass fiber cloth (Reproduced with permission from Gao, Y.; Li, H.; Liu, J., Direct writing of flexible electronics through room temperature liquid metal ink; published by PLoS ONE, 2012.) (Reproduced with permission from Zhang, Q.; Gao, Y.; Liu, J., Atomized spraying of liquid metal droplets on desired substrate surfaces as a generalized way for ubiquitous printed electronics; published by Applied Physics A: Materials Science & Processing, 2013.).

3. Printing Technologies and Apparatuses

In ancient China, Sheng Bi, a brilliant inventor who lived around 1041 AD, invented a new type of printing method: movable-type printing, the convenience of which led to the spread of knowledge and culture; the basic principle and method of modern letterpress stamping is the same as movable-type printing in spite of its different equipment and technical conditions. Thus, the invention of an excellent printing method and the selection of appropriate printing ink can have a huge impact on human life. Table 1 shows a conductivity comparison of several typical electroconductive inks [61]. The corrosivity of the liquid metal ink for different substrates (Table 2) prevents liquid gallium from being printed on specific flexible substrates.

Table 2. The corrosion between liquid gallium and other metals [84].

Corrosive Behavior	Easy to React with Gallium	Better Resistance for Corrosion	Best Resistance for Corrosion
Metal	Iron, nickel, cadmium, cerium, copper, aluminum, gold, manganese, platinum, silver, tin and vanadium, zirconium	Titanium, tantalum, niobium, molybdenum, beryllium	Tungsten, rhenium, sintering BeO, Al_2O_3, and fused quartz, graphite
Temperature range	Appearing corrosion at 200 °C or even lower	Appearing corrosion above 400 °C	No corrosion even at 800 °C

3.1. Printing Electronic Circuit

Nowadays, due to the high surface tension and good conductive ability of liquid metal, printing circuit using liquid metal ink has attracted more and more attention. The concept of DREAM ink (direct writing or printing of electronics based on alloy and metal ink) [62], as defined by Liu et al., can lead to alternative electronic devices in different fields [63,85,86]. So far, these are the different methods of printing a personal electronic circuit: screen printing [70], atomized spraying [75], microcontact printing (μCP) [87], masked deposition [88], and inkjet printing [89].

3.1.1. Direct Painting or Writing

Because of its good wettability with multifarious substrates, room-temperature liquid metal (RTLM) can be directly painted or written on paper, glass, or cloth as liquid metal ink [63]. Sheng et al. directly painted the liquid metal ink on the surface of VHB 4905 acrylic films to manufacture a capacitor sensor (Figure 5A) [90]. Because of its gravity and adhesion to substrates, the liquid metal alloy can be directly installed into the refill as a kind of ink to write conductive texts or lines [91,92]. Zheng et al. developed the liquid metal roller-ball pen (LMRP), which can write conductive lines or electronic devices on soft plates with the tip diameters ranging from 200 μm to 1000 μm (Figure 5B) [93]. Boley et al. illustrated a direct writing system, using the liquid metal alloy to fabricate small-scale stretchable electronics [73].

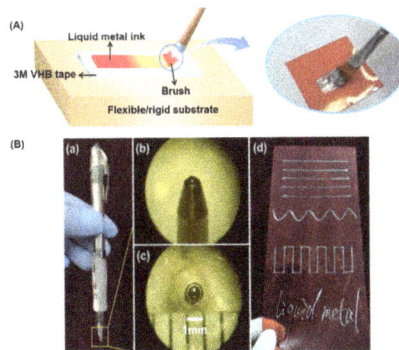

Figure 5. Printing methods of direct painting or writing without machine. (**A**) Direct painting of the liquid metal ink on the surface of VHB 4905 acrylic films to manufacture capacitor sensor [90]; (**B**) the liquid metal ink roller-ball pen and conductive tracks written by it [93] (Reproduced with permission from Zheng, Y.; Zhang, Q.; Liu, J., Pervasive liquid metal based direct writing electronics with roller-ball pen; published by AIP Advances, 2013.).

3.1.2. Mechanical Printing Methods

Compared to traditional direct painting or writing, mechanical printing methods can realize digital control, which makes the liquid metal flexible electronic circuits more accurate (Figure 6).

Zheng et al. have manufactured and demonstrated a versatile desktop liquid metal printer that can print either a plane circuit or a 3D metal object. The machine is made up of a syringe, a nitrogen gas tube, a pressure controller, a teach pendant, a stage, the *X* axis, the *Y* axis, and the *Z* axis (Figure 6B) [69]. In order to achieve a match between different material substrate and printing ink, they have designed a brush-like porous needle to print files of different types. A desktop printer can print a variety of electrical patterns via computer software. In particular, paper is low-cost and recyclable, which makes it a common substrate material and allows the concept of printed-circuits-on-paper (PCP) [69]. Along with that, the same group created another fully automatic liquid metal printer for pervasive electronic circuit printing [18]. Its theory of tapping mode printing is fixing an ink storage tube on the printing driver, clamping the driver on the guide rail using a sliding wheel, which can make the printer head move along the guide rail in the *X* direction, and driving the base plate along in the *Y* direction using a motor (Figure 6A). Consequently, one can control the position of the printer head in the printing area of the base plate. The tapping mode means that the nib can move a small distance in the *Z* direction. When printing, the printer head falls to contact the base, uplifts a certain height over the printing substrate, so as to avoid the printed line. In the process of printing, the printing header bead rotates due to the basal friction, which makes the liquid metal ink in the pen container run downwards. One can use the tapping mode method to manufacture various printed-circuit-on-board (PCB) circuits, antennae, and so on. In addition, other mechanical system printing methods to print personal electronic circuits using liquid metal include inkjet printing and coelectrospinning (Figure 6C,D) [89,94,95].

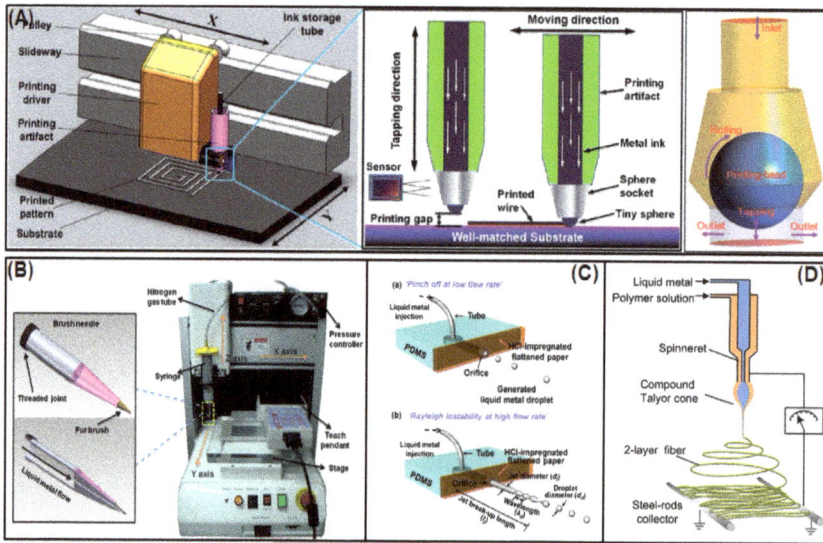

Figure 6. Mechanical printing methods. (**A**) The tapping mode printer and its printing process [18] (Reproduced with permission from Zheng, Y.; He, Z.-Z.; Yang, J.; Liu, J., Personal electronics printing via tapping mode composite liquid metal ink delivery and adhesion mechanism; published by Scientific Reports, 2014.); (**B**) The desktop printer to directly print circuits on paper [69] (Reproduced with permission from Zheng, Y.; He, Z.; Gao, Y.; Liu, J., Direct desktop printed-circuits-on-paper flexible electronics; published by Scientific Reports, 2013.); (**C**) inkjet printing of the liquid metal droplet based on pinch-off and Rayleigh instability [94] (Reproduced with permission from Kim, D.; Yoo, J.H.; Lee, Y.; Choi, W.; Yoo, K.; Lee, J.B.J., Gallium-Based Liquid Metal Inkjet Printing; published by IEEE Proceedings, 2014.); (**D**) coelectrospinning to fabricate a light-emitting coaxial nanofiber with a liquid metal core and polymer shell [95] (Reproduced with permission from Yang, H.; Lightner, C.R.; Dong, L., Light-emitting coaxial nanofibers; published by I ACS NANO, 2012.).

3.1.3. Mask-Based Printing Methods

Printing methods based on masks are the most commonly used method to manufacture liquid metal flexible electronics in the laboratory. Kramer et al. provided a masked deposition method to fabricate liquid metal hyperelastic electronic circuits [88] (Figure 7A). At the same time, Gozen et al. used the masked deposition method to prepare micro-scale high-density soft-matter electronics [66]. Wang et al. proposed a rapid fabrication of flexible functional circuits based on liquid metal dual-trans printing, which can be applied to quickly fabricate a flexible functional electronic device fitting to any complex surface shape (Figure 7B) [58]. First one can print out a liquid metal circuit on PVC membrane, then use polydimethylsiloxane (PDMS) solution to cover the circuit. After a period of time, PDMS solution can be cured and when it is liquid, arbitrary shape objects can be immersed into PDMS solution. Finally, cool the whole object to make liquid metal solid and one can completely transfer the initial liquid metal circuit to a PDMS flexible substrate. In the process, after PDMS curing, one can peel off the PVC membrane and target object to obtain PDMS device embedded liquid metal flexible circuits. Because the shape of the PDMS substrate completely fits with objects, one can achieve highly conformal electronic devices. This technology has significant implications for sensing, monitoring in health care, and households as a corresponding device to conduct specific functions. Common methods mentioned in the literature also include microcontact printing [87], atomized spraying [75], screen printing [70], and CO_2 laser ablation [96] (Figure 8D). The theory of the atomized spraying and the screen printing methods is that the gun sprays the liquid metal ink onto the substrate through a specific mask to form a circuit which has the same shape as the mask (Figure 8B,C).

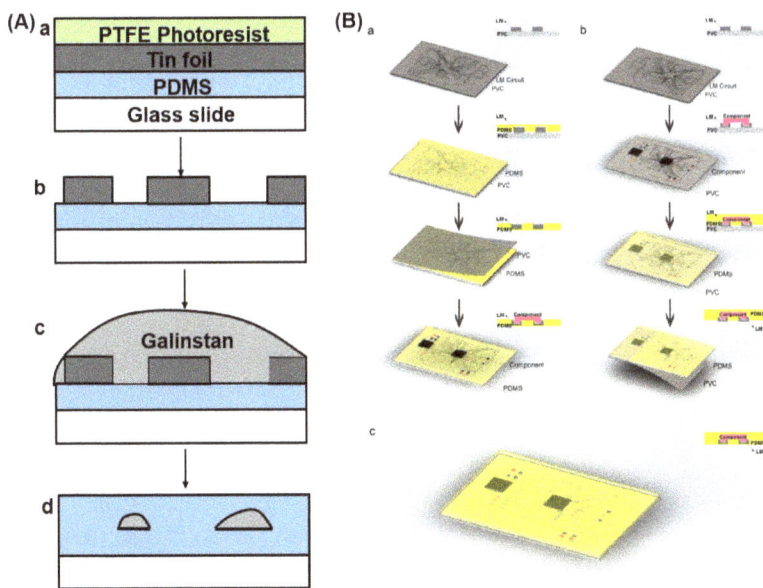

Figure 7. Mask-based printing methods. (**A**) The masked deposition process of gallium–indium alloys for liquid-embedded elastomer conductors, (**a**) painting photoresist onto tin foil; (**b**) micro-channel on the elastomer surface; (**c**) flooding the surface with galinstan; (**d**) removing excessgalinstan with a thin film, cooling the liquid metal in a freezer and encapsulating it by coating (Modified from [88]); (**B**) the process of dual-trans printing fabrication [58] (Reproduced with permission from Wang, Q.; Yu, Y.; Yang, J.; Liu, J., Fast fabrication of flexible functional circuits based on liquid metal dual-trans printing; published by Advanced Material, 2015.).

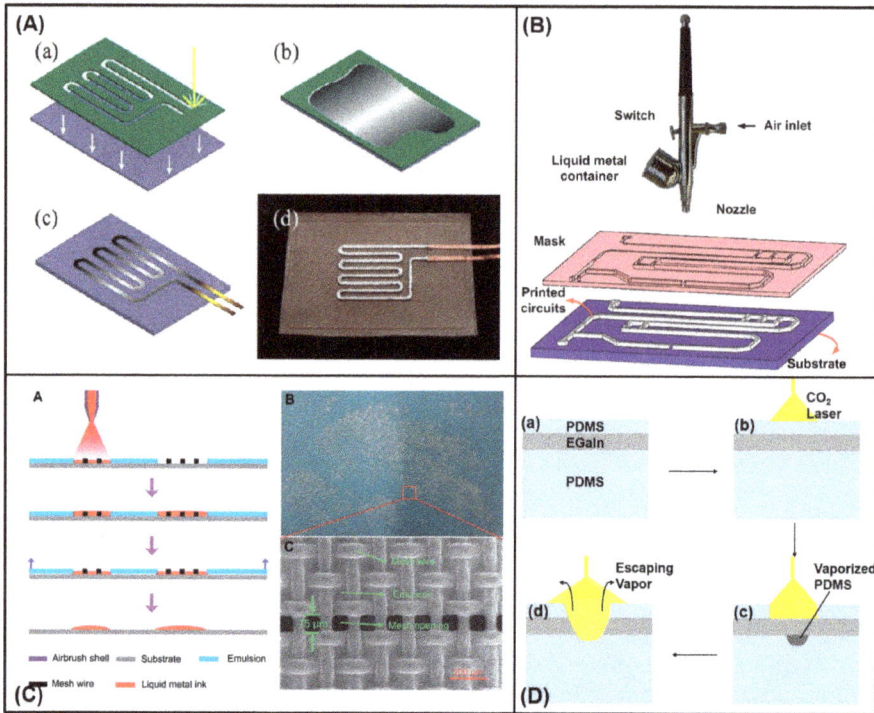

Figure 8. Mask-based printing methods. (**A**) The microcontact printing process to fabricate the liquid metal soft-matter circuit [87], (**a**) laser engraving a thin film; (**b**) spreading GaIn alloy over the stencil; (**c**) removing the stencil and inserting the copper wires before sealing; (**d**) unsealed galinstan heater on VHB elastomer produced with stencil lithography (Reproduced with permission from Tabatabai, A.; Fassler, A.; Usiak, C.; Majidi, C., Liquid-phase gallium–indium alloy electronics with microcontact printing; published by Langmuir, 2013.); (**B**) atomized spraying of liquid metal droplets to rapid prototype circuits [75] (Reproduced with permission from Zhang, Q.; Gao, Y.; Liu, J., Atomized spraying of liquid metal droplets on desired substrate surfaces as a generalized way for ubiquitous printed electronics; published by Applied Physics A: Materials Science & Processing, 2013.); (**C**) the screen printing process and scanning electron microscopy image of the screen mesh [70] (Reproduced with permission from Wang, L.; Liu, J., Pressured liquid metal screen printing for rapid manufacture of high resolution electronic patterns; published by RSC Advances, 2015.); (**D**) proposed CO_2 laser ablation mechanism of the liquid metal alloy on the PDMS substrate (Modified from [96]).

The common method for preparation of liquid metal microfluidic electronics is pouring the liquid metal alloy into a specific shape of channel to form the functional circuit. Therefore, from the strict definition, the preparation method of microfluidic electronics cannot be classified as fabrication of liquid metal flexible printed electronics. However, Figure 9 demonstrates the processing steps of manufacturing liquid alloy microfluidic wireless power transfer, in which the mask layer and atomization printing technology are applied and the power transfer displays large flexibility in its rolled state and can attach to the human arm [97].

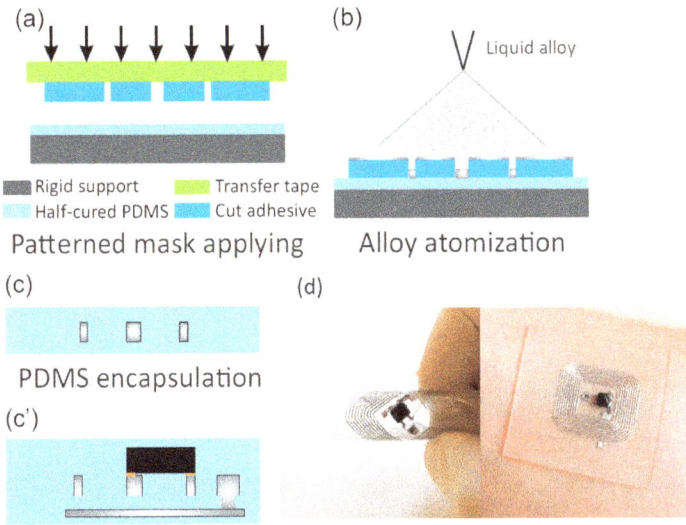

Figure 9. Processing steps of manufacturing liquid alloy microfluidic wireless power transfer (**a**–**c**), an active component mounted in the circuit followed by encapsulation (**c′**) and in rolled state and attached to human arm (**d**) [97] (Reproduced with permission from Jeong, S.H.; Hjort, K.; Wu, Z., Tape transfer atomization patterning of liquid alloys for microfluidic stretchable wireless power transfer; published by Scientific Reports, 2015.).

3.1.4. High-Resolution Nanoimprint Lithography

Unlike the relatively large liquid metal droplets, one can also prepare liquid metal nanoparticles to print micro-scale or nano-scale wires. John et al. produced small features through mechanically sintering gallium–indium nanoparticles to realize high-resolution nanoimprint lithography [98]. As shown in Reference [98], 1 μm coalesced EGaIn line arrays were formed by reducing the size of the sintering tool. Via inkjet printing EGaIn nanoparticles dispersions on the surface of nitrile glove, the functionalized elastomer glove with liquid metal micro-wire arrays represents excellent stretchability when holding a tennis ball.

3.2. 3D Printing

3D printing is a technology to construct an object by means of printing layer by layer; in other words, it is a kind of typical additive manufacturing technology. Different from traditional subtractive technology, additive manufacturing is a top-down, layer-by-layer manufacturing process. It can not only avoid corrosion and contribute to environmental protection, but also saves a lot of raw materials. This significantly reduces environmental pollution and the cost of materials. Figure 10A shows the difference between subtractive and additive manufacturing technologies [99]. Next we will introduce several 3D printing technologies using low melting point metal ink based on the additive mode.

Wang et al. first put forward the concept of liquid phase 3D printing to employ gallium-, bismuth-, and indium-based alloys whose melting points are higher than room temperature and lower than 300 °C to accomplish the printing process [100]. The liquid phase 3D printing is achieved in a liquid environment, and the liquid can be water, ethanol, kerosene, electrolyte solution, etc. However, the melting point of liquid metal ink must be lower than that of the liquid environment to ensure the printed item run. In the experiments, Wang et al. introduced $Bi_{35}In_{48.6}Sn_{16}Zn_{0.4}$ as printing ink because it does not absorb/release as much heat as normal metal in the process of phase change (Figure 11B) [100]. Shortly after, they originated a late-model 3D printing technology, hybrid 3D printing, which means interactive printing of various inks or combination of multiple printing methods. In Figure 10B, $Bi_{35}In_{48.6}Sn_{16}Zn_{0.4}$ and 705 silicone rubber (nonmetal) inks were adopted to accomplish hybrid 3D printing [64]. The 705 silicone rubber serves as a kind of waterproof, anticorrosive, transparent, and insulating adhesive, and can be solidified when absorbing water vapor at room temperature. Therefore, it is often used as an electrical packaging material. First, one can print the bottom layer on the substrate using 705 silicone rubber and print the middle layer on the 705 silicone rubber layer using $Bi_{35}In_{48.6}Sn_{16}Zn_{0.4}$. After that, one can print the top layer using 705 silicone rubber. When adequately solidified, a sandwich structure can be obtained once taking off the printed item from the basement. Compatible hybrid 3D printing of metal and nonmetal inks fully embodies the good mechanical strength and electrical/thermal conductive of metal ink, and excellent insulation of nonmetal ink, which makes the printing circuit applicable in a harsh environment. Fassler et al. applied the common freeze-casting method to manufacture various liquid metal based 3D structures (Figure 11A) [101]. Ladd et al. tested the direct-writing microcomponents of the liquid metal alloy for 3D free-standing liquid metal microstructures at room temperature [102]. Liquid metal can also be employed as an injectable 3D bio-electrode to measure the electrocardiograph (ECG) or electroencephalogram (EEG) [54,57,103].

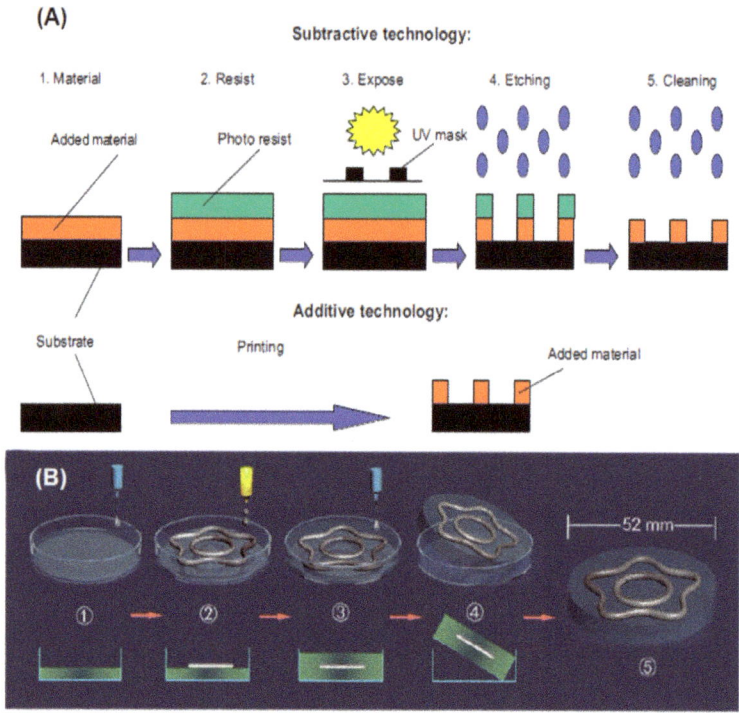

Figure 10. (**A**) Difference between subtractive and additive manufacturing technologies [99] (Reproduced with permission from Kunnari, E.; Valkama, J.; Keskinen, M.; Mansikkamäki, P., Environmental evaluation of new technology: Printed electronics case study.; published by Journal of Cleaner Production, 2009.); (**B**) hybrid 3D printing process with $Bi_{35}In_{48.6}Sn_{16}Zn_{0.4}$ and 705 silicone rubber inks [64] (Reproduced with permission from Wang, L.; Liu, J., Compatible hybrid 3D printing of metal and nonmetal inks for direct manufacture of end functional devices; published by Science China Technological Sciences, 2014.).

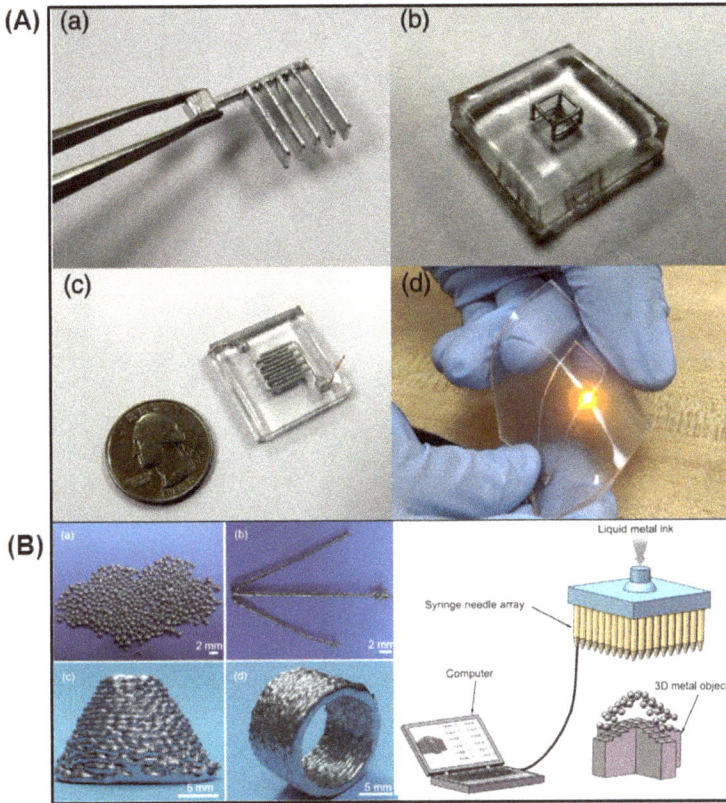

Figure 11. (**A**) The electronical components based on the liquid metal alloy [101] (Reproduced with permission from Fassler, A.; Majidi, C., 3D structures of liquid-phase gain alloy embedded in pdms with freeze casting; published by Lab on a Chip, 2013.); (**B**) liquid phase 3D printing method [100], (**a**) liquid metal balls; (**b**) liquid metal rods; (**c**) cone structure; (**d**) cylinder structure (Reproduced with permission from Wang, L.; Liu, J., Liquid phase 3D printing for quickly manufacturing conductive metal objects with low melting point alloy ink; published by Science China Technological Sciences, 2014.)

3.3. Printing Devices

Liu et al. have delivered printers that have now been put into practical use (Table 3) [18,100]. Liquid metal circuit printers have completely changed the traditional mode, breaking the technical bottleneck on personal electronics manufacturing, which makes it a reality to quickly fabricate electronic circuits with extremely low costs. Through computer control, even people with no electronics experience can download a circuit diagram from the Internet and print it directly. Inkjet printing equipment is based on the liquid metal screen printing process. Through a USB cable, the printer can connect with a computer to realize online printing, and can also print offline via an SD card. Before printing one needs to prepare .gcode files to control the nozzle movement. A 3D Metal Printer is a desktop fused deposition modeling (FDM) device; it substitutes traditional metal parts processing to prepare metal parts with a complex spatial structure.

Table 3. Three kinds of printers [69] (Reproduced with permission from Zheng, Y.; He, Z.; Gao, Y.; Liu, J., Direct desktop printed-circuits-on-paper flexible electronics; published by Scientific Reports, 2013.).

	Liquid Metal Printer	Personal Circuit Printer	Ubiquitous Inkjet Printer	3D Metal Printer
Device type				
Printed out items				
Application Fields		circuit; art; devices; rapid prototyping.	electronic circuit; art; radio frequency identification label; antenna.	circuit; radar; antenna; molds.

4. Typical Applications

As a novel electronic ink material, the flexibility and shape-preserving of the liquid metal alloy allows it to play an important role in fabricating components and soft circuit. Its non-toxicity and benign biocompatibility make it an implant material in the body, with uses such as the selected low melting point metal skeleton material [51], high-resolution angiography contrast [52], biological micro-droplets [95], nerve connection [35,36], human exoskeleton [104], and electronic skin [55,56,105].

The technology of directly printing liquid metal electronic circuit transforms conventional rigid circuits process to flexible desktop manufacturing [106]. With a way to rapidly manufacture various flexible household electronics with different functions, one can envisage that daily life will undergo big changes in the near future. Imagine a quilt with flexible functional circuits containing temperature or pressure sensors that can monitor human sleep quality and adjust the quilt temperature according to body temperature to achieve intelligent heating or cooling. In fact, all of these needs can be implemented by liquid metal flexible circuits. Liquid metal flexible printed electronics can be directly printed on many soft substrates, and could be used to produce lots of complex artwork. Zhang et al. demonstrated that even leaves can become a circuit board, which provides a solid theoretical basis for the universality of liquid metal atomized spraying manufacturing flexible electronics [75].

Through the direct-writing method, Li et al. drew a liquid metal thermocouple on paper [85]. The liquid metal thermocouple has many special advantages, such as small contact resistance and thermal resistance, no need for welding, and a wide working temperature zone. The thermoelectric effect is the theoretical basis of thermocouple temperature measurement. The thermoelectric power produced by the thermocouple has nothing to do with its shape or size, but relates to the composition of the thermoelectric materials and the temperature difference between the two ends. The thermocouple composed of Ga and EGaIn not only has temperature sensing ability but also bending capacity.

Fassler et al. prepared two kinds of capacitors—the square wave capacitor and the spiral capacitor—and a square planar spiral inductor [107]. They calculated the values of the capacitors or inductor of various shapes via formula derivation. A soft-matter diode, a new type of flexible functional equipment component with liquid metal electrodes, was developed by So et al. [29]. Because the liquid metal EGaIn electrodes' sandwich layers are conductive forward from Ga_xO_y to Ga while nonconductive backward from Ga to Ga_xO_y, the soft-matter diode has the function of limiting current and can pass the current one-way through.

At the same time, because of its flexibility and fluidity, liquid alloy is widely used in sensors and functional devices; Wu's group made important contributions to the field [108]. They revealed a liquid alloy microfluidic stretchable large-area wireless strain antenna, consisting of two layers of liquid metal alloy filled microfluidic channels in a silicone elastomer [109]. A stretchable ultra-high frequency radio frequency identity tag printed on human skin demonstrated that liquid alloy microfluidic electronics can be applied to electronic skin [56]. Liquid alloy microfluidic electronics can also be fabricated as antennae with strong stretchability, rolled space [110,111], and liquid alloy microfluidic wireless power transfer [97]. Matsuzaki et al. designed a wearable data glove with local strain sensing using GaInSn elestrodes (Figure 12A) [59]; the data glove can judge the degree of finger bending according to the change in resistance at different monitoring points. Krupenkin et al. modeled a footwear-embedded microfluidic energy harvester, which is a wearable mechanical-to-electrical energy conversion based on liquid metal microdroplets [60]. Meanwhile, the liquid metal Galinstan can fabricate a stretchable loudspeaker to apply audio frequency electric signal (Figure 13A) [112]. When the liquid metal circuit is cut off, once reconnected, it can work well without additional repairmen (Figure 13B) [113].

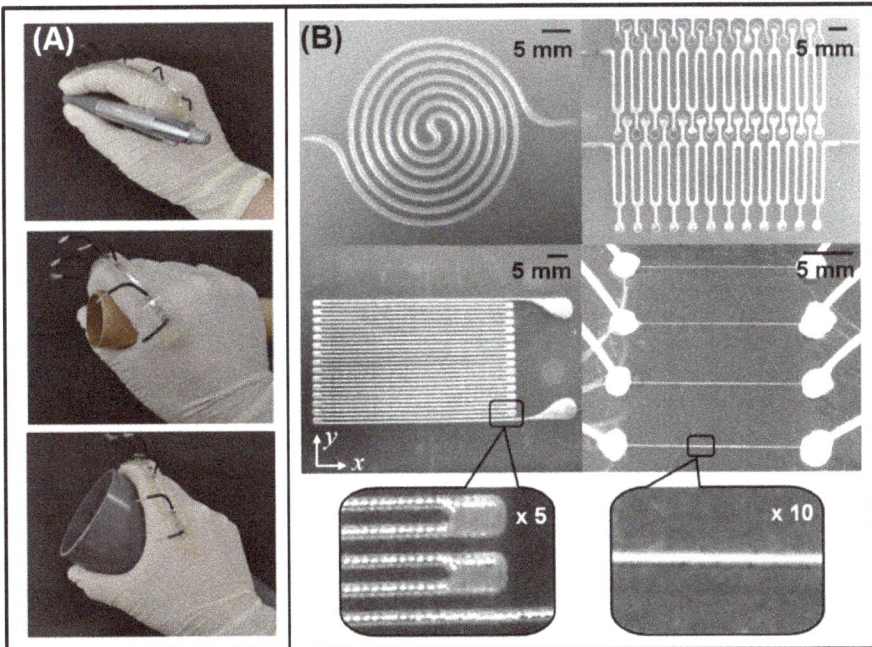

Figure 12. Sensors based on liquid metal deformation. (**A**) A wearable data glove with local strain sensing using GaInSn elestrodes [59] (Reproduced with permission from Matsuzaki, R.; Tabayashi, K., Highly stretchable, global, and distributed local strain sensing line using gainsn electrodes for wearable electronics; published by Advanced Functional Material, 2015.); (**B**) hyperelastic pressure sensors with microchannels of conductive liquid metal alloy [114] (Reproduced with permission from Park, Y.-L.; Majidi, C.; Kramer, R.; Berard, P.; Wood, R.J., Hyperelastic pressure sensing with a liquid-embedded elastomer; published by Journal of Micromechanics and Microengineering, 2010.).

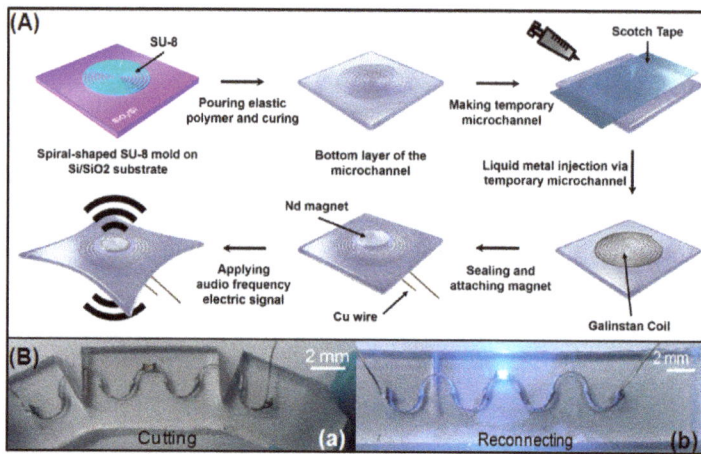

Figure 13. The liquid metal flexible printed electronics functional devices. (**A**) The fabrication process of the liquid metal stretchable loudspeaker [112] (Reproduced with permission from Jin, S.W.; Park, J.; Hong, S.Y.; Park, H.; Jeong, Y.R.; Park, J.; Lee, S.-S. ; Ha, J.S., Stretchable loudspeaker using liquid metal microchannel; published by Scientific Reports, 2015.); (**B**) Physical separation (a) and reconnection (b) of the liquid metal circuit with an LED [113] (Reproduced with permission from Li, G.; Wu, X.; Lee, D.-W., A galinstan-based inkjet printing system for highly stretchable electronics with self-healing capability; published by Lab on a Chip, 2016.).

5. Discussion and Conclusions

As new-generation inks for flexible printed electronics, liquid metals possess outstanding versatility and specific merits of fluidity, conductivity, metallicity, and electromagnetic properties. They thus offer tremendous opportunities for making future flectional electronic devices. However, there are still challenges in preparing, processing, and utilizing high-performance inks. For example, the low adhesion to many substrates, easy oxidation, and high surface tension of liquid alloy inks still restrict the realization of high-precision, stable circuits. To solve such key issues, we can mix the liquid metal with materials that are metal or non-metal. Therefore, new hybrid functional liquid metal materials can be made with desired properties of semiconductivity, semi-flexibility, enhanced adhesion, antioxidant ability, etc. This would enable the inks to adapt well to practical situations and have more functions. For example, the liquid metal nanocrystallization can improve printing precision, and one can modify the surface of liquid metal nanoparticles, e.g., by combining with the luminous substance through metal bond to achieve the self-luminous colorful printing.

So far, liquid metal ink has already been printed on a wide range of soft substrates, such as plate, paper, cloth, skin, leaves, and so on, which opens up many opportunities for making flexible electronical components like sensors, actuators, wearable bioelectronics, electrical skin, etc. The use of such diverse flexible substrates significantly enriches the application field of liquid metal flexible printed electronics, and one can thus expand related practices from human beings to other organisms such as animals and plants. For example, one can even print flexible circuits with features of energy harvesting and light emission on leaves to capture the energy of swinging leaves to supply light at night.

Various kinds of liquid metal printing technologies are enabling efficient circuit fabrication. At this stage, a series of technological problems still need to be tackled, such as precise and reliable connection between the printed liquid metal circuits with the traditional integrated circuits. High precision printing is critical to achieve the liquid metal integrated circuits. Regarding the specific fabrication, hybrid flexible printing with planar printing and 3D printing are worth trying. Furthermore, cooling of complex liquid metal circuits should be carefully monitored to avoid damage from the "thermal barrier." Meanwhile, encapsulation is often a necessity for making ending user devices. For instance, long-lasting liquid metal skin electronics mainly depend on packaging technology. PDMS is often applied to package liquid metal flexible circuits. However, for various substrates in different environments, one needs to consider diverse encapsulation technologies to make the circuit more secure in order to achieve the desired functionality. For applications in organisms, the non-toxicity of packaging materials especially needs to be considered. Sometimes, liquid metal's movement and shape-shifting under the electric field can produce more self-configurational operations of the circuits [106,107]. These unconventional, dynamically changing or self-destructive circuits are also worth pursuing.

In summary, liquid metal flexible printed electronics break through the limitations of traditional flexible electronics. The process of circuits manufacturing is straightforward, fast, stable, and relatively compatible with existing integrated circuit technology. Liquid metal flexible printed electronics would allow us to make high-quality products such as wearable devices, electronic skin, medical implants, flexible display, and solar panels that will meet the coming demands. The existing liquid metal flexible printed technology is a combination of a liquid metal soft circuit and rigid components or chips, so not completely flexible electronics in the larger sense. At the same time, this kind of semi-flexible circuit may partially limit its application. A basic significance of liquid metal flexible printed electronics lies in the popularity of liquid metal printing technology: people can discretionarily print their personalized flexible functional devices on any surface, which can significantly expand traditional electronics engineering. By overcoming the problems in obtaining a high rate of good product, self-repair of damaged circuits, long service life, low environmental pollution, and waste recycling, liquid metal flexible printed electronics can contribute to a bright future.

Acknowledgments: This work was partially supported by Beijing Municipal Science and Technology Funding (Under Grant No. Z151100003715002) and Key Project Funding of the Chinese Academy of Sciences.

Author Contributions: Xuelin Wang wrote the whole manuscript. Jing Liu supervised the work and wrote part of the manuscript.

Conflicts of Interest: The authors declare no conflict of interest.

References

1. Cui, Z. *Printed Electronics: Materials, Technologies and Applications*; Higher Education Press: Beijing, China, 2012.
2. Lin, Y.; Gritsenko, D.; Liu, Q.; Lu, X.; Xu, J. Recent advancements in functionalized paper-based electronics. *ACS Appl. Mater. Interfaces* **2016**, *8*, 20501–20515. [CrossRef] [PubMed]
3. Yin, Z.; Huang, Y.; Bu, N.; Wang, X.; Xiong, Y. Inkjet printing for flexible electronics: Materials, processes and equipments. *Chin. Sci. Bull.* **2010**, *55*, 3383–3407. [CrossRef]
4. Wong, W.S.; Salleo, A. *Flexible Electronics: Materials and Applications*; Springer: New York, NY, USA, 2009.
5. Reuss, R.H.; Chalamala, B.R.; Moussessian, A.; Kane, M.G.; Kumar, A.; Zhang, D.C.; Rogers, J.A.; Hatalis, M.; Temple, D.; Moddel, G.; et al. Macroelectronics: Perspectives on technology and applications. *Proc. IEEE* **2005**, *93*, 1239–1256. [CrossRef]
6. Kim, D.-H.; Ahn, J.-H.; Choi, W.M.; Kim, H.-S.; Kim, T.-H.; Song, J.; Huang, Y.Y.; Liu, Z.; Lu, C.; Rogers, J.A. Stretchable and foldable silicon integrated circuits. *Science* **2008**, *320*, 507–511. [CrossRef] [PubMed]
7. Crawford, G.P. *Flexible Flat Panel Displays*; John Wiley & Sons, Ltd.: New York, NY, USA, 2005.
8. Stoppa, M.; Chiolerio, A. Wearable electronics and smart textiles: A critical review. *Sensors* **2014**, *14*, 11957–11992. [CrossRef] [PubMed]
9. Koo, M.; Park, K.-I.; Lee, S.H.; Suh, M.; Jeon, D.Y.; Cho, J.W.; Kang, K.; Lee, K.J. Bendable inorganic thin-film battery for fully flexible electronic systems. *NANO Lett.* **2012**, *12*, 4810–4816. [CrossRef] [PubMed]
10. Reeder, J.; Kaltenbrunner, M.; Ware, T.; Arreaga-Salas, D.; Avendano-Bolivar, A.; Yokota, T.; Inoue, Y.; Sekino, M.; Voit, W.; Sekitani, T.; et al. Mechanically adaptive organic transistors for implantable electronics. *Adv. Mater.* **2014**, *26*, 4967–4973. [CrossRef] [PubMed]

11. Kang, D.; Pikhitsa, P.V.; Choi, Y.W.; Lee, C.; Shin, S.S.; Piao, L.; Park, B.; Suh, K.-Y.; Kim, T.-I.; Choi, M. Ultrasensitive mechanical crack-based sensor inspired by the spider sensory system. *Nature* **2014**, *516*, 222–226. [CrossRef] [PubMed]

12. Russo, A.; Ahn, B.Y.; Adams, J.J.; Duoss, E.B.; Bernhard, J.T.; Lewis, J.A. Pen-on-paper flexible electronics. *Adv. Mater.* **2011**, *23*, 3426–3430. [CrossRef] [PubMed]

13. Jung, I.; Shin, G.; Malyarchuk, V.; Ha, J.S.; Rogers, J.A. Paraboloid electronic eye cameras using deformable arrays of photodetectors in hexagonal mesh layouts. *Appl. Phys. Lett.* **2010**, *96*, 021110. [CrossRef]

14. Schwartz, G.; Tee, B.C.K.; Mei, J.; Appleton, A.L.; Kim, D.H.; Wang, H.; Bao, Z. Flexible polymer transistors with high pressure sensitivity for application in electronic skin and health monitoring. *Nat. Commun.* **2013**, *4*, 54–56. [CrossRef] [PubMed]

15. Wang, X.; Dong, L.; Zhang, H.; Yu, R.; Pan, C.; Wang, Z.L. Recent progress in electronic skin. *Adv. Sci.* **2015**, *2*, 10. [CrossRef]

16. Kim, D.-H.; Lu, N.; Ma, R.; Kim, Y.-S.; Kim, R.-H.; Wang, S.; Wu, J.; Won, S.M.; Tao, H.; Islam, A.; et al. Epidermal electronics. *Science* **2011**, *333*, 838–843. [CrossRef] [PubMed]

17. Choi, W.M.; Song, J.; Khang, D.-Y.; Jiang, H.; Huang, Y.Y.; Rogers, J.A. Biaxially stretchable "wavy" silicon nanomembranes. *NANO Lett.* **2007**, *7*, 1655–1663. [CrossRef] [PubMed]

18. Zheng, Y.; He, Z.-Z.; Yang, J.; Liu, J. Personal electronics printing via tapping mode composite liquid metal ink delivery and adhesion mechanism. *Sci. Rep.* **2014**, *4*, 101–104. [CrossRef] [PubMed]

19. Ghosh, D.; Maki, S.P.; Lyons, C.; Theiss, S.D.; Owings, R.R. Hybrid dielectric thin films on flexible substrates for embedded capacitor applications. *IEEE Trans. Compon. Packag. Manuf. Technol.* **2016**, *6*, 941–945. [CrossRef]

20. Singh, M.; Haverinen, H.M.; Dhagat, P.; Jabbour, G.E. Inkjet printing-process and its applications. *Adv. Mater.* **2010**, *22*, 673–685. [CrossRef] [PubMed]

21. MicroPen Technologies. Available online: http://www.Micropen.com/ (accessed on 1 September 2016).

22. Kim, S.S.; Na, S.I.; Jo, J.; Tae, G.; Kim, D.Y. Efficient polymer solar cells fabricated by simple brush painting. *Adv. Mater.* **2007**, *19*, 4410–4415. [CrossRef]

23. Jang, J. Displays develop a new flexibility. *Mater. Today* **2006**, *9*, 46–52. [CrossRef]

24. Krebs, F.C. Fabrication and processing of polymer solar cells: A review of printing and coating techniques. *Sol. Energy Mater. Sol. Cells* **2009**, *93*, 394–412. [CrossRef]

25. Mayer, A.C.; Scully, S.R.; Hardin, B.E.; Rowell, M.W.; McGehee, M.D. Polymer-based solar cells. *Mater. Today* **2007**, *10*, 28–33. [CrossRef]

26. Someya, T.; Kato, Y.; Sekitani, T.; Iba, S.; Noguchi, Y.; Murase, Y.; Kawaguchi, H.; Sakurai, T. Conformable, flexible, large-area networks of pressure and thermal sensors with organic transistor active matrixes. *Proc. Natl. Acad. Sci. USA* **2005**, *102*, 12321–12325. [CrossRef] [PubMed]

27. Madden, P.G.A. *Development and Modeling of Conducting Polymer Actuators and the Fabrication of a Conducting Polymer Based Feedback Loop*; Massachusetts Institute of Technology: Cambridge, MA, USA, 2003.

28. Hammock, M.L.; Chortos, A.; Tee, B.C.K.; Tok, J.B.H.; Bao, Z. 25th anniversary article: The evolution of electronic skin (e-skin): A brief history, design considerations, and recent progress. *Adv. Mater.* **2013**, *25*, 5997–6038. [CrossRef] [PubMed]

29. So, J.-H.; Koo, H.-J.; Dickey, M.D.; Velev, O.D. Ionic current rectification in soft-matter diodes with liquid-metal electrodes. *Adv. Funct. Mater.* **2012**, *22*, 625–631. [CrossRef]

30. Niu, S.; Wang, X.; Yi, F.; Zhou, Y.S.; Wang, Z.L. A universal self-charging system driven by random biomechanical energy for sustainable operation of mobile electronics. *Nat. Commun.* **2015**, *6*, 8975. [CrossRef] [PubMed]

31. Zheng, Q.; Zhang, H.; Shi, B.; Xue, X.; Liu, Z.; Jin, Y.; Ma, Y.; Zou, Y.; Wang, X.; An, Z.; et al. In vivo self-powered wireless cardiac monitoring via implantable triboelectric nanogenerator. *ACS Nano* **2016**, *10*, 6510–6518. [CrossRef] [PubMed]

32. Rogers, J.A.; Bao, Z.; Baldwin, K.; Dodabalapur, A.; Crone, B.; Raju, V.R.; Kuck, V.; Katz, H.; Amundson, K.; Ewing, J.; et al. Paper-like electronic displays: Large-area rubber-stamped plastic sheets of electronics and microencapsulated electrophoretic inks. *Proc. Natl. Acad. Sci. USA* **2001**, *98*, 4835–4840. [CrossRef] [PubMed]

33. Wang, X.; Tian, H.; Mohammad, M.A.; Li, C.; Wu, C.; Yang, Y.; Ren, T.-L. A spectrally tunable all-graphene-based flexible field-effect light-emitting device. *Nat. Commun.* **2015**, *6*, 37–42. [CrossRef] [PubMed]

34. Lessing, J.; Morin, S.A.; Keplinger, C.; Tayi, A.S.; Whitesides, G.M. Stretchable conductive composites based on metal wools for use as electrical vias in soft devices. *Adv. Funct. Mater.* **2015**, *25*, 1418–1425. [CrossRef]

35. Zhang, J.; Sheng, L.; Jin, C.; Liu, J. Liquid Metal as Connecting or Functional Recovery Channel for the Transected Sciatic Nerve. Available online: https://arxiv.org/abs/1404.5931 (accessed on 28 September 2016).

36. Liu, F.; Yu, Y.; Yi, L.; Liu, J. Liquid metal as reconnection agent for peripheral nerve injury. *Sci. Bull.* **2016**, *61*, 939–947. [CrossRef]

37. Wang, X.; Zhang, H.; Yu, R.; Dong, L.; Peng, D.; Zhang, A.; Zhang, Y.; Liu, H.; Pan, C.; Wang, Z.L. Dynamic pressure mapping of personalized handwriting by a flexible sensor matrix based on the mechanoluminescence process. *Adv. Mater.* **2015**, *27*, 2324–2331. [CrossRef] [PubMed]

38. Tee, B.C.K.; Wang, C.; Allen, R.; Bao, Z. An electrically and mechanically self-healing composite with pressure- and flexion-sensitive properties for electronic skin applications. *Nat. Nano* **2012**, *7*, 825–832. [CrossRef] [PubMed]

39. Kim, J.; Lee, M.; Shim, H.J.; Ghaffari, R.; Cho, H.R.; Son, D.; Jung, Y.H.; Soh, M.; Choi, C.; Jung, S.; et al. Stretchable silicon nanoribbon electronics for skin prosthesis. *Nat. Commun.* **2014**, *5*, 5747. [CrossRef] [PubMed]

40. Tee, B.C.-K.; Chortos, A.; Berndt, A.; Nguyen, A.K.; Tom, A.; McGuire, A.; Lin, Z.C.; Tien, K.; Bae, W.-G.; Wang, H.; et al. A skin-inspired organic digital mechanoreceptor. *Science* **2015**, *350*, 313–316. [CrossRef] [PubMed]

41. Wang, S.; Li, M.; Wu, J.; Kim, D.-H.; Lu, N.; Su, Y.; Kang, Z.; Huang, Y.; Rogers, J.A. Mechanics of epidermal electronics. *J. Appl. Mech.* **2012**, *79*, 031022. [CrossRef]

42. Webb, R.C.; Ma, Y.; Krishnan, S.; Li, Y.; Yoon, S.; Guo, X.; Feng, X.; Shi, Y.; Seidel, M.; Cho, N.H.; et al. Epidermal devices for noninvasive, precise, and continuous mapping of macrovascular and microvascular blood flow. *Sci. Adv.* **2015**, *1*, e1500701. [CrossRef] [PubMed]

43. Rajan, K.; Bocchini, S.; Chiappone, A.; Roppolo, I.; Perrone, D.; Bejtka, K.; Ricciardi, C.; Pirri, C.F.; Chiolerio, A. Spin-coated silver nanocomposite resistive switching devices. *Microelectron. Eng.* **2017**, *168*, 27–31. [CrossRef]

44. Chiolerio, A.; Roppolo, I.; Bejtka, K.; Asvarov, A.; Pirri, C.F. Resistive hysteresis in flexible nanocomposites and colloidal suspensions: Interfacial coupling mechanism unveiled. *RSC Adv.* **2016**, *6*, 56661–56667. [CrossRef]

45. Tai, Y.L. Preparation and Application of Conductive Ink for Printed Electronics. Ph.D. Thesis, Fudan University, Shanghai, China, 2012.

46. Rajan, K.; Roppolo, I.; Chiappone, A.; Bocchini, S.; Perrone, D.; Chiolerio, A. Silver nanoparticle ink technology: State of the art. *Nanotechnol. Sci. Appl.* **2016**, *9*, 1–13. [PubMed]

47. Xiong, Z.; Liu, C. Optimization of inkjet printed PEDOT:PSS thin films through annealing processes. *Org. Electron.* **2012**, *13*, 1532–1540. [CrossRef]

48. Chiolerio, A.; Bocchini, S.; Scaravaggi, F.; Porro, S.; Perrone, D.; Beretta, D.; Caironi, M.; Pirri, C.F. Synthesis of polyaniline-based inks for inkjet printed devices: Electrical characterization highlighting the effect of primary and secondary doping. *Semicond. Sci. Technol.* **2015**, *30*, 104001. [CrossRef]

49. Petukhov, D.I.; Kirikova, M.N.; Bessonov, A.A.; Bailey, M.J.A. Nickel and copper conductive patterns fabricated by reactive inkjet printing combined with electroless plating. *Mater. Lett.* **2014**, *132*, 302–306. [CrossRef]

50. Plevachuk, Y.; Sklyarchuk, V.; Yakymovych, A.; Svec, P.; Janickovic, D.; Illekova, E. Electrical conductivity and viscosity of liquid Sn–Sb–Cu alloys. *J. Mater. Sci. Mater. Electron.* **2011**, *22*, 631–638. [CrossRef]

51. Yi, L.; Jin, C.; Wang, L.; Liu, J. Liquid-solid phase transition alloy as reversible and rapid molding bone cement. *Biomaterials* **2014**, *35*, 9789–9801. [CrossRef] [PubMed]

52. Wang, Q.; Yu, Y.; Pan, K.; Liu, J. Liquid metal angiography for mega contrast X-ray visualization of vascular network in reconstructing in vitro organ anatomy. *IEEE Trans. Biomed. Eng.* **2014**, *61*, 2161–2166. [CrossRef] [PubMed]

53. Lu, Y.; Hu, Q.; Lin, Y.; Pacardo, D.B.; Wang, C.; Sun, W.; Ligler, F.S.; Dickey, M.D.; Gu, Z. Transformable liquid-metal nanomedicine. *Nat. Commun.* **2015**, *6*, 10066. [CrossRef] [PubMed]

54. Jin, C.; Zhang, J.; Li, X.; Yang, X.; Li, J.; Liu, J. Injectable 3-D fabrication of medical electronics at the target biological tissues. *Sci. Rep.* **2013**, *3*, 3442. [CrossRef] [PubMed]

55. Guo, C.; Yu, Y.; Liu, J. Rapidly patterning conductive components on skin substrates as physiological testing devices via liquid metal spraying and pre-designed mask. *J. Mater. Chem. B* **2014**, *2*, 5739–5745. [CrossRef]

56. Jeong, S.H.; Hagman, A.; Hjort, K.; Jobs, M.; Sundqvist, J.; Wu, Z. Liquid alloy printing of microfluidic stretchable electronics. *Lab Chip* **2012**, *12*, 4657–4664. [CrossRef] [PubMed]

57. Yu, Y.; Zhang, J.; Liu, J. Biomedical implementation of liquid metal ink as drawable ecg electrode and skin circuit. *PLoS ONE* **2013**, *8*, e58771. [CrossRef] [PubMed]

58. Wang, Q.; Yu, Y.; Yang, J.; Liu, J. Fast fabrication of flexible functional circuits based on liquid metal dual-trans printing. *Adv. Mater.* **2015**, *27*, 7109–7116. [CrossRef] [PubMed]

59. Matsuzaki, R.; Tabayashi, K. Highly stretchable, global, and distributed local strain sensing line using gainsn electrodes for wearable electronics. *Adv. Funct. Mater.* **2015**, *25*, 3806–3813. [CrossRef]

60. Krupenkin, T.; Taylor, J.A. Reverse electrowetting as a new approach to high-power energy harvesting. *Nat. Commun.* **2011**, *2*, 73–86. [CrossRef] [PubMed]

61. Wang, L.; Liu, J. Research advancement of liquid metal printed electronics ink (in Chinese). *Imaging Sci. Photochem.* **2014**, *32*, 382–392.

62. Zhang, Q.; Zheng, Y.; Liu, J. Direct writing of electronics based on alloy and metal (dream) ink: A newly emerging area and its impact on energy, environment and health sciences. *Front. Energy* **2012**, *6*, 311–340. [CrossRef]

63. Gao, Y.; Li, H.; Liu, J. Direct writing of flexible electronics through room temperature liquid metal ink. *PLoS ONE* **2012**, *7*, e45485. [CrossRef] [PubMed]

64. Wang, L.; Liu, J. Compatible hybrid 3D printing of metal and nonmetal inks for direct manufacture of end functional devices. *Sci. China Technol. Sci.* **2014**, *57*, 2089–2095. [CrossRef]

65. Tripathi, V.; Loh, Y.L. Thermal conductivity of a granular metal. *Phys. Rev. Lett.* **2006**, *96*, 046805. [CrossRef] [PubMed]

66. Gozen, B.A.; Tabatabai, A.; Ozdoganlar, O.B.; Majidi, C. High-density soft-matter electronics with micron-scale line width. *Adv. Mater.* **2014**, *26*, 5211–5216. [CrossRef] [PubMed]

67. Sivan, V.; Tang, S.-Y.; O'Mullane, A.P.; Petersen, P.; Eshtiaghi, N.; Kalantar-zadeh, K.; Mitchell, A. Liquid metal marbles. *Adv. Funct. Mater.* **2013**, *23*, 144–152. [CrossRef]

68. Du, K.; Glogowski, E.; Tuominen, M.T.; Emrick, T.; Russell, T.P.; Dinsmore, A.D. Self-assembly of gold nanoparticles on gallium droplets: Controlling charge transport through microscopic devices. *Langmuir* **2013**, *29*, 13640–13646. [CrossRef] [PubMed]

69. Zheng, Y.; He, Z.; Gao, Y.; Liu, J. Direct desktop printed-circuits-on-paper flexible electronics. *Sci. Rep.* **2013**, *3*, 1786. [CrossRef]

70. Wang, L.; Liu, J. Pressured liquid metal screen printing for rapid manufacture of high resolution electronic patterns. *RSC Adv.* **2015**, *5*, 57686–57691. [CrossRef]

71. Kim, D.; Lee, Y.; Lee, D.-W.; Choi, W.; Yoo, K.; Lee, J.-B. Hydrochloric acid-impregnated paper for gallium-based liquid metal microfluidics. *Sens. Actuators B Chem.* **2015**, *207*, 199–205. [CrossRef]

72. Kubo, M.; Li, X.; Kim, C.; Hashimoto, M.; Wiley, B.J.; Ham, D.; Whitesides, G.M. Stretchable microfluidic radiofrequency antennas. *Adv. Mater.* **2010**, *22*, 2749–2752. [CrossRef] [PubMed]

73. Boley, J.W.; White, E.L.; Chiu, G.T.C.; Kramer, R.K. Direct writing of gallium-indium alloy for stretchable electronics. *Adv. Funct. Mater.* **2014**, *24*, 3501–3507. [CrossRef]

74. Kim, D.; Thissen, P.; Viner, G.; Lee, D.-W.; Choi, W.; Chabal, Y.J.; Lee, J.-B. Recovery of nonwetting characteristics by surface modification of gallium-based liquid metal droplets using hydrochloric acid vapor. *ACS Appl. Mater. Interfaces* **2013**, *5*, 179–185. [CrossRef] [PubMed]

75. Zhang, Q.; Gao, Y.; Liu, J. Atomized spraying of liquid metal droplets on desired substrate surfaces as a generalized way for ubiquitous printed electronics. *Appl. Phys. A* **2014**, *116*, 1091–1097. [CrossRef]

76. Xu, Q.; Oudalov, N.; Guo, Q.; Jaeger, H.M.; Brown, E. Effect of oxidation on the mechanical properties of liquid gallium and eutectic gallium-indium. *Phys. Fluids* **2012**, *24*, 063101. [CrossRef]

77. Schrader, M.E. Young-dupre revisited. *Langmuir* **1995**, *11*, 3585–3589. [CrossRef]

78. Liu, Y.; Gao, M.; Mei, S.; Han, Y.; Liu, J. Ultra-compliant liquid metal electrodes with in-plane self-healing capability for dielectric elastomer actuators. *Appl. Phys. Lett.* **2013**, *103*, 064101. [CrossRef]

79. Tiberto, P.; Gupta, S.; Bianco, S.; Celegato, F.; Martino, P.; Chiolerio, A.; Tagliaferro, A.; Allia, P. Morphology and magnetic properties of island-like Co and Ni films obtained by de-wetting. *J. Nanopart. Res.* **2011**, *13*, 245–255. [CrossRef]

80. Doudrick, K.; Liu, S.; Mutunga, E.M.; Klein, K.L.; Damle, V.; Varanasi, K.K.; Rykaczewski, K. Different shades of oxide: From nanoscale wetting mechanisms to contact printing of gallium-based liquid metals. *Langmuir* **2014**, *30*, 6867–6877. [CrossRef] [PubMed]

81. Li, G.; Parmar, M.; Kim, D.; Lee, J.-B.; Lee, D.-W. Pdms based coplanar microfluidic channels for the surface reduction of oxidized galinstan. *Lab Chip* **2014**, *14*, 200–209. [CrossRef] [PubMed]

82. Furmidge, C.G.L. Studies at phase interfaces. I. The sliding of liquid drops on solid surfaces and a theory for spray retention. *J. Colloid Sci.* **1962**, *17*, 309–324. [CrossRef]

83. Kramer, R.K.; Boley, J.W.; Stone, H.A.; Weaver, J.C.; Wood, R.J. Effect of microtextured surface topography on the wetting behavior of eutectic gallium–indium alloys. *Langmuir* **2014**, *30*, 533–539. [CrossRef] [PubMed]

84. Yi, L.T. Injectable Bone Cement Based on Low-Melting-Point Alloy with Liquid-Solid Phase Switchable Capability. Ph.D. Thesis, Tsinghua University, Beijing, China, 2015.

85. Li, H.; Yang, Y.; Liu, J. Printable tiny thermocouple by liquid metal gallium and its matching metal. *Appl. Phys. Lett.* **2012**, *101*, 073511. [CrossRef]

86. Liu, J.; Li, H. A Liquid Metal Based Printed Circuit Board and Its Fabrication Method. China Patent 201110140156, 14 October 2011.

87. Tabatabai, A.; Fassler, A.; Usiak, C.; Majidi, C. Liquid-phase gallium–indium alloy electronics with microcontact printing. *Langmuir* **2013**, *29*, 6194–6200. [CrossRef] [PubMed]

88. Kramer, R.K.; Majidi, C.; Wood, R.J. Masked deposition of gallium-indium alloys for liquid-embedded elastomer conductors. *Adv. Funct. Mater.* **2013**, *23*, 5292–5296. [CrossRef]

89. Jeong, J.-A.; Kim, H.-K. Characteristics of inkjet-printed nano indium tin oxide particles for transparent conducting electrodes. *Curr. Appl. Phys.* **2010**, *10*, e105–e108. [CrossRef]

90. Sheng, L.; Teo, S.; Liu, J. Liquid-metal-painted stretchable capacitor sensors for wearable healthcare electronics. *J. Med. Biol. Eng.* **2016**, *36*, 265–272. [CrossRef]

91. Wang, L.; Liu, J. Printing low-melting-point alloy ink to directly make a solidified circuit or functional device with a heating pen. *Proc. R. Soc. A Math. Phys. Eng. Sci.* **2014**, *470*, 0609. [CrossRef] [PubMed]

92. Yang, J.; Liu, J. Direct printing and assembly of fm radio at the user end via liquid metal printer. *Circ. World* **2014**, *40*, 134–140. [CrossRef]

93. Zheng, Y.; Zhang, Q.; Liu, J. Pervasive liquid metal based direct writing electronics with roller-ball pen. *AIP Adv.* **2013**, *3*, 6459–6463. [CrossRef]

94. Kim, D.; Yoo, J.H.; Lee, Y.; Choi, W.; Yoo, K.; Lee, J.B.J. Gallium-Based Liquid Metal Inkjet Printing. In Proceedings of the 2014 IEEE 27th International Conference on Micro Electro Mechanical Systems (MEMS), San Francisco, CA, USA, 26–30 January 2014; pp. 967–970.

95. Yang, H.; Lightner, C.R.; Dong, L. Light-emitting coaxial nanofibers. *ACS Nano* **2012**, *6*, 622–628. [CrossRef] [PubMed]

96. Lu, T.; Finkenauer, L.; Wissman, J.; Majidi, C. Rapid prototyping for soft-matter electronics. *Adv. Funct. Mater.* **2014**, *24*, 3351–3356. [CrossRef]

97. Jeong, S.H.; Hjort, K.; Wu, Z. Tape transfer atomization patterning of liquid alloys for microfluidic stretchable wireless power transfer. *Sci. Rep.* **2015**, *5*, 8419. [CrossRef] [PubMed]

98. Boley, J.W.; White, E.L.; Kramer, R.K. Mechanically sintered gallium–indium nanoparticles. *Adv. Mater.* **2015**, *27*, 2355–2360. [CrossRef] [PubMed]

99. Kunnari, E.; Valkama, J.; Keskinen, M.; Mansikkamäki, P. Environmental evaluation of new technology: Printed electronics case study. *J. Clean. Prod.* **2009**, *17*, 791–799. [CrossRef]

100. Wang, L.; Liu, J. Liquid phase 3D printing for quickly manufacturing conductive metal objects with low melting point alloy ink. *Sci. China Technol. Sci.* **2014**, *57*, 1721–1728. [CrossRef]

101. Fassler, A.; Majidi, C. 3D structures of liquid-phase gain alloy embedded in pdms with freeze casting. *Lab Chip* **2013**, *13*, 4442–4450. [CrossRef] [PubMed]

102. Ladd, C.; So, J.-H.; Muth, J.; Dickey, M.D. 3D printing of free standing liquid metal microstructures. *Adv. Mater.* **2013**, *25*, 5081–5085. [CrossRef] [PubMed]

103. Liu, R.; Yang, X.; Jin, C.; Fu, J.; Chen, W.; Liu, J. Development of three-dimension microelectrode array for bioelectric measurement using the liquidmetal-micromolding technique. *Appl. Phys. Lett.* **2013**, *103*, 193701. [CrossRef]

104. Deng, Y.; Liu, J. Flexible mechanical joint as human exoskeleton using low-melting-point alloy. *J. Med. Devices* **2014**, *8*, 044506. [CrossRef]

105. Li, J.; Guo, C.; Wang, Z.; Gao, K.; Shi, X.; Liu, J. Electrical stimulation towards melanoma therapy via liquid metal printed electronics on skin. *Clin. Transl. Med.* **2016**, *5*, 1–7. [CrossRef] [PubMed]

106. Wang, L.; Liu, J. Ink spraying based liquid metal printed electronics for directly making smart home appliances. *ECS J. Solid State Sci. Technol.* **2015**, *4*, 3057–3062. [CrossRef]

107. Fassler, A.; Majidi, C. Soft-matter capacitors and inductors for hyperelastic strain sensing and stretchable electronics. *Smart Mater. Struct.* **2013**, *22*, 955–966. [CrossRef]

108. Cheng, S.; Wu, Z. Microfluidic electronics. *Lab Chip* **2012**, *12*, 2782–2791. [CrossRef] [PubMed]

109. Cheng, S.; Wu, Z. A microfluidic, reversibly stretchable, large-area wireless strain sensor. *Adv. Funct. Mater.* **2011**, *21*, 2282–2290. [CrossRef]

110. Wu, Z.; Hjort, K.; Jeong, S.H. Microfluidic stretchable radio-frequency devices. *Proc. IEEE* **2015**, *103*, 1211–1225. [CrossRef]

111. Jeong, S.; Hjort, K.; Wu, Z. Tape transfer printing of a liquid metal alloy for stretchable RF electronics. *Sensors* **2014**, *14*, 16311–16321. [CrossRef] [PubMed]

112. Jin, S.W.; Park, J.; Hong, S.Y.; Park, H.; Jeong, Y.R.; Park, J.; Lee, S.-S.; Ha, J.S. Stretchable loudspeaker using liquid metal microchannel. *Sci. Rep.* **2015**, *5*, 11695. [CrossRef] [PubMed]

113. Li, G.; Wu, X.; Lee, D.-W. A galinstan-based inkjet printing system for highly stretchable electronics with self-healing capability. *Lab Chip* **2016**, *16*, 1366–1373. [CrossRef] [PubMed]

114. Park, Y.-L.; Majidi, C.; Kramer, R.; Berard, P.; Wood, R.J. Hyperelastic pressure sensing with a liquid-embedded elastomer. *J. Micromech. Microeng.* **2010**, *20*, 125029–125034. [CrossRef]

MDPI AG

St. Alban-Anlage 66

4052 Basel, Switzerland

Tel. +41 61 683 77 34

Fax +41 61 302 89 18

http://www.mdpi.com

Micromachines Editorial Office

E-mail: micromachines@mdpi.com

http://www.mdpi.com/journal/micromachines

www.ingramcontent.com/pod-product-compliance
Lightning Source LLC
Chambersburg PA
CBHW051857210326
41597CB00033B/5934